Emerging Contaminants and One Health Approach

Emerging contaminants, or contaminants of concern, which range from microplastics, industrial chemicals, cosmetics, and pesticides, to pharmaceuticals and pathogens, present unique issues and challenges to communities across the globe. The risk assessment and management of these potential hazards require an integrated, multidisciplinary approach in connection with the holistic framework of One Health. This book explains the need for collaboration between researchers, policymakers, public health professionals, and other stakeholders to address the serious gaps between emerging contaminants and One Health. It offers practical strategies and solutions for managing emerging contaminants effectively.

Features

- Discusses an important and timely concept of contaminants of concern from the perspective of One Health
- Focuses on the existing research gaps in managing emerging contaminants
- Emphasizes the urgent need for collaboration between field experts in biotechnology, bioresources, bioeconomics, environmental engineering, and environmental sciences
- Provides state-of-the-art technological developments, their advantages, and challenges associated with emerging contaminants

Professionals and researchers in Environmental Engineering and Sciences, Biosciences, and Public Health will benefit from the practical approach, strategies, and solutions featured in this book. It is also great supplementary reading for students taking courses in areas such as Environmental Health, Pollution and Environmental Assessment, and Environmental Policy and Regulation.

Emerging Contaminants and One Health Approach

Management and Future Directions

Edited by
Ramsha Khan, Saurabh Shukla,
Preeti Chaturvedi Bhargava, and Paolo Roccaro

CRC Press
Taylor & Francis Group
Boca Raton London New York

CRC Press is an imprint of the
Taylor & Francis Group, an **informa** business

Designed cover image: © Getty Images

First edition published 2026
by CRC Press
2385 NW Executive Center Drive, Suite 320, Boca Raton FL 33431

and by CRC Press
4 Park Square, Milton Park, Abingdon, Oxon, OX14 4RN

CRC Press is an imprint of Taylor & Francis Group, LLC

ISBN: 978-1-032-62378-8 (hbk)
ISBN: 978-1-032-62382-5 (pbk)
ISBN: 978-1-032-62380-1 (ebk)

DOI: 10.1201/9781032623801

Typeset in Times
by Newgen Publishing UK

Contents

Preface

Environmental pollution is an incurable disease. It can only be prevented. – Barry Commoner

In the past few years, emerging contaminants (ECs) have become a challenge for scientists, policymakers, and public health professionals across the world. ECs, which include pharmaceuticals, personal care products, microplastics, industrial chemicals, pathogens, and so forth, can persist in the environment for very long periods. The presence of these ECs in different environmental matrices like water, soil, and air poses several health risks to humans, animals, and environment. There are significant research gaps in understanding the fate, pathways, and impacts of these ECs on human, animal, and environmental health. The One Health approach, which recognizes the connection between humans, animals, and the environment, provides a holistic framework for addressing these challenges. One Health is a multidisciplinary approach. It can help in creating efficient strategies toward monitoring, managing, and mitigating the risks of emerging contaminants.

This book, titled ***Emerging Contaminants and One Health Approach: Management and Future Directions***, comprises 12 chapters, written by 39 contributors from different countries. These chapters discuss the impacts of ECs within the holistic framework of the One Health approach. It begins with an introduction to the concept and its relevance, followed by its relevance in managing different types of contaminants. The third chapter focuses on existing policies and regulatory measures, highlighting the gaps. Subsequent chapters discuss several ECs such as PAHs, PFAS, microplastics, antibiotics, pesticides, endocrine disruptors, and persistent organic pollutants, with a focus on their health impacts on the ecosystem. The book also explores innovative strategies like cradle-to-cradle management, technological tools such as machine learning and blockchain, and concludes with future directions for further research and the need for policy reforms.

The editors extend their heartfelt thanks to all the contributing authors for their efforts in producing timely and high-quality chapters. This book would not have been possible without the assistance of several colleagues and friends. From selecting contributors to reviewing chapters, they have helped in many ways. Finally, we would like to thank the staff of Taylor & Francis for their continuous support and professional editing of the publication.

Ramsha Khan
Saurabh Shukla
Preeti Chaturvedi Bhargava
Paolo Roccaro

About the Editors

Ramsha Khan is currently working as a Postdoctoral Researcher at the University of Catania, Italy. A key focus of her research work involves promoting sustainability in agricultural systems through integration of circular economy approaches, precision farming through remote sensing, water quality monitoring, detection and analysis of heavy metals, microplastics, and waste valorization.

Saurabh Shukla is currently working as a Postdoctoral Researcher at University of Catania, Italy. His research interests include monitoring surface and ground water resources, monitoring agricultural systems through precision farming and waste valorization, and carbon-based materials for wastewater remediation. His research interests align with the Sustainable Development Goal (SDG 6).

Preeti Chaturvedi Bhargava is a Senior Principal Scientist at the CSIR-Indian Institute of Toxicology Research (CSIR-IITR), Lucknow. Her major areas of research are Industrial and Environmental Biotechnology, Bioremediation, Antimicrobial resistance, Wastewater Treatment & Process Engineering, Metagenomics, and Waste Management. Currently, her work involves biochar synthesis using different catalysts for enhancing adsorption capacity for remediation and adsorption of pollutants from wastewater. She is also involved in the development of green technologies for effluent treatment for sustainable environment and health. She has published over 90 research/review papers in peer-reviewed reputed journals. She has also edited three peer-reviewed books, chapters and magazine articles. Her work won Best Work and Most Cited Paper Awards in Bioresource Technology, Elsevier in 2020, 2021, and 2022. She has worked as visiting scientist at EPFL, Lausanne, Switzerland; North A & F University, Yangling, China; and the Department of Bioprocess Engineering and Biotechnology, UFPR, Curitiba, Brazil. She is a life member of many renowned international and national scientific societies. She is also a gazette member of the State Ganga committee, U.P., and a member of subcommittees in bodies like the Bureau of Indian Standard (BIS), the Central Pollution Control Board (CPCB), the Food Safety and Standards Authority of India (FSSAI), and so on. She has served as associate editor of the *Indian Journal of Microbiology*, as a guest editor for special issues of *Bioresource Technology* and the *Indian Journal of Experimental Biology*, as a member of the editorial board of the *Journal of Innovation in Applied Research*, and so forth.

Paolo Roccaro is Full Professor of Sanitary and Environmental Engineering at the Department of Civil Engineering and Architecture, University of Catania, Italy. He has been a Fulbright Scholar at the University of Arizona in Tucson and a Visiting Scientist at the University of Washington, Seattle, WA. He has coordinated projects on Microwave Regeneration of PFAS-Exhausted Granular Activated Carbon as co-PI (grant from The Water Research Foundation, USA); on innovative tools to control organic matter and disinfection byproducts in drinking water into DBP as WP Leader (grant from EU – Horizon Europe), for the SiciliAn MicronanOTecH Research and Innovation CEnter (SAMOTHRACE) as Task Leader (grant from MUR Ecosistema dell'innovazione PNRR); on smart and ecofriendly systems for Sicilian agriculture as co-PI (grant from the Sicilian Region); on advanced processes for the removal of emerging contaminants from water as PI (grant from the University of Catania); on the recovery and use of volcanic ash from Mt. Etna – REUCET as PI (grant from the Italian Ministry of Environment); and on wastewater reuse as co-PI (grant from the Italian Ministry of Instruction University and Research (MIUR). He has also worked on other projects funded by the European Commission, and the U.S. Environmental Protection Agency (USEPA). His research is focused on water quality, removal of micropollutants from water, formation and control of disinfection by-products (DBP) in water, removal and control of contaminants of emerging concern (CEC) in water and wastewaters systems, wastewater treatment and reuse, advanced treatment processes, monitoring and modeling trace contaminants in water by spectroscopy, real-time monitoring, soil and aquifer remediation, water quality management in precision farming, waste recovery and valorization, circular economy and sustainability in water and waste management, and human exposure to airborne contaminants.

Contributors

Orhue Osayamen Amengialue
Department of Biological Science
College of Science and Computing
Wellspring University
Benin City, Nigeria

M. Sai Anjani
Center for Informetrics and Statistics
SR University
Warangal, Telangana, India

Thilini U. Ariyadasa
Department of Chemical and Process Engineering
University of Moratuwa
Katubedda, Moratuwa, Sri Lanka

Meghna Arya
Aquatic Toxicology Laboratory
Environmental Toxicology Group
Food, Drug and Chemical, Environment and Systems Toxicology (FEST) Division
CSIR-Indian Institute of Toxicology Research (CSIR-IITR), Vishvigyan Bhawan
Lucknow, Uttar Pradesh, India

Luca Baccini
Department of Civil Engineering and Architecture
University of Catania
Catania, Italy

Shirisha Balle
Center for Informetrics and Statistics
SR University
Telangana, India

Preeti Chaturvedi Bhargava
Aquatic Toxicology Laboratory
Environmental Toxicology Group
Food, Drug and Chemical, Environment and System Toxicology (FEST) Division
CSIR-Indian Institute of Toxicology Research (CSIR-IITR), Vishvigyan Bhawan
Lucknow, Uttar Pradesh, India
and
Academy of Scientific and Innovative Research (AcSIR)
Ghaziabad, Uttar Pradesh, India

Deepshi Chaurasia
Aquatic Toxicology Laboratory
Environmental Toxicology Group
Food, Drug and Chemical, Environment and System Toxicology (FEST) Division
CSIR-Indian Institute of Toxicology Research (CSIR-IITR), Vishvigyan Bhawan
Lucknow, Uttar Pradesh, India

Damini
Graphic Era Hill University
Dehradun, Uttarakhand, India

Palashpriya Das
Dr D Y Patil Biotechnology Institute Pune
Maharashtra, India
and
Guru Nanak Institute of Pharmaceutical Science and Technology
Kolkata, India

Antonino Di Bella
Department of Civil Engineering and Architecture
University of Catania
Catania, Italy

Philip Idemudia Edogun
Department of Biological Science
College of Science and Computing
Wellspring University
Benin City, Nigeria

Precious Ahosegbamen Egharevba
Department of Biological Science
College of Science and Computing
Wellspring University
Benin City, Nigeria

Filippo Fazzino
Department of Civil Engineering and Architecture
University of Catania
Catania, Italy

Ravindran Gobinath
Professor and Director, SCRI
Symbiosis International (Deemed) University
Pune

Francis Aibuedefe Igiebor
Department of Microbiology
Faculty of Science
Delta State University
Abraka, Nigeria

Aishuehien Imas Iyoha
Department of Biological Science
College of Science and Computing
Wellspring University
Benin City, Nigeria

Ramsha Khan
Department of Civil Engineering and Architecture
University of Catania
Catania, Italy

Alaa El Din Mahmoud
Environmental Sciences Department
Faculty of Science
Alexandria University
Alexandria, Egypt
and
Green Technology Group
Faculty of Science
Alexandria University
Alexandria, Egypt

Eduardo Alberto Lopez Maldonado
Faculty of Chemical Sciences and Engineering
Autonomous University of Baja California
Tijuana, Baja California, Mexico

Luigi Marino
Department of Civil Engineering and Architecture
University of Catania
Catania, Italy

Muthukumar P
Department of Geology
V.O. Chidambaram College
Thoothukudi, Tamil Nadu, India

Uzoamaka Nneka Ngwoke
Department of Plant Biology and Biotechnology
Faculty of Life Sciences
University of Benin
Benin City, Nigeria

Henrietta Omokhele Obueh
Department of Biological Science
College of Science and Computing
Wellspring University
Benin City, Nigeria

Haruna Ochoyama
Department of Science Laboratory Technology
Faculty of Life Sciences
University of Benin
Benin City, Nigeria

Philip Oseikhuenmhen Ogboye
Department of Biological Science
College of Science and Computing
Wellspring University
Benin City, Nigeria

Sakshi Ranjan
Dr D Y Patil Biotechnology Institute Pune
Maharashtra, India

R. Archana Reddy
Dept of Computer Science and Artificial Intelligence
SR University
Warangal, Telangana, India

Paolo Roccaro
Department of Civil Engineering and Architecture
University of Catania
Catania, Italy

Selvam S
Department of Geology
V.O. Chidambaram College
Thoothukudi, Tamil Nadu, India

Saurabh Shukla
Department of Civil Engineering and Architecture
University of Catania
Catania, Italy

G. Shyamala
Department of Civil Engineering
SR University
Telangana, India

Anuradha Singh
Aquatic Toxicology Laboratory
Environmental Toxicology Group
Food, Drug & Chemical, Environment and System Toxicology (FEST) Division
CSIR-Indian Institute of Toxicology Research (CSIR-IITR), Vishvigyan Bhawan
Lucknow, Uttar Pradesh, India

Rahul Vikram Singh
Academy of Scientific and Innovative Research (AcSIR)
Ghaziabad, India

Shagufta Tanveer
Aquatic Toxicology Laboratory
Environmental Toxicology Group
Food, Drug and Chemical, Environment and Systems Toxicology (FEST) Division
CSIR-Indian Institute of Toxicology Research (CSIR-IITR), Vishvigyan Bhawan
Lucknow, Uttar Pradesh, India

Nosa Jude Uwuigiaren
Department of Biological Science
College of Science and Computing
Wellspring University
Benin City, Nigeria

Ulrich Vasconcelos
Departamento de Biotecnologia
Universidade Federal da Paraíba, UFPB
Brazil

1 Introduction to Emerging Contaminants and One Health

Deepshi Chaurasia, Anuradha Singh, Thilini U. Ariyadasa, and Preeti Chaturvedi Bhargava

1.1 INTRODUCTION

The unprecedented rise in population, industrialization, and urbanization has caused serious effects on the ecosystem. Such kinds of developmental activities have overexploited natural resources and shifted attention toward synthetic compounds. Emerging contaminants (ECs) are one such class of compounds that have recently gained recognition owing to their adverse effects on the environment. ECs are mainly defined as natural or synthetic chemical/biological agents that are not commonly detected in biomes but may potentially be harmful and pose ecological health risks. ECs are also termed contaminants of emerging concern as they were previously unknown and not generally monitored. The class includes personal care products (PCPs), pesticides, pharmaceuticals, per- and poly-fluoroalkyl substances (PFAS), nanomaterials, plastics microplastics, and other exogenous compounds (Puri et al., 2023). ECs enter the biosphere through various routes such as effluent discharge from industries, hospitals, in run-off from agricultural fields, from landfill sites, poultry farms, and so forth. These compounds have the capacity to bio-accumulate and go through additional transformation to sustain their long-term existence in the environment (Machado et al., 2016). Farming, being totally dependent on freshwater availability and soil productivity, is drastically affected by ECs (Bayabil et al., 2022). Irrigation of farmland with untreated water contaminates the land and crops with toxic metals and chemicals that have pernicious effects on the environment (Al-Hazmi et al., 2023). Other sources of the influx of ECs include: excavation & mining activities, construction works, mineral processing, and energy production (Naidu et al., 2021). The induction of such puissant pollutants into the global surroundings has triggered the need for exhaustive approaches to address the health of biomes and livingbeings.

The concept of "One Health" involves a holistic and transdisciplinary approach that entails the designing of a conceptual framework for mitigating global health challenges to achieve optimal health of humans, animals, and the environment. By deducing experience and know-how from various fields: public health, ecological studies, and veterinary science, integrated strategies can be designed to reduce ECs'

DOI: 10.1201/9781032623801-1

related risks and tocomplement the well-being of organisms (Wang et al., 2024). Thisrequires combined efforts and multisectoral expertise at the local, national, and international levels to attain "One Health". The implementation of a "One Health" approach in tackling EC-related issues will enhance global health by adopting robust and fair environmental policies at regional and intercontinental levels (Wang et al., 2024).

1.2 SOURCES OF ECS

Due tothe pace of global development, the quality of the environment has been compromised,whichhas led to the introduction of several alternate pollutants in the environment. One such type includes ECs, which are now included as common ingredients in general household chemicals, food additives, cosmetics, and so forth. Overexploitation of these compounds has led to their presence in soil and water domains, consequently causing adverse effects on the health of living beings. Contaminants can get into the environment from point sources and non-point/diffuse sources (Figure 1.1).

Point sources mainly include distinct locations such as discharges from industries, treatment plants, and so on. Non-point sources includeland run-off, drainage, seepage, and so forth that originate from diffuse sources and are difficult to trace, control, and treat. As point sources, landfill sites are prime donors of ECs in surface

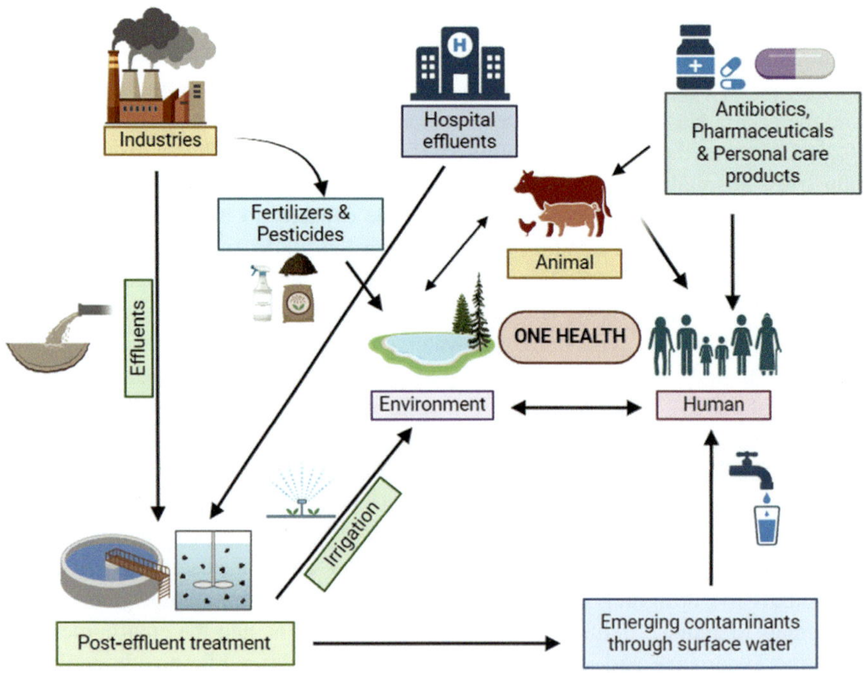

FIGURE 1.1 Sources and routes of emerging contaminants in ecosystem.

water, through which these contaminants leach and enter groundwater aquifers. The primary contributing non-point/diffuse sources of ECs is the application of manure and bio-solids to agricultural land. After permeating the environs via leaching, the contaminants percolate into the groundwater, pass through treatment plants, and finally reach the aquatic reservoirs,namely rivers. The prime routes through which ECs enter aquatic systems are the discharge of municipal wastewater, effluents from hospitals and industries, sewer overflow, leachate, and run-off from irrigational lands (Kumar et al., 2022). The source of ECs determines the attributes of the concerned substance such as its transmission and transformation, which depends on factors such as water solubility, temperature, polarity, pH, volatility, organic matter content, and so on (Lapworth et al., 2012) (Table 1.1).

1.3 CLASSIFICATION/TYPES OF ECS

The following types of ECs have been reported in water (surface water, groundwater) and soil: pharmaceuticals, pesticides, PCPs, plastics and microplastics, food additives, synthetic dyes, PFAS, and so on. Adetailed description of each type is described in the following section.

1.3.1 PHARMACEUTICALS

Recent industrialization has boosted the global pharmaceuticals industry with high-value products such as antibiotics, drugs, analgesics, steroids, and so forth that are mainly utilized for human and veterinary medication. These therapeutic agents are mainly utilized for the cure of infectious diseases in humans, plants, and animals. The easy access to these pharmaceutical agents through over-the-counter (OTC) or other means has resulted in the emergence of an underrated threat, namely antibiotic resistance. Once consumed, the antibiotics are partially metabolized and get excreted into the environment. The antibiotic residues persist in the surroundings owing to incomplete removal through wastewater treatment plants. Moreover, bio-accumulations of products like active pharmaceutical ingredients (APIs) also cause adverse effects on the ecosystem (Gogoi et al., 2018). Biomagnification of these organic compounds hasalso been reported in the food web (Singh et al., 2023). Although these compounds have been entering the ecosystem from a long time, the study and knowledge of their adverse effects have only recently been taken into consideration. The existence of antibiotics even at ng/L doses in the environment may affect the living population (Richardson and Kimura, 2017). The manure obtained from animal farms is loaded with these synthetic organic contaminants (SOCs) that are administered to increase production capacity. Application of such manure on fields can transfer the resistant microbes and genes indirectly to living beings via the food chain (Rathi et al., 2021). Commonly reported pharmaceuticals in water include: antibiotics; triclosan and diclofenac; gemfibrozil; hormones: 17-estradiol, 17β-estradiol, estrone, and estriol; antacids; steroids; antidepressants; analgesics; lipid-lowering drugs; and tranquilizers (Kamal et al., 2023).

TABLE 1.1
Some common types of ECs with examples, applications and environmental occurrence

S. no	EC Types	Examples	Applications	Environmental occurrence	References
1.	Pharmaceutical and Personal care products (PPCPs)	Benzophenone (BP); 2,4-dihydroxy benzophenone (BP-1); Miconazole; triclocarbon; carbamazepine; bisphenols; parbenes; phthalates; triclosan musk xylene; galaxolide; octocrylene	Cosmetics & Fragrance products	Seawater and river water; soil	Kung et al., 2018;
2.	Pesticides	Dichlorodiphenyltrichloro-ethane (DDT); Dichlorodiphenyldichloro-ethylene (DDE); Dicofol; Benzene hexachloride (BHC) Lindane; Pentachlorophenol; 1,4-Dichlorobenzene; Chloropropylate; Chlorbenzylate; Methoxychlor Heptachlor; Dimefox; Mipafox; Methyl parathion; Fenitrothion	Insecticides; Herbicides	Soil	Jayaraj et al., 2016;
3.	Antibiotics	Cephalosporin; Amoxicillin; Sulfamethoxazole; codein; Penicillin G; antipyrine; diclofenac; naproxen; ibuprofen; acetaminophen; Norfloxacin;paracetamol; metronidazole; Tetracycline; azithromycin;ketorolac; methylparaben Ciprofloxacin	Drugs	Ground water and agriculture	Chaturvedi et al., 2021; Rathi et al., 2021

4.	Synthetic dyes	Azo dye (Tartrazine; sudan 1; allura red)	Food products	Effluents	Monisha et al., 2023
5.	Food Additives	Triethyl citrate; terpineol; butylatedhydroxyanisole; camphor; butylatedhydroxytoluene; cis-3-hexenol; hexanoic acid; triacetin; menthol; parabens	Food preservative and ingredients	Ground water	Stuart et al., 2012
6.	Other ECs	Perfluorooctanesulfonate; organo-phosphate phthalate perfluorooctanoic acid; synthetic muscles; perchlorate and polycyclic siloxanes	Industries; cosmetics; plastic; industrial actuators; powered exoskeletons; fire work; airbag	Effluents and wastewater	Barcelo et al., 2012; Lei et al., 2015

1.3.2 Pesticides

Pesticides are essentially used during farming activities to protect crops and plants from pests, weeds, insects, and microbes. The group mainly encompasses four classes: insecticides, herbicides, fungicides, and bactericides. Thesemainly enter the environment through run-off from application sites, and on the basis of theirsolubility, the compound gets bio-accumulated in water, soil, and plants (Gani and Kazmi, 2017). Pesticides invade the human body through different routes,namelythrough the spraying of pesticides on plants and crops; by consumption of raw fruits and vegetables, or by direct skin contact. These compounds enter water streams by erosion, irrational chemical discharge, and aerial scattering. The unmanaged application of pesticides on land leads to their percolation in soil,where they get absorbed by the roots and as a result of biomagnification, their proportion increases. Biopesticides are targeted only toward specific organisms, whereas chemical pesticides have an impact on non-targeted organisms as well (Intisar et al., 2022). The most frequently encountered pesticides in the environment are: Endosulfan, hexachlorocyclohexane, and Dichlorodiphenyltrichloroethane (DDT) (Khan et al., 2022). Exposure to pesticides weakens the human lungs by enhancing epithelial permeability, lowering mucociliary clearance and restricting the activity of macrophages (Shetty et al., 2023).

1.3.3 Personal Care Products (PCPs)

With advancesin people'slifestyle, there has been an explosion of SOCs in the market. Thisclass basically involves chemicals used for beauty, health, nutrition, cleaning, cosmetics, perfumes, skin & hair related products, and so on. Regular use of such compounds has accelerated their distribution in the environment. Owing to the retention and bio-accumulative nature of theirbioactive ingredients, these compounds pose a threat to living beings (Juliano and Magrini, 2017).Some frequently detected PCPs are: polydimethylsiloxane, nano titanium dioxide (Fang et al., 2015), UV filters, and triclosan (Jagini et al., 2019);fragrances pollutants like ethanol, pinene, acetone, limonene, and acetaldehyde galaxolide; and preservativeslike diethyl phthalate, nano zinc oxide, benzophenone, parabens, and methoxycinnamate (Mabitla et al., 2023). Similarly, polycyclic musks (PCMs), which belong to the class of synthetic fragrance compounds,areubiquitously applied in PCPs (body lotions, perfumes, toothpastes, soaps, shampoo, cosmetics) and household cleaners (laundry detergents, fabric softeners, room fresheners) to render a pleasing odor (Liu et al., 2021). The presence of UV-filters/sunscreens, such as BP-4 (benzophenone-4) and PBSA (phenylbenzimidazole-5-sulphonic acid), has been reported in wastewater from treatment plants (Wilkinson et al., 2017).

1.3.4 Plastic and Microplastics

Plastics as an entity include synthetic and semi-synthetic materials that can be molded into any shape. They are further categorized into various types based on the building material and side-chain (acrylics, polyesters, silicones), production process (condensation, poly-addition), and physical properties (tensile strength, density) (Mafuta et al., 2021). Afterusage, plastics are either sorted for additional applications (plastic pellets, flakes) as a step towardtheirmanagement, but ordinarily theyaredisposed of in the open environment or incinerated (Geyer et al., 2020). Plastic particles ofsize <5mm are termed microplastics (MPs). This group is further categorized into primary and secondary MPs depending on the source. Primary MPs are derived from common household items such as toothpaste, skin cleansing products, plastic pellets, and so forth. In contrast, secondary MPsare generated from the disintegration of large plastics as a result of physical, chemical, and biological processes (Laskar et al., 2019). MPs have been reported to cause oxidative stress, intestinal damage, and to lower the nutrient supply in soil-dwelling organisms (Zhang et al., 2022b). Their ubiquitous presence in the environs, help these particles to accumulate in the food chain and serve as carriers for toxic chemicals and pathogens to invade the living body (Angnunavuri et al., 2020).These contaminants are reported to affect the biological organization at successive levels namely (sub)cellular, individual, population, and ecosystem, and to cause effects such as: reducingorgan regeneration capacity, causing apoptosis, alteringmetabolism, and so on (Kumar et al., 2021). MPs also trigger the soil structure by lowering its nutrient and water retention capacity,which hasgrave repercussions forcrop growth (Kumar et al., 2023).

1.3.5 Food Additives

Food additives are basically used to augment the physical appearance, such as the color, taste, and shelf life, of food. The European Union (EU) has categorized food additives into the following types: sweeteners, preservatives, colors, antioxidants, pH adjusters, emulsifiers, thickeners, gels,and so forth. Based on their origin, they are further classified as natural food additives, modified natural additives, and synthetic food additives (Paşca et al., 2018). Other than economic and commercial benefits, the unregulated consumption of food additives has caused adverse health impacts such as cancer, allergic problems, and so forth. In comparison to other age groups, children have been found to be at highest risk of such impacts. Studies have reported that long-term consumption of food additives may cause attention deficit hyperactivity disorder (ADHD) among children (Pluym et al., 2023). Similarly, emulsifiers such as polysorbate, carboxymethyl cellulose can affect gut microbiota (causing ulcerative colitis), and alter immune, inflammatory response and mental health. Synthetic antioxidants such as butylatedhydroxytoluene (BHT) and butylatedhydroxyanisole (BHA), which are used as preservatives in fat-rich products like vegetable oils, animal fats, dairy products, are reported to impart carcinogenic effects, cause skin allergies, and raisecholesterol levels in blood (Jideani et al., 2021).Another food additive of prime concern is BisphenolA (BPA), asynthetic chemical that is widely utilized to manufacture polycarbonate plastics and epoxy resins. These polycarbonate plastics are mainly used to store foods and beverages. Epoxy resinsare usedprimarilyto protect the inside of food cans. BPA mainly enters the human body through the consumption of packaged food. BPA is responsible for causing critical health conditions. It is a potent endocrine-disrupting chemical that affects hormonal and enzyme synthesis. Moreover, it also alters the nervous, reproductive, and immune systems, and can also induce carcinogenesis (Lalani et al., 2024).

1.3.6 Synthetic Dyes

Dyes find applications in various sectors like the food, paper, cosmetics, textiles, and leather industries. The shortage of natural dyes has led to increased dependency on synthetic dyes, which are now being produced at a rate of 8 x 10^5 tons per year globally. These synthetic dyes are either soluble (acidic, basic, or coordinated with metal ions) or insoluble (azodyes, sulfur dye) (Slama et al., 2021). Major contaminants present in these dye-rich effluents are: bleaching agents, acids and bases, salts, and heavy metals like arsenic, chromium, and cadmium. These contaminated effluents affect vegetation by reducing photosynthesis and the ability to fix nitrogen and carbon dioxide by plants (Chandanshive et al., 2018). The presence of dyes in surface water reduces the concentration of dissolved oxygen, thereby affecting aquatic life. Also, the metal ions present in these dyes gets boundto fish gills, and when humans consume these fish, they get exposed to lethal health effects. Other health implications caused by subjection to dyes include: irritation of eyes and skin, and allergic reactions (Intisar et al., 2023).

1.3.7 PFAS

PFAS represents a group of synthetic chemicals characterized by the presence of a chain of linked carbon and fluorine atoms. As the carbon-fluorine bond is the strongest, these compounds do not degrade easily in the environment. They are popular for their heat-, water-, stain-, oil-, and grease-resistantproperties. Perfluorooctanesulfonic acid (PFOS), perfluorohexanesulfonate (PFHxS), perfluorooctanoic acid (PFOA), GenX chemicals, perfluorononanoic acid (PFNA), perfluorodecanoic acid (PFDA), and perfluorobutanesulfonate (PFBS) are some of the common PFAS known. PFAS serve as components of everyday products: food packaging, non-stick cookware, fire-fighting foams, textiles, and water-repellant coatings (Li et al., 2024). Due to their ubiquitous applications, PFAS have been detected in surface waters, groundwater, drinking water, sediments, soil, and air (Evich et al., 2022), and render multiple adverse outcomes on living beings and the environment. These include obesogenic effects in children; impact of steroidogenesis in the gonads, affecting the male reproductive system (Daugherty et al., 2023); high cholesterol; elevated blood pressure; reduced vaccine efficacy; kidney and testicular cancer; hormonal imbalance; andhypertension (Eze et al., 2024). Owing to these serious health effects, PFAS are regarded as contaminants of emerging concern by the US Environmental Protection Agency (US EPA) (Podder et al., 2021).

1.4 IMPACT OF ECS ON ENVIRONMENT AND HUMAN HEALTH

ECs are reported to cause deleterious effects to flora, fauna, and human beings. Attributes such as bioaccumulation, biomagnification, persistence, and mobility account for their unregulated presence in the biosphere (Yadav et al., 2021). The long-term persistence of ECs in the environment is responsible for their movement in various matrices such as air, water, and soil, thereby posing serious risks to the associated biota (Figure 1.2).

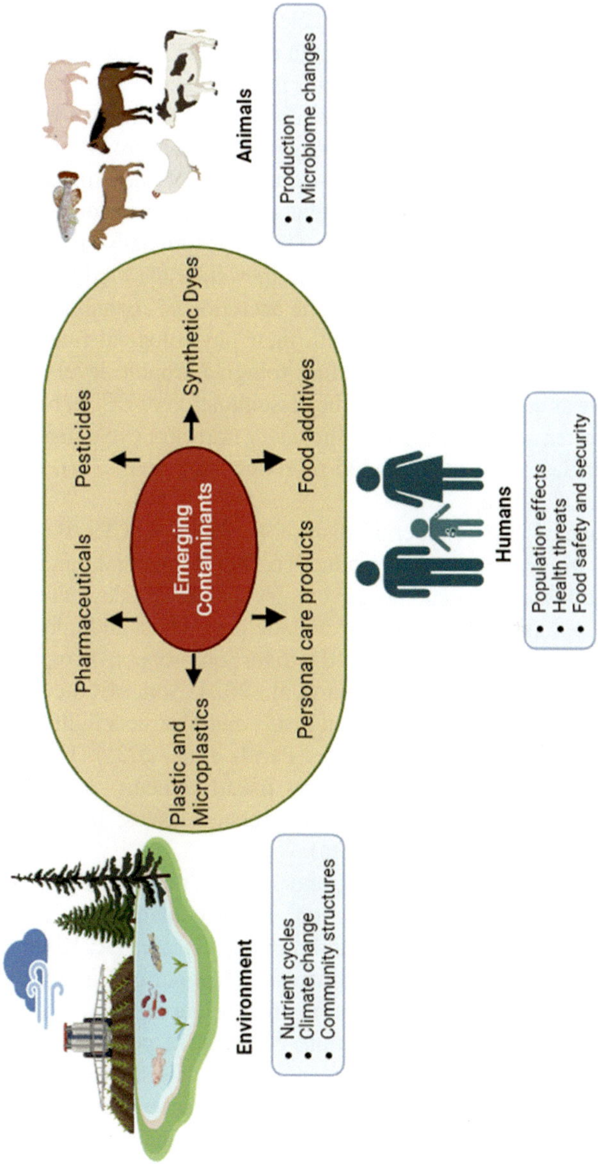

FIGURE 1.2 Types of emerging contaminants and theireffects on the environment, animals, and human health.

The presence of ECs in **soil** interrupts the physiological, behavioral, and reproductive abilities of soil organisms (earthworms, nematodes, arthropods). These pollutants alter the metabolic activities of soil microbes,which further affect the soil fertility, nutrient cycling, and cause disparity in the distribution of essential elements responsible for maintaining soil productivity (Khan et al., 2023a). PPCPs, pesticides, PFAS, and MPs have been commonly detected in soil ecosystems and have caused shifts in microbial community structures (Ma et al., 2023). This affects soil's ability to inhibit pathogens and sustain its properties in the presence of stressors and varying climatic conditions. Similar effects are observed when plants get exposed to ECs. Soils contaminated with ECs encounter obstructed plant growth and development that influence overall plant biodiversity. An outcome of such incidence involves a decrease in the efficiency of processes related to nutrient absorption that impacts plant health (Moreno-Jiménez et al., 2022). The bioaccumulative nature of ECs helps them to persist in plant tissues, through which they enter the food chain (Wang et al., 2022). This feature is also responsible for the accretion of contaminants in tissues of soil-dwelling organisms that interrupt with their physiological functions and complicate the soil food webs, thereby leading tohigher trophic levels. Such infected organisms behave as carriers, leading to the dissemination of ECs in biomes (Hopkins et al., 2023). Further, some ECs disrupt the water transport cycle inside plant tissues resulting in poor development, modified reproductive patterns, and imperiled plant strength (Wang et al., 2022).

The occurrence of ECs (endocrine disruptors, POPs, PPCPs, PFAS) in **surface water** systems triggers several reproductive, developmental, and physiological changes in aquatic organisms and poses ecological risks (Krithiga et al., 2022). MPs are typically treated as food particles by aquatic organisms, and upon ingestion can damage organs, cause blockages, inhibit the digestion process, disrupt the endocrine system, and suppress immunity (Harmon et al., 2024).Some ECs cause foaming in water bodies, which decreases the oxygen levels and obstructs light penetration that poses serious consequences on marine life (Parida et al., 2021). Therelease of such toxicants intothe open environment provokes risks like neuro- and nephro-toxicity, cancer, liver damage, dermal toxicity, heart-related issues, immunotoxicity, and genotoxicity in humans (Singh et al., 2024).

ECs like PAHs, pesticides, parabens, and so forth thrive in the **air** and majorly affect the respiratory system via inhalation. These contaminants are also responsible for causing dermal, heart-related, and immune diseases (Al-Kindi et al., 2020). Studies have reported that these air pollutants can trigger cell injury by the formation of reactive oxygen species (ROS) in organelles like mitochondria, endoplasmic reticulum, and cell membrane (Zhang et al., 2022a). The pollutants in a vapor state develop threatening effects on life-forms due to their persistence, toxicity, and long-range transport (Barroso et al., 2019). The fumes released from coking plants are filled with a mixture of aromatic compounds (chloro and nitro-phenols) that may damage DNA and cause lipid peroxidation, thereby contributing to increased risks (Wang et al., 2024). Bioaerosol is another important carrier of ECs that spread across transboundary regions and can cause health complications such as: allergies, respiratory diseases, neurological disorders, and can even lead to life-threatening diseases like cancer (King et al., 2020) (Table 1.2).

TABLE 1.2
Type of ECs and their effects on OneHealth

S. no	Type of ECs	Effect on	Toxicity	References
1.	Microplastics	*Crassostreagigas; Chlorella pyrenoidosa; Daphnia magna*	Gene expression for enzymes (antioxidant); reduced growth; reduced photosynthetic activity; mortality	Revel et al., 2020;Mao et al., 2018
2.	Pesticides	Humans	Immunotoxicity, effect on growth, damage liver; kidney; enhance birth defects, cancer; neurotoxicity	Jayaraj et al., 2016
3.	Pharmaceuticals and Personal care products	Humans	Hypersensitivity, dermatitis, skin sensitization and allergic contact, lymphomatoid, effect on autoimmunity and thyroid homeostasis	Skarha et al., 2019
4.	Plastic	Humans	Endocrine disruption; hepatocellular carcinomas in rodents; ROS production; apoptosis effect on endocrine function	Benjamin et al., 2015
5.	Antibiotics	Humans	Diarrhoea; hypersensitivity type I & III; fever perioperative anaphylaxis; maculopapular; urticaria and serum sickness-typereaction	Hutchings et al., 2019;
6.	Synthetic dyes	Humans, Plants	Carcinogenic and mutagenic effects, oxidative stress in plants; inhibit photosynthesis and carbon-dioxide assimilation rate inhibit biodegradability by aerobic microorganism and toxic effect on food chain	Manzoor and Sharma, 2020;
7.	PFAS	Humans, animals, plants	Liver effects, increased serum cholesterol, lower vaccination response, thyroid disease, cancer (testicular, kidney), reproductive effects, ROS generation, damage to plant cell structure and functions	ITRC 2023; Li et al., 2022

1.5 DETECTION AND REMOVAL TECHNOLOGIES FOR ECS

1.5.1 Detection

The detection of such complicated compounds in open environments requires the use of high-end instrumentation for accurate measurement even at trace levels (Table 1.3). This demands a high range of selectivity and sensitivity to monitor the presence of desired ECs in a mixture of compounds. The detection process starts with a sample preparation method. As the collected samples contain impurities and are not in suitable form to be directly analyzed, they are processed through a series of steps involving pH adjustments, extraction, filtration, cleaning, and so on. This is required to ensure that the desired compound is suspended in a suitable matrix and is present at a specific concentration that enhances its detection (Khan et al., 2023b). Extraction is performed using solid-liquid, solid-phase, liquid-liquid, ultrasound-assisted, and supercritical fluid methods (Martín-Pozo et al., 2019). Post-sample preparation and techniques like chromatography coupled with mass spectrometry (MS) areusually employed for quantification. Other detection methods like capillary electrophoresis, sensor-based techniques, and nuclear magnetic resonance spectroscopy areused for comprehensive analysis (non-target compounds) (Ismail and Mokhtar, 2020).

TABLE 1.3
Methods for detection of ECs in environmental matrices

EC	Types of EC	Detection methods	Matrix	Reference
Pharmaceuticals	Antibiotics, Hormones	HPLC–MS/MS	Sludge	Kim et al., 2019
	Endocrine active compounds	LC-QTOF/MS	Lakewater	Guyader et al., 2019
	-	FTIR	Wastewater	Quintelas et al., 2019
PCPs	Parabens, Triclosan, bisphenol A	GC-MS with derivatization	Water samples	Arismendi et al., 2019
PFAS	-	LC-HRMS	Food	Nobile et al., 2020
	-	NMR	Environmental samples	Camdzic et al., 2021
Food additives	Artificial sweeteners	LC-HRMS/MS	Surface waters	Osorio et al., 2018
Pesticides	-	LC- QTOF/MS	Groundwater and stormwater run-off	Pinasseau et al., 2019
Microplastics	-	FTIR	Air	Dris et al., 2017

a) **Chromatography**
Chromatography involves the separation of compounds based on their partition coefficient between stationary and mobile phases. Gas chromatography (GC) consists of a liquid stationary phase absorbed on solid support and a mobile gas phase (helium or nitrogen). The sample is vaporized, mixed with the gaseous mobile phase, and passed through the stationary phase. GC is basically used to determine the presence of non-polar, thermostable, and volatile ECs in samples. The combination of GC with MS enhances separation capacity, offering high selectivity and sensitivity. GC-MS has been used for identification of compounds like PCPs, pesticides, andanti-inflammatory drugs (Carmona and Pico, 2018). However, the technique requires an additional derivatization reaction that makes the process tedious and can even result in loss of the desired compound (Coronado-Apodaca et al., 2023). Liquid chromatography with MS (LC-MS) is another common detection tool to analyze polar and thermolabilecompounds, namelysteroids, pesticides, dyes, phthalates, pharmaceuticals, and so forth(Khan et al., 2023b). GC or LC can be used with high-resolution mass spectrometry (HRMS) to enable selective and sensitive screening of organic pollutants (Coronado-Apodaca et al., 2023). A few other techniques like high-performance liquid chromatography (HPLC) and ultra-high-performance liquid chromatography (UHPLC) can also be engaged for detection and quantification of compounds like parabens, hormones, and anthelmintics in complex sample matrices (water, sediment) (Li et al., 2020).

b) **Spectroscopy**
Advanced spectroscopic techniques like nuclear magnetic resonance (NMR) and Fourier-transform infrared spectroscopy (FTIR) are also employed for the identification of ECs in terms of theirstructure and concentration. PFAS and nano plastics have been detected using these methods. NMR can differentiate between structurally similar compounds and can quantify trace analytes in different sample types. FTIR also helps in the identification and measurement of the contaminant. It is a rapid technique with no reagents requirement, allows reuse of sample post-analysis, and provides online monitoring of results (Ismail and Mokhtar, 2020). However, NMR is less sensitive, time-consuming, and costly than MS-based techniques, which restrict its application (Camdzic et al., 2021). Other less common spectroscopy methods involve the use of fluorescence spectroscopy (FS) and near-infrared spectroscopy (NIRS). In FS, the molecules are excited on absorption of electromagnetic radiation and while returning to the ground state, the energy is released in the form of photons. Similarly in NMR, the sample molecules are exposed to a particular wavelength (780–2526 nm), during which some rays will be absorbed and some will be scattered and reflected. These techniques have been reported for the detection of aflatoxins (Tao et al., 2018).

1.5.2 Removal Techniques

The complex nature of ECs restricts theirremoval from environmental samples through simple treatment technologies. Various refined technologies involving different approaches have been in practice to achieve a complete breakdown of these stubborn pollutants. Some of these technologies are listed below:

a) **Conventional treatment:** These methods involve stages like coagulation, flocculation, sedimentation, and disinfection. This helps to reduce turbidity, color, microbial load, suspended solids, and organic content (BOD, COD)in wastewater and drinking water (Khan et al., 2023b). However, the complex structure, low concentration, and persistent nature of ECs limit their removal through this method. Also, sometimes the disinfectants used in the process react with these compounds and form toxic by-products that are even more difficult to treat (Radwan et al., 2022).
b) **Biological treatment:** This involves approaches like the activated sludge process (ASP) and membrane bioreactors (MBR). In ASP, the organic compounds are brokendown into simple fractions using aerobic/anaerobic microbes by regulating temperature and COD (Khan et al., 2023b). But because of issues like sludge production, energy requirements, and so forth, the method is not usually applied (Hussain and Dubey, 2014). On the other hand, MBR provides better results than ASP due to its high efficacy and cost-effectiveness in treating complex compounds. Thismethod has achieved high BOD and COD reduction in pharmaceutical effluents (Hamon et al., 2018).
c) **Advanced Oxidation Process (AOP):** The process involves the production of highly reactive free radicals, commonly hydroxyl ions ,from water by activating a semiconductor (TiO_2) with artificial light or sunlight to achieve a complete breakdown of ECs. Hydroxyl ions are generated by methods like sonolysis, ozonation, photolysis, photocatalysis, Fenton, and Photo-Fenton (Khan et al., 2023b). AOPs mineralize the contaminants into carbon dioxide, water, and inorganic salts and do not release any toxic by-products in the process. Pollutants such as pesticides, antibiotics, inflammatory drugs, and industry-based chemicals can be treated by this technique. The efficiency of AOPs can be enhanced by working on certain factors such as pollutant type, sample matrix, and energy demand, developing reusable catalysts, and lowering by-product formation (Li et al., 2023).
d) **Sorption based treatment:** It is defined as the process in which a compound (sorbate) gets fixed or captured by another substance (sorbent).Depending on whether the sorbent is a solid or liquid material, the sorption process is named adsorption and absorption, respectively (Khan et al., 2023b).The separation of contaminants is mainly achieved through chemical interactions, such as van der Waals, hydrophobic, ion-exchange, hydrogen bonding, π-π interactions, and electrostatic attractions between the sorbent and analyte being studied (Nayyar et al., 2022). The phenomenon has turned out to be a promising approach for the removal of ECs from environmental samples. Also, the replacement of synthetic adsorbents by natural ones prepared from

lignocellulosic biomass has encouraged the applicability of the process in terms of its cost-effectiveness (Singh et al., 2021). Sorbent material includes zeolites, metal organic frameworks, activated carbon, nanocomposite, graphene, polymer, and biochar and carbon nanotubes (Lima et al., 2019). Among these, biochar has gained significant popularity because of its characteristic features like feedstock availability, porosity, large surface area, presence of functional groups, and environmental friendliness. Emerging pollutants like pharma compounds (paracetamol, tetracycline, diclofenac), heavy metals, fertilizers, and dyes have been successfully treated through this method (Khan et al., 2023b).

e) **Bioremediation:** The technique employs living organisms to achieve complete degradation of ECs. The involvement of more than one organism, that is, consortia in the process, improves its efficiency and efficacy. Genetically modified organisms (GMOs) are currently under study to be used as potential candidates for bioremediation. Various factors such as temperature, the presence or absence of oxygen, nutrients, and so on affect the process. The method offers an environmentally friendly, cheap, and sustainable alternative for the effective treatment of potent pollutants. Depending upon the organism involved in the remediation, the process is further termed microbial remediation (bacteria), myco-remediation (fungi), phyto-remediation (plants), and phycoremediation (algae) (Shanmuganathan et al., 2023).

1.6 MANAGEMENT OF ECS THROUGH ONE HEALTH APPROACH

The worldwide increase inproduction and consumption of ECs needs to be controlled through the adoption of stringent regulatory measures to safeguard the health of wildlife, humans, and the environment. Regulations on emissions and discharge limits should be imposed to restrict theirimpact on the biosphere. International agencies like the US EPA, the EU, the World Health Organization, and the United Nations Environment Programme (UNEP) havetakenvarious actions to regulate the maximum contaminants level in drinking water and water bodies to ensure the protection ofpublic health. These actions also cover regulating pollutant discharge levels from the point source to meet surface water quality standards (Li et al., 2024). Their collaborative efforts in formulating guidelines, recommendations, and implementing policies assist in encouraging the adoption of standards to mitigate the impact of ECs across transnational boundaries. Such a consolidated approach helps developing countries to embrace suitable strategies for monitoring and managing these complex pollutants (Puri et al., 2023). Proper waste management practices should be adopted to dispose of these pollutants in a sustainable manner. Industries should adopt sustainable practices (green technologies) to manage waste discharge from their processes, in order to reduce their ecological footprint and maintain environmental quality (Song et al., 2022). Moreover, education and public awareness can assist in decreasing the knowledge gap and hence the use of such compounds in environment. Raising awareness through campaigns, workshops, training, and outreach programs on ECs and theirimplications forliving life can assist in guiding the population to reduce its usage. An interdisciplinary approach involving insights from various scientific disciplines can facilitate individuals' understanding the mechanism of how ECs affect mankind and in developing innovative solutions for theirdetection, removal, and management (Dingemans et al., 2020).Furthermore, the integration ofpolicymakers and the scientific community promotes the sharing of knowledge and practical evidence of scientific concepts that can later be framed as regulatory policies (Trein et al., 2023).Also, discussions at the local, national, and global level can help in learning about initiatives undertaken in different regions to counteract the use of ECs, thereby promoting green development (Wang et al., 2024).These measures outline the need forinternational cooperation through partnerships and networking in terms of the exchange of ideas, information, resources, and technologies to combat the issues imposed by ECs and to protect One Health across borders (Khan et al., 2022).

1.7 CONCLUSION AND FUTURE PROSPECTS

The chapter comprehensively illustrates the types of ECs, and theirsources and toxicity impacts like neurotoxicity, reproductive toxicity, immunotoxicity, cytotoxicity, and genotoxicity on the OneHealthsystem. The concomitant effects arising from the increased overexploitation of ECs is wellknown among thc scientific community, and therefore a consolidated multisector approach is required to impede the threatening repercussions of these contaminants forflora, fauna, and human health, thus fostering the OneHealth concept. Also, a deep understanding based on the interaction between these contaminants and environmental factors is imperative to develop appropriate treatment technologies. The formulation as well as the implementation of strict policies at the national and international level will assist in the effective management of ECs worldwide.

CONFLICTS OF INTEREST

The authors declare that they have no conflicts of interest.

ACKNOWLEDGMENTS

The authors are thankful to Director, CSIR-Indian Institute of Toxicology Research (CSIR-IITR), Lucknow (India), for his valuable encouragement and support. This is CSIR-IITR manuscript no. IITR/SECC-PME/MSS/2025/037. The figures in the manuscript have been created using Biorender.com.

REFERENCES

Al-Hazmi, H.E., Mohammadi, A., Hejna, A., et al. 2023. Wastewater treatment for reuse in agriculture: Prospects and challenges. *Environmental Research*, 236, 116711.

Al-Kindi, S.G., Brook, R.D., Biswal, S., Rajagopalan, S., 2020. Environmental determinants of cardiovascular disease: Lessons learned from air pollution. *Nature Review Cardiology*, 17(10), 656–672.

Angnunavuri, P.N., Attiogbe, F, Mensah, B., 2020. Consideration of emerging environmental contaminants in Africa: Review of occurrence, formation, fate, and toxicity of plastic particles. *Scientific African*, 9, 00546.

Arismendi, D., Becerra-Herrera, M., Cerrato, I., Richter, P., 2019. Simultaneous determination of multiresidue and multiclass emerging contaminants in waters by rotating-disk sorptive extraction–derivatization-gas chromatography/mass spectrometry. *Talanta,* 201, 480–489.

Barcelo, D., 2012. *Emerging Organic Contaminants and Human Health.* Springer.

Barroso, P.J., Santos, J.L., Martín, J., et al. 2019. Emerging contaminants in the atmo sphere: Analysis, occurrence and future challenges. *Critical Reviews in Environmental Science and Technology,* 49, 104–171.

Bayabil, H.K., Teshome, F.T. and Li, Y.C., 2022. Emerging contaminants in soil and water. *Frontiers in Environmental Science,*10, 873499.

Benjamin, S., Pradeep, S., Josh, M.S., Kumar, S. and Masai, E., 2015. A monograph on the remediation of hazardous phthalates. *Journal of Hazardous Materials*, 298, 58–72.

Camdzic, D., et al. 2021. Total and class-specific analysis of per- & polyfluoroalkyl substances in environmental samples using nuclear magnetic resonance spectroscopy. *Journal of Hazardous Materials Letters*, 2, 100023.

Carmona, E.and Pico, Y., 2018. The use of chromatographic methods coupled to mass spectrometry for the study of emerging pollutants in the environment. *Critical Reviews in Analytical Chemistry,* 48, 305–316.

Chandanshive, V.V., Kadam, S.K., Khandare, R.V., et al. 2018. In situ phytoremediation of dyes from textile wastewater using garden ornamental plants, effect on soil quality and plant growth. *Chemosphere*, 210, 968–976.

Chaturvedi, P., Shukla, P., Giri, B.S., Chowdhary, P., et al. 2021. Prevalence and hazardous impact of pharmaceutical and personal care products and antibiotics in environment: A review on emerging contaminants. *Environmental Research*, 194, 110664.

Coronado-Apodaca, K.G., Rodríguez-De Luna, S.E., et al. 2023. Occurrence, transport, and detection techniques of emerging pollutants in groundwater. *Methods X*, 10, 102160.

Daugherty, S., et al. 2023. Legacy and emerging Perfluoroalkyl and Polyfluoroalkyl substances regulate Steroidogenesis in the Male Gonad. *Endocrinology,* 164, bqad142.

Dingemans, M.M., et al. 2020. Responsible water reuse needs an interdisciplinary approach to balance risks and benefits. *Water,* 12, 1264.

Evich, M.G., et al. 2022. Per-and polyfluoroalkyl substances in the environment. *Science,* 375, eabg9065.

Eze, C.G., Okeke, E.S., Nwankwo, C.E., et al. 2024. Emerging contaminants in food matrices: An overview of the occurrence, pathways, impacts and detection techniques of per-and polyfluoroalkyl substances. *Toxicology Reports*.

Fang, Q., Shi, X., Zhang, L., et al. 2015. Effect of titanium dioxide nanoparticles on the bioavailability, metabolism, and toxicity of pentachlorophenol in zebrafish larvae. *Journal of Hazardous Materials*, 283, 897–904.

Gani, K.M. and Kazmi, A.A. 2017. Contamination of emerging contaminants in Indian aquatic sources: First overview of the situation. *Journal of Hazardous, Toxic, and Radioactive Waste*, 21(3), 04016026.

Geyer, R., 2020. Production, use, and fate of synthetic polymers. In Letcher T.M. (Ed), *Plastic waste and recycling* (pp. 13–32).Academic Press.

Gogoi, A., Mazumder, P., Tyagi, et al. 2018. Occurrence and fate of emerging contaminants in water environment: A review. *Groundwater for Sustainable Development*, 6, 169–180.

Guyader, M.E., Warren, L.D., et al. 2019. Prioritizing potential endocrine active high resolution mass spectrometry (HRMS) features in Minnesota lakewater. *Science of the Total Environment,* 670, 814–825.

Hamon, P., Moulin, P., Ercolei, L. and Marrot, B., 2018. Oncological ward wastewater treatment by membrane bioreactor: Acclimation feasibility and pharmaceuticals removal performances. *Journal of Water Process Engineering,* 21, 9–26.

Harmon, S.M., Chen, Q., Ma, C., et al. 2024. The effects of microplastic pollution on aquatic organisms. In Zeng E.D. (Ed), *Microplastic contamination in aquatic environments* (pp. 355–379). Elsevier.

Hopkins, K.E., McKinney, M.A., Saini, A., et al. 2023. Characterizing the movement of per and polyfluoroalkyl substances in an avian aquatic-terrestrial food web. *Environmental Science and Technology,* 57, 20249–20260.

Hussain, A. and Dubey, S.K., 2014. Specific methanogenic activity test for anaerobic treatment of phenolic wastewater. *Desalinationand Water Treatment,* 52, 7015–7025.

Hutchings, M., Truman, A. and Wilkinson, B., 2019. Antibiotics: Past, present and future. *Current Opinionin Microbiology,* 51, 72–80.

Interstate Technology & Regulatory Council (ITRC). 2023. *PFAS Technical and Regulatory Guidance Document and Fact Sheets PFAS-1.* Washington, D.C.

Intisar, A., Ramzan, A., Hafeez, S., et al. 2023. Adsorptive and photocatalytic degradation potential of porous polymeric materials for removal of pesticides, pharmaceuticals, and dyes-based emerging contaminants from water. *Chemosphere*, 336, 139203.

Intisar, A., Ramzan, A., Sawaira, T., et al. 2022. Occurrence, toxic effects, and mitigation of pesticides as emerging environmental pollutants using robust nanomaterials–A review. *Chemosphere*, 293, 133538.

Ismail, W.N.W. and Mokhtar, S.U., 2020. *Various methods for removal, treatment, and detection of emerging water contaminants*. IntechOpen.

Jagini, S., Konda, S., Bhagawan, D. and Himabindu, V., 2019. Emerging contaminant (triclosan) identification and its treatment: a review. *SN Applied Sciences*, 1, 1–15.

Jayaraj, R., Megha, P. and Sreedev, P., 2016. Organochlorine pesticides, their toxic effects on living organisms and their fate in the environment. *Interdisciplinary Toxicology*, 9(3–4), 90–100.

Jideani, A.I., Silungwe, H., Takalani, T., et al. 2021. Antioxidant-rich natural fruit and vegetable products and human health. *International Journal of Food Properties*, 24(1), 41–67.

Juliano, C. and Magrini, G.A., 2017. Cosmetic ingredients as emerging pollutants of environmental and health concern. A mini-review. *Cosmetics*, 4(2), 11.

Kamal, N., Sindhu, R. and Bhargava, P.C., 2023. Biodegradation of emerging organic pollutant gemfibrozil: Mechanism, kinetics and pathway modelling. *Bioresource Technology*, 374, 128749.

Khan, A., Jie, Z., Wang, J., et al. 2023a. Ecological risks of microplastics contamination with green solutions and future perspectives. *Science of The Total Environment*, 899, 165688.

Khan, N.A., López-Maldonado, E.A., Majumder, A., et al. 2023b. A state-of-art-review on emerging contaminants: Environmental chemistry, health effect, and modern treatment methods. *Chemosphere*, 344, 140264.

Khan, S., et al. 2022. Emerging contaminants of high concern for the environment: Current trends and future research. *Environmental Research,* 207, 112609.

Kim, S., Chen, J., Cheng, T., Gindulyte, A., He, J., He, S., Li, Q., Shoemaker, B.A., Thiessen, P.A., Yu, B. and Zaslavsky, L., 2019. PubChem 2019 update: improved access to chemical data. *Nucleic Acids Research*, 47(D1), D1102–D1109. https://doi.org/10.1093/nar/gky1033

King, M.D., Lacey, R.E., Pak, H., et al. 2020. Assays and enumeration of bioaerosols-traditional approaches to modern practices. *Aerosol Science and Technology*, 54(5), 611–633.

Krithiga, T., Sathish, S., Renita, A.A., et al. 2022. Persistent organic pollutants in water resources: Fate, occurrence, characterization and risk analysis. *Science of The Total Environment,* 831, 154808.

Kumar, M., Chen, H., Sarsaiya, S., et al. 2021. Current research trends on micro-and nano-plastics as an emerging threat to global environment: A review. *Journal of Hazardous Materials*, 409, 124967.

Kumar, R., Qureshi, M., Vishwakarma, D.K., et al. 2022. A review on emerging water contaminants and the application of sustainable removal technologies. *Case Studies in Chemical and Environmental Engineering*, 6, 100219.

Kumar, V., Singh, E., Bhargava, P.C., et al. 2023. Micro-and nano-plastics (MNPs) as emerging pollutant in ground water: Environmental impact, potential risks, limitations and way forward towards sustainable management. *Chemical Engineering Journal*, 459, 141568.

Kung, T.A., Lee, S.H., Yang, et al. 2018. Survey of selected personal care products in surface water of coral reefs in Kenting National Park, Taiwan. *Science of The Total Environment*, 635, 1302–1307.

Lalani, A., Pouyani, N., Askari, A., et al. 2024. Food additives, benefits, and side effects: A review article. *Journal of Chemical Health Risks*, 1(1), 1–10.

Lapworth, D.J., Baran, N., Stuart, M.E. and Ward, R.S., 2012. Emerging organic contaminants in groundwater: A review of sources, fate and occurrence, *Environmental Pollution,* 163, 287–303.

Laskar, N. and Kumar, U., 2019. *Plastics and Microplastics: A Threat to Environment*, vol. 14, Environmental Technology & Innovation, 100352.

Lei, M., Zhang, L., Lei, J., et al. 2015. Overview of emerging contaminants and associated human health effects. *BioMed Research International*, 2015, p. 12.

Li, J., Sun, J. and Li, P., 2022. Exposure routes, bioaccumulation & toxic effects of per-and polyfluoroalkyl substances (PFASs) on plants: A critical review. *Environment International*, 158, 106891.

Li, L., et al. 2023. Recent progress on ruthenium-based electrocatalysts towards the hydrogen evolution reaction. *Catalysts,* 13, 1497.

Li, X., Shen, X., Jiang, W., Xi, Y. Li, S., 2024. Comprehensive review of emerging contaminants: Detection technologies, environmental impact, and management strategies. *Ecotoxicology and Environmental Safety,* 278, 116420.

Li, Y., Gan, Z., Liu, Y., et al. 2020. Determination of 19 anthelmintics in environmental water and sediment using an optimized PLE and SPE method coupled with UHPLC-MS/MS. *Scienceof The Total Environmental,* 719, 137516.

Lima, D.R., Gomes, A.A., Lima, E.C., et al. 2019. Evaluation of efficiency and selectivity in the sorption process assisted by chemometric approaches: Removal of emerging contaminants from water. *Spectrochimca Acta Part A Molecularand Biomolecular Spectroscopy,* 218, 366–373.

Liu, J., Zhang, W., Zhou, Q., et al. 2021. Polycyclic musks in the environment: A review of their concentrations and distribution, ecological effects and behavior, current concerns and future prospects. *Critical Reviews in Environmental Science and Technology*, 51(4), 323–377.

Ma, Y., Yang, K., Yu, H., et al. 2023. Effects and mechanism of microplastics on organic carbon and nitrogen cycling in agricultural soil: A review. *Soil Use Management,* 40, 12971.

Mabitla, S.S., Chiwaye, N. and Daramola, M.O., 2023. Assessment of chemical constituents of personal care products & their environmental implications: A case of South Africa. *Materials Today: Proceedings*, 105, 209–212.

Machado, K.C., Grassi, M.T., Vidal, C., et al. 2016. A preliminary nationwide survey of the presence of emerging contaminants in drinking and source waters in Brazil. *Science of the Total Environment*, 572, 138–146.

Mafuta, C., Baker, E., Rucevska, I., et al. 2021. Drowning in plastics: Marine litter and plastic waste vital graphics. UNEP, Secretariats of the Basel, Rotterdam and Stockholm Secretariats and GRID-Arendal.

Manzoor, J. and Sharma, M., 2020. Impact of textile dyes on human health and environment. In Wani, K.A., Jangid, N.K., and Bhat, A.R. (Eds), *Impact of textile dyes on public health and the environment* (pp. 162–169). IGI Global.

Mao, Y., Ai, H., Chen, Y., et al. 2018. Phytoplankton response to polystyrene microplastics: Perspective from an entire growth period. *Chemosphere,* 208, 59–68. https://doi.org/10.1016/j.chemosphere.2018.05.170

Martín-Pozo, L., et al. 2019. Analytical methods for the determination of emerging contaminants in sewage sludge samples. A review. *Talanta,* 192, 508–533.

Monisha, B., Sridharan, R., Kumar, P.S., et al. 2023. Sensing of azo toxic dyes using nanomaterials and its health effects-A review. *Chemosphere*, 313, 137614.

Moreno-Jiménez, E, Leifheit, E.F., Plaza, C., et al. 2022. Effects of microplastics on crop nutrition in fertile soils and interaction with arbuscularmycorrhizal fungi. *Journal of Sustainable Agriculture and Environment,* 1, 66–72.

Naidu, R., Biswas, B., Willett, I.R., et al. 2021. Chemical pollution: A growing peril and potential catastrophic risk to humanity. *Environmental International,* 156, 106616.

Nayyar, D., Shaikh, M.A.N., and Nawaz, T., (2022). Remediation of emerging contaminants by naturally derived adsorbents. In S. P. Singh, A. K. Agarwal, T. Gupta, and S. M. Maliyekkal (Eds.), *New trends in emerging environmental contaminants* (pp. 225–260). Springer Singapore.

Nobile, M., Arioli, F., Pavlovic, R., et al. 2020. Presence of emerging contaminants in baby food. *Food Additives and Contaminants,* 37, 131–142.

Osorio,V., Schriks, M., et al. 2018. A novel sample preparation procedure for effect-directed analysis of micro-contaminants of emerging concern in surface waters. *Talanta,* 186, 527–537.

Parida, V.K., Saidulu, D., Majumder, A., et al. 2021. Emerging contaminants in wastewater: A critical review on occurrence, existing legislations, risk assessment, and sustainable treatment alternatives. *Journal of Environmental Chemical Engineering,* 9, 105966.

Pașca, C., Coroian, A., and Socaci, S., 2018. Risks and benefits of food additives-review. *Bulletin UASVM Animal Science and Biotechnologies* 75 (2). Print ISSN 1843-5262; Electronic ISSN 1843-536X DOI: 10.15835/buasvmcn-asb: 2018.0026

Pinasseau, L., Wiest, L., and Fildier, A., et al. 2019. Use of passive sampling and high resolution mass spectrometry using a suspect screening approach to characterise emerging pollutants in contaminated groundwater and runoff. *Science of the Total Environment,* 672, 253–263.

Pluym, N., Burkhardt, T., Rögner, N., et al. 2023. Monitoring the exposure to ethoxyquin between 2000 and 2021 in urine samples from the German Environmental Specimen Bank. *Environment International*, 172, 107781.

Podder, A., Sadmani, A.A., Reinhart, D., et al. 2021. Per and poly-fluoroalkyl substances (PFAS) as a contaminant of emerging concern in surface water: A transboundary review of their occurrences and toxicity effects. *Journal of Hazardous Materials*, 419, 126361.

Puri, M., Gandhi, K., and Kumar, M.S. 2023. Emerging environmental contaminants: A global perspective on policies and regulations. *Journalof Environmental Management,* 332, 117344.

Quintelas, C., Mesquita, D.P., Ferreira, E.C. and Amaral, A.L. (2019). Quantification of pharmaceutical compounds in wastewater samples by near infrared spectroscopy (NIR). *Talanta,* 194, 507–513.

Dris, R., Gasperi, J., Mirande, C., et al. 2017. A first overview of textile fibers, including microplastics, in indoor and outdoor environments. *Environmental Pollution,* 221, 453–458.

Radwan, E.K., Abdel Ghafar, H.H., Ibrahim, M.B.M. and Moursy, A.S., 2023. Recent trends in treatment technologies of emerging contaminants. *Environmental Quality Management,* 32(3), 7–25.

Rathi, B.S., Kumar, P.S. and Show, P.L., 2021. A review on effective removal of emerging contaminants from aquatic systems: Current trends and scope for further research. *Journal of Hazardous Materials*, 409, 124413.

Revel, M., Châtel, A., Perrein-Ettajani, H., et al. 2020. Realistic environmental exposure to microplastics does not induce biological effects in the Pacific oyster Crassostreagigas. *Marine Pollution Bulletin,* 150, 110627.

Richardson, S.D. and Kimura, S.Y., 2017. Emerging environmental contaminants: Challenges facing our next generation and potential engineering solutions. *Environmental Technology & Innovation*, 8, 40–56.

Shanmuganathan, R., Kadri, M.S., Mathimani, T., Le, Q.H. and Pugazhendhi, A., 2023. Recent innovations and challenges in the eradication of emerging contaminants from aquatic systems. *Chemosphere*, 332, 138812.

Shetty, S.S., Sonkusare, S., Naik, P.B. and Madhyastha, H., 2023. Environmental pollutants and their effects on human health. *Heliyon,* 9.

Singh, A., Chaurasia, D., Khan, N., Singh, E. and Bhargava, P.C., 2023. Efficient mitigation of emerging antibiotics residues from water matrix: Integrated approaches and sustainable technologies. *Environmental Pollution*, 328, 121552.

Singh, P.K., Kumar, U., Kumar, I., et al. 2024. Critical review on toxic contaminants in surface water ecosystem: Sources, monitoring, and its impact on human health. *Environmental Science and Pollution Research*, 31(45), 56428–56462.

Singh, S., Kapoor, D., et al. 2021. Mechanism and kinetics of adsorption and removal of heavy metals from wastewater using nanomaterials. *Environmental Chemistry Letters*, 19, 2351–2381.

Skarha, J., Mínguez-Alarcón, L., Williams, P.L., et al. 2019. Cross-sectional associations between urinary triclosan and serum thyroid function biomarker concentrations in women. *Environment International*, 122, 256–262.

Slama, H.B., Chenari Bouket, A., Pourhassan, Z., et al. 2021. Diversity of synthetic dyes from textile industries, discharge impacts and treatment methods. *Applied Sciences*, 11(14), 6255.

Song, L., et al. 2022. How much is global business sectors contributing to sustainable development goals? *Sustainable Horizons,* 1, 100012.

Stuart, M., Lapworth, D., Crane, E., Hart, A., 2012. Review of risk from potential emerging contaminants in UK groundwater. *Science of The Total Environmental,* 416, 1–21.

Tao, F., Yao, H., Hruska, Z., Burger, L.W., Rajasekaran, K.and Bhatnagar, D., 2018. Recent development of optical methods in rapid and non-destructive detection of aflatoxin and fungal contamination in agricultural products. *TrAC, Trends in Analytical Chemistry,* 100, 65–81.

Trein, P., et al. 2023. Empirical research on policy integration: A review and new directions. *Policy Science,* 56, 29–48.

Wang, F., Feng, X., Liu, Y., et al. 2022. Micro(nano)plastics and terrestrial plants: Up-to date knowledge on uptake, translocation, and phytotoxicity. *Resources, Conservationand Recycling,* 185, 106503.

Wang, F., Xiang, L., Leung, K.S.Y., et al. 2024. Emerging contaminants: A one health perspective. *The Innovation,* 5(4), *100612*.

Wilkinson, J., Hooda, P.S., Barker, J., Barton, S. and Swinden, J., 2017. Occurrence, fate and transformation of emerging contaminants in water: An overarching review of the field. *Environmental Pollution*, 231, 954–970.

Yadav, D., Rangabhashiyam, S., Verma, P., et al. 2021. Environmental and health impacts of contaminants of emerging concerns: Recent treatment challenges and approaches. *Chemosphere*, 272, 129492.

Zhang, Q., Meng, X., Shi, S., et al. 2022a. Overview of particulate air pollution and human health in China: Evidence, challenges, and opportunities. *The Innovation*, 3(6).

Zhang, Y., Zhang, X., Li, X. and He, D., 2022b. Interaction of microplastics and soil animals in agricultural ecosystems. *Current Opinion in Environmental Science & Health*, 26, 100327.

2 Understanding the Conjunctive Relevance of One Health with Diversity of Emerging Contaminants

Muthukumar P and Selvam S

2.1 INTRODUCTION

The One Health approach emphasizes the deep interconnection between human, animal and environmental health, an especially relevant perspective when addressing the growing challenge of emerging contaminants (ECs). This diverse group includes pharmaceuticals, pesticides and per- and polyfluoroalkyl substances (PFAS), many of which are not fully regulated or understood. Unlike traditional contaminants such as nitrates or heavy metals, ECs often originate from novel sources such as wastewater effluent, industrial discharges and agricultural run-off, and exhibit complex chemical behaviour in the environment. Surface and groundwater systems, though often studied separately, are hydrologically linked. Surface water bodies such as rivers and lakes typically receive initial contaminant loads, while aquifers act as long-term reservoirs. These systems provide more than half of the world's drinking water, support food production and sustain ecosystems, making their contamination a serious concern.

ECs can reach surface waters through both point sources (e.g. wastewater discharge) and non-point sources (e.g. diffuse run-off from agriculture). Groundwater becomes vulnerable through recharge processes, leaching or direct anthropogenic inputs. For example, pharmaceuticals such as ibuprofen are frequently detected in rivers and wells, and pesticides such as atrazine can migrate into aquifers and PFAS, especially from firefighting foams, can contaminate both systems. This creates dual exposure pathways that affect all three pillars of One Health. These contaminants carry implications not only for public health through drinking water exposure but also for animal welfare, through irrigation and wildlife contact, and ecosystem integrity, by disrupting aquatic communities. Despite increasing awareness, much of the current research tends to focus on either surface or groundwater in isolation, overlooking their dynamic interactions. Yet, contamination in surface waters can readily infiltrate aquifers, compounding risks across the entire One Health triad.

The global dependence on surface water and groundwater highlights the pressing need to address contamination by ECs. Surface water bodies such as rivers and lakes

DOI: 10.1201/9781032623801-2

are often the first to receive pollutants due to their direct interface with human activities. Sources like urban run-off, industrial discharges and agricultural inputs introduce a complex mixture of ECs into these environments. The fate of these contaminants, whether they persist, degrade or transform, largely depends on their physicochemical characteristics. Carbamazepine, a widely detected pharmaceutical in municipal wastewater, is notably resistant to biodegradation, allowing it to stay in river systems for extended periods (Kuroda & Kobayashi, 2021). Glyphosate, an herbicide heavily used in agricultural areas, typically reaches surface waters through run-off during rainfall events, with seasonal fluctuations in its concentration patterns (Kolpin et al., 2002). PFAS, often termed 'forever chemicals' due to their environmental persistence, are introduced through industrial activities and firefighting applications, with atmospheric deposition further contributing to their widespread presence (Postigo & Barceló, 2015). These contaminants not only affect surface water quality but also serve as a conduit for groundwater contamination through infiltration.

Groundwater, considered a naturally protected source of freshwater due to its subsurface location and filtration through geological medium, is increasingly vulnerable to contamination by ECs. These contaminants enter aquifer systems through multiple pathways, including recharge from impacted surface waters, vertical leaching from agricultural zones, and direct anthropogenic inputs from septic systems, landfills and industrial waste disposal sites. Due to the low hydraulic conductivity and slow flow rates typical of many aquifer systems, ECs exhibit prolonged residence times once introduced, resulting in persistent contamination. Persistent organic pollutants such as atrazine have been detected in groundwater years following application, demonstrating significant mobility and chemical stability in the subsurface environment (Kolpin et al., 2002). Similarly, PFAS, owing to their high solubility and resistance to natural degradation, have been widely reported in aquifers proximal to industrial areas, representing a sustained risk to drinking water sources (Barnes et al., 2008). Pharmaceutical compounds, including antibiotics such as tetracycline, have also been identified in groundwater samples, particularly in urbanized regions, as a result of infiltration from inadequately treated wastewater and diffuse pollution sources (Lapworth et al., 2012).

The health impacts of ECs are not only significant but also deeply interconnected across human, animal and environmental systems. For humans, the most common exposure pathway is through drinking water drawn from contaminated surface or groundwater sources. Even at trace levels, pharmaceuticals present in drinking water may contribute to chronic health issues such as endocrine disruption and the emergence of antibiotic-resistant bacteria (Singer et al., 2016). The presence of PFAS, often referred to as 'forever chemicals', is particularly concerning. These compounds have been associated with a range of serious health outcomes including liver toxicity, immune dysfunction and even increased cancer risk, and are frequently detected in drinking water supplies at concentrations that exceed recommended safety thresholds (Postigo & Barceló, 2015).

Animals are also vulnerable. In agricultural settings, livestock exposed to pesticide-contaminated water, such as that containing atrazine, may suffer from reproductive and developmental disorders. Wildlife, particularly amphibians, are sensitive to ECs' hormonal effects, with some populations experiencing significant declines due to disrupted endocrine systems (Hayes et al., 2010). Aquatic environments face parallel threats. ECs like antibiotics can alter the structure and function of microbial communities, promoting the spread of antibiotic resistance, a problem that cascades back to both environmental and public health (Singer et al., 2016). Meanwhile, PFAS accumulate in aquatic organisms, from plankton to fish, disrupting food chains, reducing biodiversity and weakening the overall resilience of ecosystems.

Addressing the issue of ECs requires a holistic lens, which the One Health framework effectively provides by integrating human, animal and environmental health. This interdisciplinary approach encourages collaboration across fields such as environmental science, toxicology, public health and veterinary medicine to tackle complex contamination challenges. This review explores the relevance of One Health in understanding EC diversity in surface and groundwater systems. Specifically, it aims to (1) characterize EC types, sources and chemical properties; (2) assess their fate and transport; and (3) evaluate their health implications across the One Health triad. By synthesizing global research, we highlight how these interconnected water systems act as medium for ECs and stress the need for integrated, cross-sectoral solutions. The chapter proceeds with an overview of One Health, an analysis of EC diversity, a discussion on water system dynamics, health impact evaluations, and a final synthesis of challenges and strategies for mitigation.

2.2 THE ONE HEALTH FRAMEWORK: A LENS FOR EMERGING CONTAMINANTS

2.2.1 Evolution of One Health

The One Health framework has undergone a notable evolution, transitioning from a narrowly focused initiative to a comprehensive, interdisciplinary approach that addresses the interconnected health of humans, animals and the environment. Initially conceptualized as a collaborative strategy to manage zoonotic diseases, One Health has expanded to encompass a wide array of environmental stressors, including ECs such as pharmaceuticals, pesticides and PFAS in surface water and groundwater (Table 2.1). This evolution reflects a growing recognition of the interdependence of ecosystems and health outcomes, positioning One Health as a dynamic lens for addressing complex global challenges like chemical pollution (WHO, 2017).

The origins of One Health can be outlined back to the early 20th century, when veterinarians and physicians began recognizing the shared risks of zoonotic diseases such as pathogens that can transfer between animals and humans. Early efforts focused on diseases like rabies, brucellosis and tuberculosis, which posed significant threats to both human and animal populations. For instance, rabies, a viral disease primarily transmitted through animal bites, highlighted the need for coordinated efforts between medical and veterinary professionals to control outbreaks. This period marked the beginning of a collaborative approach, often referred to as 'One Medicine', which emphasized the unity of human and veterinary medicine in combating shared health threats. The term 'One Health' was formally coined in the early

TABLE 2.1
Emerging contaminants in groundwater and surface water with adverse health effects

Emerging Contaminant	Sources	Pathways to Groundwater	Pathways to Surface Water	Potential Health Effects
Pharmaceuticals (e.g. Ibuprofen, Carbamazepine)	Wastewater, hospitals, households	Recharge from treated wastewater, leaching from septic systems	Run-off from urban areas, discharge from wastewater treatment plants	Endocrine disruption, potential toxicity at high exposure levels
Pesticides (e.g. Atrazine)	Agricultural run-off, pesticide application	Leaching from cultivated fields, infiltration through soil	Run-off from agricultural lands, direct application near water bodies	Neurological effects, potential carcinogenicity
PFAS (Per- and Polyfluoroalkyl Substances)	Industrial discharges, firefighting foams	Recharge from contaminated surface water, leaching from landfills	Run-off from industrial sites, atmospheric deposition	Liver damage, immune system effects, potential cancer risk
Personal Care Products (e.g., DEET, Triclosan)	Domestic wastewater, cosmetics	Leaching from septic systems, recharge from treated wastewater	Run-off from urban areas, discharge from wastewater treatment plants	Endocrine disruption, antimicrobial resistance
Microplastics	Plastic waste, wastewater effluent	Infiltration through soil, recharge from contaminated surface water	Run-off from urban areas, direct discharge into rivers	Potential for bioaccumulation, carrier for other toxins
Nitrosamines (e.g. NDMA)	Industrial processes, water treatment byproducts	Leaching from contaminated sites, recharge from treated water	Discharge from wastewater treatment, run-off from industrial areas	Carcinogenic, potential liver toxicity

2000s, building on this foundation to advocate for a more integrated strategy that transcends disciplinary boundaries.

An essential moment in the evolution of One Health came with the global response to emerging infectious diseases in the late 20th and early 21st centuries. Outbreaks such as the 2003 SARS epidemic and the 2009 H1N1 influenza pandemic underscored the critical role of wildlife and environmental factors in disease transmission. These events prompted organizations like the World Health Organization (WHO), the Food and Agriculture Organization (FAO), and the World Organisation for Animal Health (WOAH) to formalize the One Health approach. In 2010, these agencies issued a tripartite concept note, defining One Health as a collaborative, multisectoral and transdisciplinary approach to achieve optimal health outcomes by recognizing the interconnection between people, animals, plants and their shared environment. This broader definition marked a significant shift, expanding the scope of One Health beyond zoonotic diseases to include environmental health as a core component.

The inclusion of environmental stressors like ECs in the One Health framework represents a more recent and critical development. Initially focused on biological pathogens, One Health has evolved to address chemical pollutants that threaten ecosystems and health across the triad. ECs such as pharmaceuticals, pesticides and PFAS, which contaminate surface water and groundwater, exemplify this expanded focus. For example, pharmaceuticals like ibuprofen and carbamazepine, often discharged through wastewater, persist in aquatic environments and pose risks to human health through drinking water and to aquatic ecosystems by disrupting food webs (Postigo & Barceló, 2015). Similarly, pesticides like atrazine, which leach into groundwater from agricultural fields, have been linked to endocrine disruption in wildlife and humans (Hayes et al., 2010). PFAS, known for their persistence and mobility, contaminate both surface water and groundwater, with studies detecting them in European aquifers and correlating their presence with health risks like cancer and immune suppression (Postigo & Barceló, 2015).

This evolution reflects a deeper understanding of the interdependence of ecosystems and health outcomes. Groundwater and surface water serve as critical pathways for ECs, linking environmental contamination to human and animal health. The One Health framework now integrates disciplines such as environmental science, toxicology and public health to address these challenges holistically. By recognizing the role of chemical pollutants in the health triad, One Health provides a robust platform for developing interdisciplinary solutions, such as improved wastewater treatment, sustainable agricultural practices and stricter regulations on industrial discharges. As environmental stressors like ECs continue to grow in complexity, the One Health approach will remain essential for safeguarding global health in an interconnected world (WHO, 2017).

2.2.2 ECs across the Triad

ECs disrupt the One Health triad (Human, Animal and Environmental Health) distinctly (Table 2.1). Human health is affected by contaminated surface water, and groundwater compromises drinking water safety. Pharmaceuticals like carbamazepine persist in rivers and infiltrate aquifers, with chronic exposure linked to endocrine disruption (Postigo & Barceló, 2015). PFAS in both systems are associated with cancer and immune suppression, with detections exceeding health advisories in multiple regions. For animal health, irrigation from contaminated rivers or wells exposes livestock to pesticides like neonicotinoids, which bioaccumulate and affect reproduction (Hayes et al., 2010). Wildlife drinking from affected surface waters or springs faces similar risks, with PFAS detected in fish and birds. Regarding environmental health, ECs alter aquatic ecosystems – antibiotics in rivers foster microbial resistance (Singer et al., 2016), while PFAS in groundwater disrupt food webs, reducing biodiversity.

2.2.3 Literature Review

Global studies have increasingly highlighted the hydrological connectivity between surface water and groundwater, the spread of ECs, and their cascading effects on health and ecosystems, focusing on the distribution, fate and impacts of ECs, and underscored the relevance of One Health in addressing these multifaceted challenges.

Singer et al. (2016) investigated antibiotic residues in Asian rivers, finding significant concentrations of antibiotics like tetracycline and sulfamethoxazole in the Yangtze and Pearl Rivers. These residues were linked to the proliferation of antibiotic resistance genes (ARGs), which reduce the efficacy of human treatments and impact livestock health through contaminated irrigation water. This study highlights the role of surface water as a conduit for ARGs, affecting both human and animal health, and emphasizes the need for integrated water management strategies.

Postigo and Barceló (2015) conducted a comprehensive analysis of PFAS in European surface waters and groundwater, detecting perfluorooctanoic acid (PFOA) and perfluoro octane sulfonate (PFOS) at concentrations exceeding health advisories in regions like the Netherlands and Germany. They correlated these levels with health risks, including liver damage and immune suppression, and noted ecosystem shifts such as reduced biodiversity in aquatic systems due to bioaccumulation in fish. Their findings underscore the hydrological connectivity between surface water and groundwater, as PFAS in rivers were found to recharge aquifers, amplifying exposure risks.

Hayes et al. (2010) examined the impact of the pesticide atrazine in U.S. streams, particularly in the Midwest, where concentrations in surface water reached 1–3 μg/L during peak agricultural seasons. They linked atrazine exposure to amphibian declines, noting a 50% reduction in frog populations due to endocrine disruption, which caused reproductive abnormalities. This animal-environmental feedback loop demonstrates how ECs in surface water can disrupt ecosystems, with potential downstream effects on human health through food chain contamination.

In India, Kumar et al. (2019) detected pharmaceuticals such as ibuprofen and ciprofloxacin in the Ganges River, with concentrations ranging from 0.5 to 2 μg/L. Downstream groundwater samples showed similar profiles, with 70% of wells containing detectable levels of these compounds, highlighting hydrological connectivity through recharge processes. This study emphasizes the vulnerability of groundwater as a long-term reservoir for ECs, posing risks to drinking water supplies and agricultural systems.

Kolpin et al. (2002) studied ECs in U.S. water systems, detecting pesticides like atrazine and pharmaceuticals like carbamazepine in both surface water and groundwater across the Midwest. They found that surface water concentrations of atrazine peaked at 10 μg/L during spring run-off, with 30% of groundwater samples showing detectable levels due to recharge. This connectivity amplifies risks across the One Health triad, as contaminated water affects human health, livestock and aquatic ecosystems.

Lapworth et al. (2012) investigated groundwater contamination in the UK, finding antibiotics like tetracycline in 20% of rural wells at concentrations up to 0.1 μg/L. They attributed this to leaching from agricultural fields and septic systems, noting that groundwater acts as a reservoir, prolonging exposure risks. This study highlights the need for One Health approaches to address diffuse contamination sources impacting human and animal health.

Ricolfi et al. (2025) analysed PFAS in China's Poyang Lake, reporting total PFAS concentrations ranging from 23 to 1,000 ng/L in surface water. PFOA was the dominant compound, with groundwater samples showing 10–50 ng/L due to recharge. They noted bioaccumulation in fish, with PFAS levels in carp reaching 5 ng/g, posing risks to human health through dietary exposure and to ecosystems through food web disruptions.

DeLuca et al. (2021) reviewed global PFAS distribution, finding PFOA and PFOS in 90% of surface water samples worldwide, with concentrations in industrial areas like the U.S. and China reaching 500 ng/L. They highlighted PFAS persistence, noting their role in ecosystem shifts, such as reduced microbial diversity in contaminated rivers, which affects nutrient cycling and water quality. Sunderland et al. (2019) explored human exposure to PFAS, estimating that drinking water contributes 60% of total PFAS intake in contaminated regions. They reported PFOS levels in U.S. tap water at 20–100 ng/L, linked to health risks like decreased fertility and increased cancer incidence, emphasizing the human health dimension of One Health.

Wang et al. (2017) estimated global PFOS emissions from 1958 to 2015 at 1,228–8,740 tons, with significant contributions from industrial discharges into surface water. They noted that PFOS persists in groundwater for decades, with concentrations in Chinese aquifers reaching 50 ng/L, illustrating the long-term environmental health risks of ECs. Kuroda and Kobayashi (2021) detected carbamazepine in Japanese rivers at 0.2–1 μg/L, with 40% of groundwater samples showing traces due to recharge from treated wastewater. This persistence in water systems poses risks to human health through drinking water and to aquatic ecosystems by altering microbial communities.

Campanale et al. (2022) studied karst aquifers in Italy, finding that surface water contaminated with PFAS (50 ng/L) rapidly recharged groundwater, with 80% of aquifer samples showing detectable levels. This rapid transport underscores the hydrological connectivity that amplifies EC risks across the One Health framework. Barnes et al. (2008) reported PFAS in U.S. groundwater near industrial sites, with PFOA levels at 100–200 ng/L. They noted bioaccumulation in wildlife, with PFAS concentrations in fish reaching 10 ng/g, affecting animal health and posing indirect risks to humans through consumption. Zushi and Masunaga (2009) analysed seasonal patterns of ECs in Japanese stormwater, finding pesticides like atrazine at 1 μg/L in summer and PFAS at 50 ng/L in winter. They highlighted stormwater as a key pathway for ECs into surface water, with subsequent recharge into groundwater, impacting all One Health components. Tran et al. (2019) studied stormwater in Australia, detecting pharmaceuticals like ibuprofen at 0.5 μg/L and ARGs, which were 30% more prevalent in urban run-off. This study links urban activities to environmental contamination, with implications for human and animal health through water use. Sharma et al. (2016) reviewed PFAS in Asian water systems, finding PFOA in Indian rivers at 20–80 ng/L, with groundwater levels at 10–30 ng/L due to leaching. They noted ecosystem impacts, such as reduced fish populations, and health risks like thyroid disorders in humans.

Lopez-Espinosa et al. (2016) found PFAS in Spanish tap water at 15–40 ng/L, linked to developmental effects in children, such as delayed growth. This study emphasizes the human health risks of ECs in drinking water, a key exposure pathway in the One Health framework.

Ahrens et al. (2023) detected PFAS in Arctic water systems at 5–20 ng/L, illustrating long-range transport and persistence. They noted bioaccumulation in polar bears, with PFOS levels at 50 ng/g, affecting animal health and highlighting global environmental health concerns. Kotthoff et al. (2015) analysed PFAS in European consumer products, finding leaching into surface water at 10–50 ng/L. They linked this to groundwater contamination through recharge, with implications for drinking water safety and ecosystem stability. Zhang et al. (2022) reported PFAS in South Korean rivers at 30–100 ng/L, with 60% of groundwater samples showing traces. They noted impacts on aquatic biodiversity, such as reduced algae growth, affecting ecosystem services like water purification. Karásková et al. (2016) studied PFAS in Czech surface waters, finding PFOA at 20–60 ng/L, with groundwater levels at 5–15 ng/L due to recharge. They highlighted bioaccumulation in fish, with PFAS levels at 3 ng/g, posing risks to human and animal health.

Fenton et al. (2021) reviewed PFAS toxicity, noting their lipophilic properties lead to bioaccumulation in humans and wildlife, with PFOS levels in human blood at 10–50 ng/mL. They linked this to health risks like kidney disorders, emphasizing the need for One Health approaches. These studies collectively illustrate the global scope of EC contamination, their movement through hydrologically connected systems, and their impacts across the One Health triad. Surface water acts as an initial receptor for ECs, while groundwater serves as a long-term reservoir, amplifying exposure risks. The findings underscore the relevance of One Health in integrating these disparate impacts into a cohesive framework for EC management, advocating for interdisciplinary strategies to mitigate their threats to human, animal and environmental health.

2.3 DIVERSITY OF EMERGING CONTAMINANTS IN SURFACE WATER AND GROUNDWATER

2.3.1 Types and Sources

ECs in surface water and groundwater include pharmaceuticals, pesticides and PFAS, each with distinct sources that contribute to their widespread distribution. Pharmaceuticals (e.g. tetracycline, ibuprofen) enter surface water through wastewater treatment plant effluents, and groundwater through septic leakage or recharge, with up to 90% passing treatment unchanged due to their resistance to conventional treatment processes like biodegradation or adsorption (Barnes et al., 2008). For example, studies in the U.S. have detected antibiotics like sulfamethoxazole in rivers at concentrations of 0.1–1 μg/L, originating from hospital and domestic wastewater (Kolpin et al., 2002). In developing regions, untreated sewage discharge exacerbates this issue, with rivers like the Yamuna in India showing pharmaceutical levels as high as 5 μg/L (Mutiyar & Mittal, 2014). Pesticides (e.g. glyphosate, neonicotinoids) contaminate rivers from agricultural run-off and aquifers through leaching, with seasonal pulses driven by rainfall (Kolpin et al., 2002). Neonicotinoids, such as imidacloprid, have been found in European rivers at 0.01–0.5 μg/L, particularly during planting seasons, due to their high solubility and mobility (Morrissey et al., 2015). PFAS (e.g. PFOA, PFOS) reach surface water from industrial discharges and firefighting foams, infiltrating groundwater through landfill leachates or surface recharge (Postigo & Barceló, 2015). In Australia, PFAS contamination near firefighting training sites has led to surface water concentrations of 50–200 ng/L, with subsequent groundwater contamination at 10–50 ng/L due to leaching (Baduel et al., 2017). Additionally, personal care products (PCPs) like triclosan, used in soaps and cosmetics, are emerging as significant ECs, entering surface water through urban wastewater and groundwater through septic systems, with triclosan detected in U.S. streams at 0.14–2.3 μg/L (Halden & Paull, 2005). These diverse sources highlight the complexity of EC contamination, necessitating targeted mitigation strategies to address both point and diffuse inputs across different regions.

2.3.2 Chemical Diversity

ECs exhibit varied chemical properties that influence their behaviour in surface water and groundwater, complicating their management. Pharmaceuticals range from polar (e.g. ciprofloxacin, zwitterionic) to hydrophobic (e.g. ibuprofen, log Kow 3.97), affecting their mobility and fate – polar compounds like ciprofloxacin are highly mobile in surface flows, while hydrophobic ones like ibuprofen tend to sorb to aquifer sediments, reducing their transport but increasing persistence (Ternes et al., 2004). For example, diclofenac, with a log Kow of 4.51, has been shown to accumulate in river sediments, with concentrations up to 0.5 μg/kg, impacting benthic organisms (Schwaiger et al., 2004). Pesticides vary from water-soluble atrazine (Kd ~1–5 L/kg) to persistent DDT metabolites, with differing transport rates. Atrazine's solubility (33 mg/L) allows it to move quickly through surface water, while DDT metabolites, with a log Kow of 6.91, persist in groundwater for decades due to their low solubility and high sorption to soil particles (Mackay et al., 2006). PFAS, with fluorinated structures, are highly mobile and resistant to degradation (log Kow ~4.8 for PFOA), persisting across systems (Lapworth et al., 2012). Their surfactant properties enable PFAS to migrate through both aqueous and soil phases, with studies showing PFOA half-lives exceeding 90 years in groundwater (Hekster et al., 2003). Additionally, ECs like PCPs exhibit diverse properties, such as triclosan, with a log Kow of 4.76, is moderately hydrophobic, leading to bioaccumulation in aquatic organisms, while DEET (log Kow 2.18) is more mobile, detected in surface water at 0.01–0.1 μg/L (Brausch & Rand, 2011). This chemical diversity affects not only transport but also toxicity, as polar compounds may be more bioavailable, while hydrophobic ones pose long-term risks through accumulation. Understanding these properties is crucial for predicting EC behaviour and designing effective remediation strategies, such as tailored filtration systems or source control measures.

2.3.3 Pathways to Water Systems

ECs enter surface water through wastewater outfalls, urban run-off and agricultural drainage, with rivers showing concentrations in the ng/L–µg/L range (Kuroda & Kobayashi, 2021). For example, in Japan, urban rivers like the Tama River have shown pharmaceutical levels of 0.2–1 µg/L, primarily from wastewater treatment plant discharges, with seasonal variations driven by dilution during rainy periods (Nakada et al., 2008). Stormwater run-off in urban areas also introduces ECs, with studies in Australia detecting PFAS at 20–100 ng/L in run-off from industrial zones (Nguyen et al., 2018). Agricultural drainage contributes significantly, with pesticides like glyphosate reaching surface water at 1–5 µg/L during peak application periods in the U.S. Midwest (Battaglin et al., 2014). Groundwater contamination occurs through recharge from polluted surface water, direct leaching or wastewater infiltration. Karst aquifers enhance rapid transport, as seen in Italy, where PFAS from surface water infiltrated groundwater within weeks, reaching 50 ng/L (Campanale et al., 2022). Sandy systems, conversely, favour gradual infiltration, with studies in the Netherlands showing atrazine in groundwater at 0.1–0.5 µg/L years after application due to slow leaching through sandy soils (Hantush et al., 2000). Hydrological connectivity, e.g., river-aquifer exchange amplifies EC spread. In the Mississippi River Basin, river-aquifer interactions have led to pharmaceutical detections in groundwater at 0.01–0.1 µg/L, mirroring surface water profiles (Bradley et al., 2017). Additionally, atmospheric deposition plays a role, particularly for PFAS, which can deposit onto surface water and infiltrate groundwater through precipitation, with global studies estimating 10–20% of PFAS in remote lakes originates from atmospheric sources (Muir et al., 2019). These pathways highlight the need for integrated monitoring and management strategies that account for both surface and subsurface dynamics.

2.3.4 Detection Advances

Techniques like liquid chromatography-mass spectrometry (LC-MS) detect pharmaceuticals, while gas chromatography-mass spectrometry (GC-MS) targets pesticides and PFAS, offering high sensitivity for trace-level analysis (Marino et al., 2024). LC-MS has been pivotal in detecting antibiotics like tetracycline in surface water at 1–10 ng/L, with tandem mass spectrometry (MS/MS) improving specificity by identifying molecular fragments (Hao et al., 2007).

GC-MS is effective for PFAS, with studies in Europe using it to quantify PFOA at 5–20 ng/L in groundwater, despite challenges with matrix effects (Loos et al., 2010). Surface water's dynamic flow requires frequent sampling, often monthly or event-based, to capture seasonal peaks, as seen in U.S. rivers where pesticide levels spike post-rainfall (Gilliom et al., 2006). Groundwater's low concentrations demand high sensitivity (ng/L detection limits), with techniques like solid-phase extraction (SPE) coupled with LC-MS achieving detection limits of 0.1 ng/L for pharmaceuticals

(Focazio et al., 2008). Matrix interference and cost remain challenges, particularly in complex matrices like groundwater, where organic matter can suppress signals, requiring advanced cleanup methods (Lapworth et al., 2012). Emerging technologies, such as high-resolution mass spectrometry (HRMS), are improving detection by identifying unknown ECs through non-targeted analysis, with studies identifying over 50 new PFAS compounds in U.S. rivers (Strynar et al., 2015). Passive sampling devices, like polar organic chemical integrative samplers (POCIS), are also gaining traction, providing time-weighted average concentrations of ECs in surface water, with applications in the UK detecting pharmaceuticals at 0.01–0.5 μg/L over 30-day periods (Lohmann et al., 2012). These advances are critical for monitoring ECs, but high costs and the need for skilled personnel limit their widespread adoption, particularly in developing regions.

2.4 SURFACE WATER AND GROUNDWATER AS CONJUNCTIVE PATHWAYS FOR ECs

2.4.1 Hydrological Context

Surface water and groundwater are intricately linked through hydrological processes (Figure 2.1) such as recharge (surface to ground) and discharge (ground to surface), creating a dynamic system that facilitates the movement of ECs. Rivers recharge aquifers during high-flow periods, as seen in the Mississippi River Basin, where up to 30% of aquifer recharge originates from river infiltration, carrying ECs like pesticides into groundwater (Bradley et al., 2017). Springs, conversely, feed streams by discharging groundwater, often reintroducing ECs into surface water systems; for example, in Florida, springs discharging into the Suwannee River have been found to contain pharmaceuticals like carbamazepine at 0.01–0.1 μg/L, reflecting upstream groundwater contamination (Katz et al., 2011). Wetlands bridge both systems, acting as transitional zones where surface water and groundwater interact, making them particularly vulnerable to EC accumulation. In the Everglades, wetlands have shown PFAS concentrations of 10–50 ng/L, attributed to both surface run-off and groundwater discharge (Loos et al., 2010). Surface water acts as an initial sink for ECs, rapidly receiving pollutants from point sources like wastewater outfalls and diffuse sources like agricultural run-off, with rivers often showing EC concentrations in the ng/L–μg/L range (Kuroda & Kobayashi, 2021). Groundwater, however, serves as a long-term reservoir due to its slow movement and limited natural attenuation, with contaminants like PFAS persisting for decades (Campanale et al., 2022). In karst systems, such as those in Italy, rapid recharge through fractures allows ECs like PFAS to move from surface water to groundwater within weeks, with concentrations reaching 50 ng/L (Campanale et al., 2022). In contrast, sandy aquifers, like those in the Netherlands, exhibit slower infiltration, with pesticides like atrazine detected at 0.1–0.5 μg/L years after surface application due to gradual recharge (Hantush et al., 2000). This hydrological connectivity amplifies the vulnerability of both systems to ECs, as contaminants can cycle between surface water and groundwater, creating a feedback loop that exacerbates exposure risks across the One Health triad such as human, animal, and environmental health.

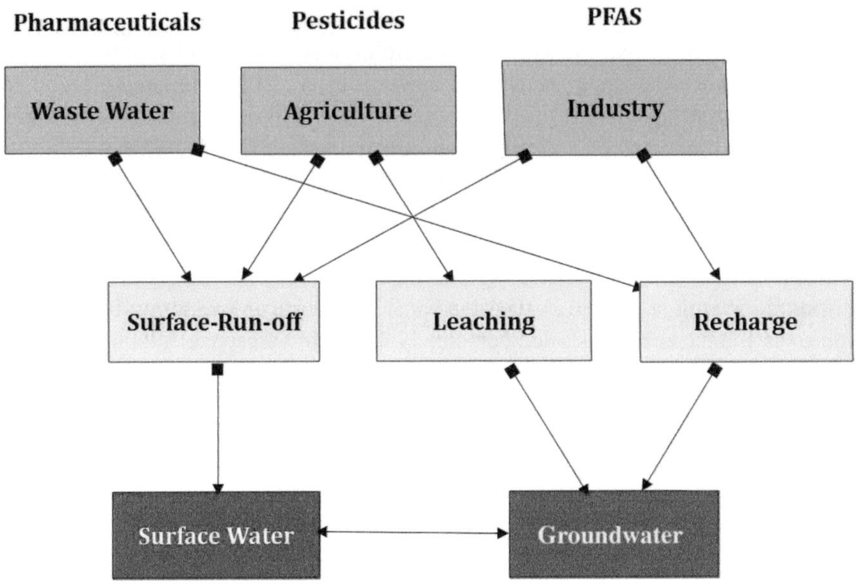

FIGURE 2.1 Source and pathways to the water systems.

2.4.2 Fate and Transport

The fate and transport of ECs vary significantly between surface water and groundwater, influenced by system-specific hydrological and geochemical conditions. In surface water, turbulent flow dilutes but disperses ECs, enabling rapid transport over large distances; for instance, in the Danube River, pharmaceuticals like ibuprofen have been detected 100 km downstream from wastewater outfalls at concentrations of 0.05–0.2 μg/L (Loos et al., 2010). Sorption to sediments retains hydrophobic compounds like PFAS, with studies showing PFOA sorption coefficients (Kd) of 2–10 L/kg in river sediments, reducing their mobility but increasing their persistence in depositional zones (Higgins & Luthy, 2006). Polar pharmaceuticals, such as ciprofloxacin, travel freely in surface water due to their high solubility, with concentrations in Asian rivers reaching 1 μg/L during low-flow periods (Tran et al., 2019). In groundwater, sorption to organic matter and slow biodegradation prolong residence times, with antibiotics like tetracycline degrading 50% faster in oxygenated rivers (half-life ~10 days) than in anoxic aquifers (half-life ~30 days) due to limited microbial activity (Singer et al., 2016). This slow degradation allows ECs to persist, as seen with PFAS in U.S. groundwater, where PFOA concentrations of 100–200 ng/L have remained stable for over a decade (Barnes et al., 2008). Dispersion in fractures enhances groundwater transport, particularly in karst or fractured bedrock systems; for example, in the Chalk aquifer of the UK, PFAS were found to migrate 1–2 km/year through fractures, reaching concentrations of 20–50 ng/L (Sorensen et al., 2015). Additionally, photodegradation plays a role in surface water, breaking down certain ECs like diclofenac (half-life ~1 day under sunlight), but is negligible in groundwater, where lack of light and oxygen limits natural attenuation (Liu et al., 2011). Temperature and pH also influence EC fate; in tropical rivers, higher temperatures (25–30°C) accelerate the degradation of pesticides like atrazine (half-life ~20 days), while in colder groundwater systems (10–15°C), degradation slows significantly (half-life ~60 days) (Gilliom et al., 2006). These differences highlight the need for system-specific models to predict EC behaviour, as their transport and persistence directly impact exposure risks across the One Health triad.

2.4.3 Global Evidence

Global studies provide compelling evidence of EC contamination in surface water and groundwater, illustrating their hydrological connectivity and impacts across the One Health framework (Figure 2.2, Table 2.2). Kolpin et al. (2002) found pesticides like atrazine in U.S. rivers and wells, with surface water concentrations peaking at 10 μg/L during spring run-off and groundwater levels at 0.1–1 μg/L, linked by recharge processes in the Midwest. They noted that 60% of wells in agricultural areas showed pesticide residues, reflecting the role of leaching and recharge in contaminant transport. Barnes et al. (2008) detected PFAS in surface water and groundwater near industrial sites in the U.S., with concentrations up to 1 μg/L; for example, near a manufacturing facility in West Virginia, PFOA levels in groundwater reached 500 ng/L, with surface water showing similar profiles due to discharge, posing risks to drinking water and

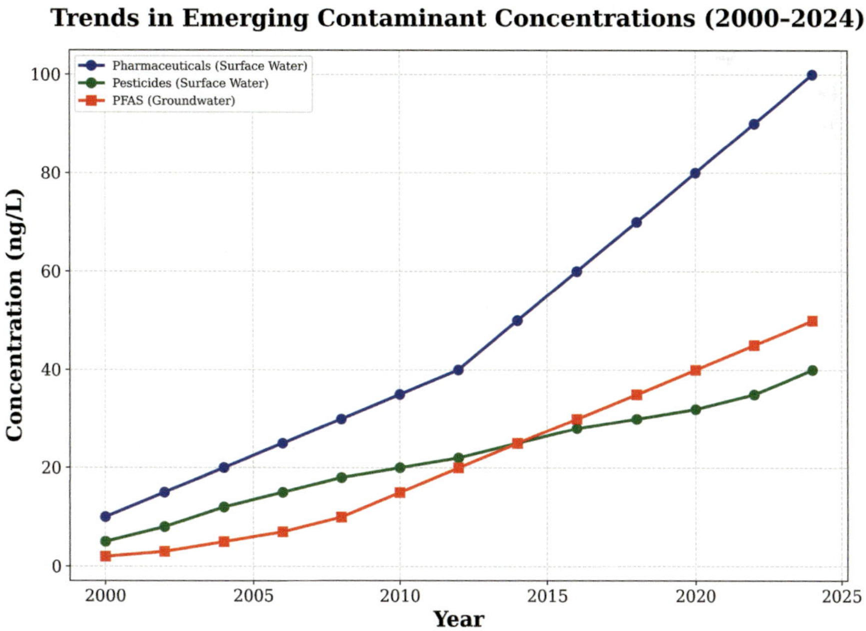

FIGURE 2.2 Trends in emerging contaminant concentrations (2000–2024). (Data compiled from published studies cited in the manuscript.)

TABLE 2.2
Comparison of emerging contaminants in surface water and groundwater across countries

Country	EC Type	Surface Water Concentration	Groundwater Concentration	References
United States	Pesticides (Atrazine)	10 μg/L (rivers, spring run-off)	0.1–1 μg/L (wells in agricultural areas)	Kolpin et al. (2002); Bradley et al. (2017); Hayes et al. (2010); Gilliom et al. (2006)
United States	PFAS (PFOA)	1 μg/L (near industrial sites)	500 ng/L (near industrial sites)	Barnes et al. (2008)
Kenya	Pharmaceuticals (Ibuprofen)	0.5–2 μg/L (rivers near Nairobi)	0.01–0.1 μg/L (aquifers near urban areas)	Lapworth et al. (2012)
Australia	PFAS (PFOA)	200–500 ng/L (near airports)	50–100 ng/L (near firefighting bases)	Postigo & Barceló (2015)
India	Pharmaceuticals (Ciprofloxacin)	10– 130ng/L (Ganges River)	0.01–0.05 μg/L (downstream wells)	Bhagat et al. (2020); Jassal et al. (2023)
India	PFAS (PFOA)	20–80 ng/L (rivers)	10–30 ng/L (groundwater)	Sharma et al. (2016)
China	PFAS (PFOA)	23–1,000 ng/L (Poyang Lake)	10–50 ng/L (nearby groundwater)	Ricolfi et al. (2025)
Italy	PFAS (PFOA)	50 ng/L (surface water in karst regions)	50 ng/L (karst aquifers)	Campanale et al. (2022)
South Korea	PFAS (PFOA)	30–100 ng/L (rivers)	5–15 ng/L (60% of groundwater samples)	Zhang et al. (2022)
Japan	Pharmaceuticals (Carbamazepine)	0.2–1 μg/L (Tama River)	0.01–0.1 μg/L (40% of groundwater samples)	Kuroda & Kobayashi (2021)
UK	PFAS (PFOA)	20–50 ng/L (rivers)	20–50 ng/L (Chalk aquifer)	Sorensen et al. (2015)

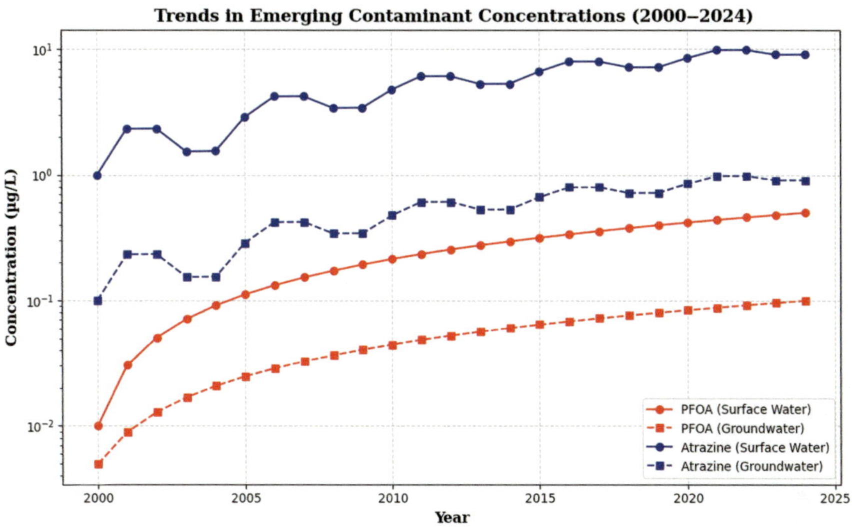

FIGURE 2.3 Contaminant concentration trends in 2000–2024. (Data compiled from published studies cited in the chapter.)

aquatic ecosystems. Lapworth et al. (2012) reported pharmaceuticals in African rivers and aquifers, reflecting urban-agricultural gradients in Kenya, rivers near Nairobi contained ibuprofen at 0.5–2 µg/L, with nearby aquifers showing 0.01–0.1 µg/L due to recharge, highlighting the urban impact on water quality. In Australia, PFAS from firefighting bases contaminated both systems, with surface water near Sydney airports showing PFOA at 200–500 ng/L and groundwater at 50–100 ng/L, prompting health advisories for drinking water (Postigo & Barceló, 2015). In China, Ricolfi et al. (2025) found PFAS in Poyang Lake at 23–1,000 ng/L, with groundwater levels at 10–50 ng/L due to recharge, and noted bioaccumulation in fish (5 ng/g), affecting both human and animal health. Bhagat et al. (2020) detected pharmaceuticals in the Ganges River, India, with concentrations of ciprofloxacin at 0.5–2 µg/L, and downstream groundwater showed similar profiles (0.01–0.05 µg/L), illustrating river-aquifer exchange. In Europe, Campanale et al. (2022) reported PFAS in Italian karst aquifers at 50 ng/L, linked to surface water recharge, with rapid transport through fractures amplifying exposure risks. Sharma et al. (2016) found PFAS in Indian rivers at 20–80 ng/L, with groundwater at 10–30 ng/L, noting ecosystem impacts like reduced fish populations. In South Korea, Zhang et al. (2022) detected PFAS in rivers at 30–100 ng/L, with 60% of groundwater samples showing traces, affecting aquatic biodiversity. These findings show the global trends in EC contamination concentration in water systems (Figure 2.3) and the need for One Health approaches to address their impacts on human, animal, and environmental health.

2.5 ONE HEALTH IMPLICATIONS OF ECs IN SURFACE WATER AND GROUNDWATER

2.5.1 Human Health

ECs in drinking water sourced from rivers and wells pose significant risks to human health, as both surface water and groundwater are contaminated with persistent pollutants. PFAS, such as PFOA and PFOS are linked to severe health outcomes, including liver damage, immune system suppression, and increased cancer risk, with concentrations (Figure 2.4) in drinking water often exceeding the U.S. EPA guideline of 70 ng/L (Postigo & Barceló, 2015). For instance, in the Veneto region of Italy, PFAS levels in tap water reached 100–200 ng/L, correlating with elevated rates of kidney cancer in exposed populations (Mastrantonio et al., 2018). Pharmaceuticals, particularly antibiotics like tetracycline, contribute to antimicrobial resistance (AMR), a global health crisis that complicates the treatment of infections. Singer et al. (2016) found that antibiotic residues in Asian rivers, such as the Pearl River, at 0.5–2 µg/L, foster the proliferation of resistance genes, reducing the efficacy of treatments like amoxicillin for humans. Additionally, pharmaceuticals like carbamazepine, detected in U.S. wells at 0.01–0.1 µg/L, pose risks of chronic exposure, potentially leading to neurological effects (Focazio et al., 2008). These findings highlight the urgent need for improved water treatment and monitoring to mitigate human health risks from ECs.

2.5.2 Animal Health

ECs in water systems impact animal health through agricultural and environmental exposure pathways. Irrigation transfers pesticides like atrazine to livestock, with

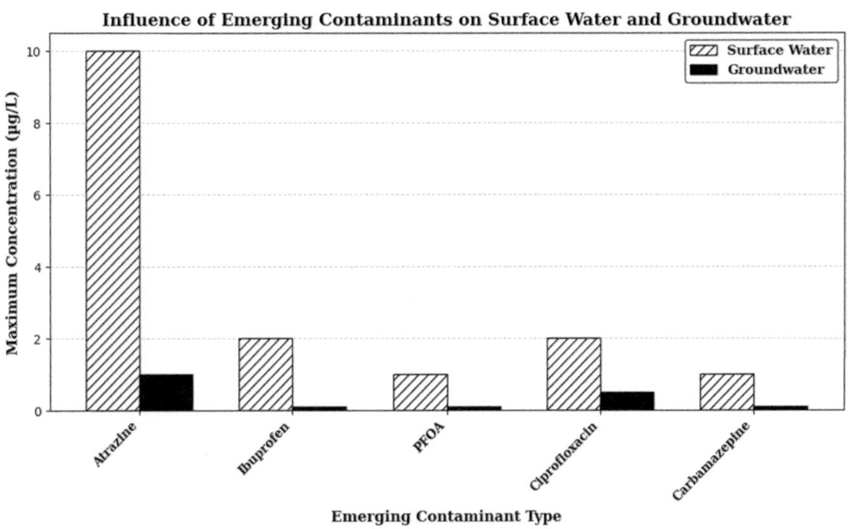

FIGURE 2.4 Influence of emerging contaminants in water system. (Data compiled from published studies cited in the chapter.)

Hayes et al. (2010) reporting endocrine disruption in cattle, where atrazine levels of 0.1–1 μg/L in irrigation water led to reduced fertility rates. Wildlife drinking from contaminated streams or springs bioaccumulates PFAS, affecting reproduction; for example, in the Great Lakes, fish exposed to PFOS at 10–50 ng/L exhibited impaired reproductive success, with a 30% reduction in egg production (Ankley et al., 2005). In Australia, PFAS from firefighting bases contaminated surface water used by grazing animals, leading to PFOS accumulation in cattle livers at 5–20 ng/g, posing risks to both animal health and human consumption (Death et al., 2021). These impacts underscore the vulnerability of animals to ECs across diverse exposure routes.

2.5.3 Environmental Health

ECs disrupt ecosystems, with cascading effects on environmental health. Antibiotics in rivers, such as tetracycline at 0.5–2 μg/L in the Yangtze River, alter microbial diversity by promoting resistant strains, which disrupts nutrient cycling and water purification processes (Singer et al., 2016). PFAS in groundwater bioaccumulate in fish, reducing biodiversity; Postigo & Barceló (2015) reported PFOS levels of 5–10 ng/g in European fish, leading to a 20% decline in species richness in contaminated lakes due to food web disruptions. Pesticides like atrazine, detected in U.S. streams at 1–10 μg/L, cause algal blooms by eliminating competing species, further degrading aquatic ecosystems (Kolpin et al., 2002). These environmental impacts highlight the broader ecological consequences of EC contamination.

2.5.4 Interconnections

Contaminated water cycles through humans (drinking), animals (irrigation), and ecosystems (food webs), creating feedback loops that amplify risks across the One Health triad. For example, PFAS in drinking water (human exposure) bioaccumulate in fish (environmental impact), which are consumed by wildlife and humans, perpetuating exposure (Postigo & Barceló, 2015). Irrigation water containing atrazine affects livestock health, enters the human food chain via dairy products, and leaches into groundwater, impacting aquatic ecosystems (Hayes et al., 2010). Antibiotic resistance genes from rivers spread to soil through irrigation, affecting microbial communities, which in turn influence human and animal health through food production (Singer et al., 2016). These interconnections demand a One Health approach to break the cycle of contamination through integrated strategies like advanced water treatment, sustainable agriculture, and ecosystem restoration.

2.6 CHALLENGES AND OPPORTUNITIES

2.6.1 Analytical Challenges

Detecting ECs in surface water and groundwater presents significant analytical challenges due to the distinct characteristics of each system and the complexity of the contaminants involved. Surface water's variability, driven by seasonal flow changes, precipitation events, and dilution, complicates detection, requiring frequent and event-based sampling to capture peak concentrations. For instance, in the U.S., pesticide levels in rivers like the Mississippi spike to 10 μg/L during spring run-off but drop to 0.1 μg/L in dry periods, necessitating high-frequency monitoring (Gilliom et al., 2006). Groundwater's low EC levels, often in the ng/L range, demand high-sensitivity techniques, as seen with PFAS in European aquifers where concentrations of 5–15 ng/L require detection limits below 1 ng/L (Loos et al., 2010). Techniques like liquid chromatography-mass spectrometry (LC-MS) and gas chromatography-mass spectrometry (GC-MS) are widely used but require refinement for complex matrices. In surface water, organic matter and suspended solids cause matrix interference, suppressing signals and reducing accuracy; for example, LC-MS analysis of pharmaceuticals in the Danube River showed a 20% signal suppression due to humic acids (Loos et al., 2010). In groundwater, low concentrations and co-occurring ions like calcium complicate detection, often requiring pre-concentration steps like solid-phase extraction (SPE), which increases analysis time and cost (Verlicchi et al., 2023; Focazio et al., 2008). The high costs of these methods such as LC-MS systems can cost $100,000–$300,000, with additional expenses for maintenance and skilled personnel limit their widespread use, particularly in developing regions where monitoring infrastructure is scarce (Marino et al., 2025; Lapworth et al., 2012). Emerging techniques like high-resolution mass spectrometry (HRMS) offer promise for non-targeted analysis, identifying unknown ECs such as novel PFAS compounds in U.S. rivers (Strynar et al., 2015), but their adoption is hindered by cost and complexity. These challenges highlight the need for affordable, robust analytical methods to improve EC monitoring globally to identify the hotspots (Figure 2.5).

2.6.2 Regulatory Gaps

Regulatory frameworks for ECs in surface water and groundwater remain inadequate, with few standards addressing their growing presence. For example, PFAS guidelines lag behind detections while the U.S. EPA set a health advisory of 70 ng/L for PFOA and PFOS combined, concentrations in European tap water often exceed 100 ng/L, with no enforceable limits in many countries (Postigo & Barceló, 2015). However, the revised European Drinking Water Directive (Directive (EU) 2020/2184) has recently introduced, for the first time, binding limits for PFAS in drinking water, marking a significant regulatory step forward within the EU. In Australia, PFAS levels in groundwater near firefighting bases reached 50–100 ng/L, yet national standards were only introduced in 2018, years after contamination was identified (Baduel et al., 2017). Pharmaceuticals like carbamazepine, detected in U.S. wells at 0.01–0.1 μg/L, lack regulatory limits despite their potential to contribute to antimicrobial resistance (Focazio et al., 2008). Pesticides like atrazine are regulated in some regions. e.g. the EU bans atrazine, but the U.S. allows up to 3 μg/L in drinking water yet detections in groundwater often exceed these limits, reaching 0.1–1 μg/L in the Midwest (Kolpin et al., 2002). The lack of harmonized regulations across countries exacerbates the problem, as ECs move through transboundary water systems like the Ganges River, where ciprofloxacin levels of 0.5–2 μg/L in India lack regulatory oversight (Bhagat et al., 2020). This regulatory fragmentation hinders effective management, as

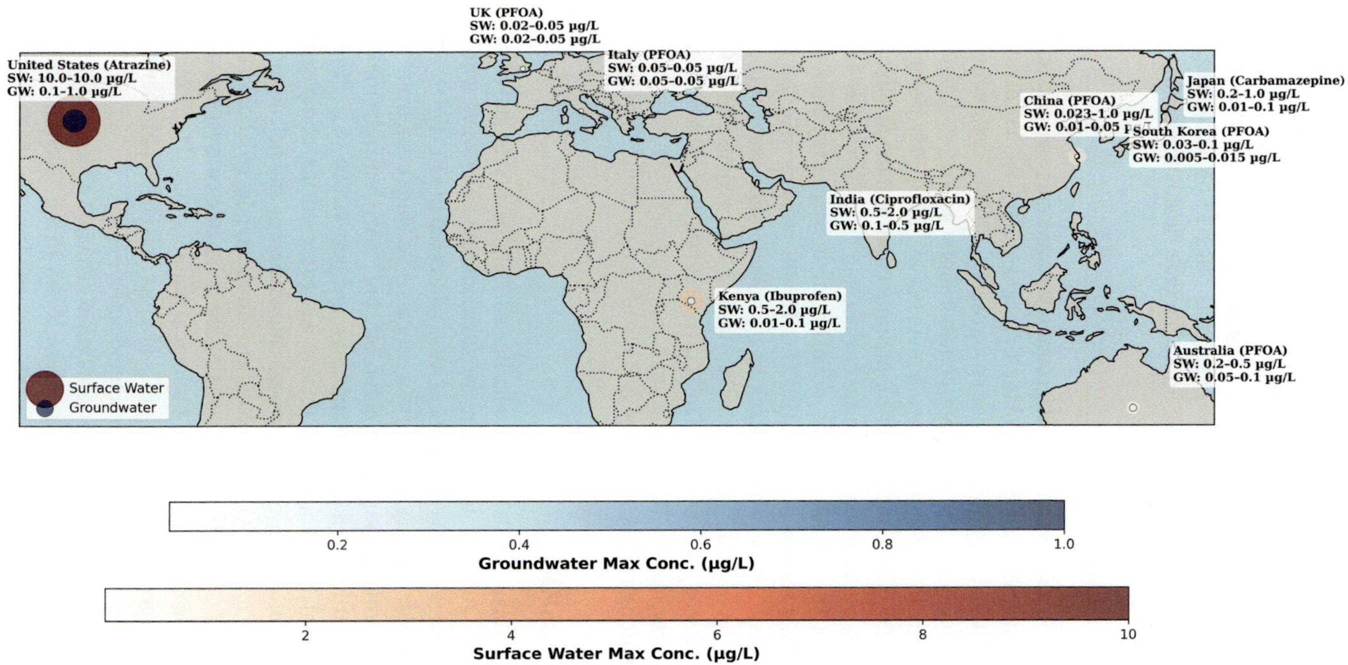

FIGURE 2.5 Global hotspots of emerging contaminants in surface water and groundwater. (Data compiled from published studies cited in the manuscript.)

countries with stricter standards may still face contamination from upstream sources. Harmonized regulations are needed to set science-based limits, improve monitoring and enforce mitigation measures like advanced wastewater treatment. International frameworks, such as the Stockholm Convention which already includes PFOS, its salts, and related compounds as persistent organic pollutants, have taken important steps toward regulation. However, further expansion may be necessary to include a broader range of emerging contaminants such as other PFAS and pharmaceuticals, ensuring a coordinated global response under the One Health approach.

2.6.3 Opportunities

Addressing ECs in surface water and groundwater presents opportunities through innovative technologies, integrated modelling and policy frameworks. Real-time sensors, such as those using fluorescence spectroscopy, enable continuous monitoring of ECs like pharmaceuticals, detecting concentrations as low as 1 ng/L in rivers, which improves early warning systems (Carstea et al., 2016). Hydrological modelling, such as MODFLOW coupled with contaminant transport models like MT3DMS, can predict EC movement between surface water and groundwater, as demonstrated in the Mississippi River Basin, where models accurately forecast atrazine transport with 85% accuracy (Nolan et al., 2012). One Health policies, like the EU's Water Framework Directive, promote interdisciplinary collaboration, integrating health and environmental data to set EC standards (European Commission, 2020). Community engagement enhances awareness and facilitates action through which local initiatives in India, such as the Ganges River clean-up campaigns, have reduced pharmaceutical inputs by 30% through public education and waste management (Sharma et al., 2016).

2.7 CONCLUSION

Emerging contaminants (ECs) such as pharmaceuticals, pesticides and PFAS threaten surface water and groundwater, which are interconnected pathways within the One Health triad. This review has comprehensively addressed the EC types, sources, and chemical properties, identifying pharmaceuticals like ibuprofen from wastewater, pesticides like atrazine from agricultural run-off, and PFAS from industrial discharges, with diverse properties such as high mobility and persistence (e.g. PFOA) driving their environmental behaviour. This review has explored the complexity of ECs, highlighting their diverse origins, chemical characteristics and behaviours in aquatic environments. The impacts extend beyond environmental degradation. Chronic, low-level exposure to ECs affects human health, alters aquatic ecosystems and disrupts the delicate balance of microbial and biological communities. These interconnected effects show the urgency of addressing ECs not just as isolated pollutants, but as part of a systemic challenge requiring coordinated action.

Although progress has been made in analytical detection and modelling, persistent gaps remain in regulation, in public awareness and in our scientific understanding of the full extent of these contaminants. This review demands integrated research to further characterize ECs' diversity and policy development to mitigate their threats, leveraging the One Health framework to safeguard water systems and health across the triad.

REFERENCES

Ahrens, L., Rakovic, J., Ekdahl, S., & Kallenborn, R. (2023). Environmental distribution of per-and polyfluoroalkyl substances (PFAS) on Svalbard: Local sources and long-range transport to the Arctic. *Chemosphere*, *345*, 140463. https://doi.org/10.1016/j.chemosphere.2023.140463

Ankley, G. T., Kuehl, D. W., Kahl, M. D., Jensen, K. M., Linnum, A., Leino, R. L., & Villeneuve, D. A. (2005). Reproductive and developmental toxicity and bioconcentration of perfluorooctanesulfonate in a partial life-cycle test with the fathead minnow (Pimephales promelas). *Environmental Toxicology and Chemistry*, *24*(9), 2316–2324. https://doi.org/10.1897/04-634R.1

Baduel, C., Mueller, J. F., Rotander, A., Corfield, J., & Gomez-Ramos, M. J. (2017). Discovery of novel per-and polyfluoroalkyl substances (PFASs) at a fire fighting training ground and preliminary investigation of their fate and mobility. *Chemosphere*, *185*, 1030–1038. https://doi.org/10.1016/j.chemosphere.2017.06.096

Barnes, K. K., Kolpin, D. W., Furlong, E. T., Zaugg, S. D., Meyer, M. T., & Barber, L. B. (2008). A national reconnaissance of pharmaceuticals and other organic wastewater contaminants in the United States—I) Groundwater. *Science of The Total Environment*, *402*(2–3), 192–200. https://doi.org/10.1016/j.scitotenv.2008.04.028

Battaglin, W. A., Meyer, M. T., Kuivila, K. M., & Dietze, J. E. (2014). Glyphosate and its degradation product AMPA occur frequently and widely in U.S. rivers, streams, and groundwater. *Journal of the American Water Resources Association*, *50*(2), 275–290. https://doi.org/10.1111/jawr.12159

Bhagat, C., Kumar, M., Tyagi, V.K. and Mohapatra, P.K., (2020). Proclivities for prevalence and treatment of antibiotics in the ambient water: a review. *Npj Clean Water*, *3*(1), 42. https://doi.org/10.1038/s41545-020-00087-x

Bradley, P. M., Barber, L. B., Duris, J. W., Foreman, W. T., Furlong, E. T., Hubbard, L. E., ... & Verplanck, P. L. (2017). Riverbank filtration potential of pharmaceuticals in a wastewater-impacted stream. *Environmental Pollution, 230*, 321–332. https://doi.org/10.1016/j.envpol.2014.06.028

Brausch, J. M., & Rand, G. M. (2011). A review of personal care products in the aquatic environment: Environmental concentrations and toxicity. *Chemosphere*, *82*(11), 1518–1532. https://doi.org/10.1016/j.chemosphere.2010.11.018.

Campanale, C., Losacco, D., Triozzi, M., Massarelli, C., & Uricchio, V. F. (2022). An overall perspective for the study of emerging contaminants in karst aquifers. *Resources*, 11(11), *105*. https://doi.org/10.3390/resources11110105

Carstea, E. M., Bridgeman, J., Baker, A., & Banicescu, I. (2016). Fluorescence spectroscopy for wastewater monitoring: A review. *Water Research*, 95, 205–219. https://doi.org/10.1016/j.watres.2016.03.021

Death, C., Bell, S., Champness, D., Milne, R., & Reichman, S. (2021). Per- and polyfluoroalkyl substances (PFAS) in livestock and game species: A review. *Science of The Total Environment*, *774*, 145–156. https://doi.org/10.1016/j.scitotenv.2020.144795

DeLuca, N. M., Angrish, M., Wilkins, A., Thayer, K., & Hubal, E. A. C. (2021). Human exposure pathways to poly-and perfluoroalkyl substances (PFAS) from indoor media: A systematic review protocol. *Environment International*, *146*, 106308. https://doi.org/10.1016/j.envint.2020.106308

European Commission. (2020). Directive 2000/60/EC of the European Parliament and of the Council establishing a framework for Community action in the field of water policy. *Official Journal of the European Union*. http://data.europa.eu/eli/dir/2000/60/oj

Fenton, S. E., Ducatman, A., Boobis, A., DeWitt, J. C., Lau, C., Ng, C., Smith, J. S., & Roberts, S. M. (2021). Per-and polyfluoroalkyl substance toxicity and human health review: Current state of knowledge and strategies for informing future research. *Environmental Toxicology and Chemistry*, *40*(3), 606–630. https://doi.org/10.1002/etc.4890

Focazio, M. J., Kolpin, D. W., Barnes, K. K., Furlong, E. T., Meyer, M. T., Zaugg, S. D., ... & Thurman, E. M. (2008). A national reconnaissance for pharmaceuticals and other organic wastewater contaminants in the United States—II) Untreated drinking water sources. *Science of The Total Environment*, *402*(2–3), 201–216. https://doi.org/10.1016/j.scitotenv.2008.02.021

Gilliom, R. J., Barbash, J. E., Crawford, C. G., Hamilton, P. A., Martin, J. D., Nakagaki, N., ... & Wolock, D. M. (2006). The quality of our nation's waters: Pesticides in the nation's streams and ground water, 1992–2001. *U.S. Geological Survey Circular, 1291*. https://doi.org/10.3133/cir1291

Halden, R. U., & Paull, D. H. (2005). Co-occurrence of triclocarban and triclosan in U.S. water resources. *Environmental Science & Technology*, *39*(6), 1420–1426. https://doi.org/10.1021/es049071e

Hantush, M. M., Marino, M. A., & Islam, M. R. (2000). Models for leaching of pesticides in soils and groundwater. *Journal of Hydrology*, *227*(1–4), 66–83. https://doi.org/10.1016/S0022-1694(99)00166-3

Hao, C., Lissemore, L., Nguyen, B., Kleywegt, S., Yang, P., & Solomon, K. (2007). Determination of pharmaceuticals in environmental waters by liquid chromatography/electrospray ionization/tandem mass spectrometry. *Analytical and Bioanalytical Chemistry*, *387*(4), 1247–1257 https://doi.org/10.1007/s00216-005-0199-y

Hayes, T. B., Anderson, L. L., Beasley, V. R., de Solla, S. R., Iguchi, T., Ingraham, H., ... & Willingham, E. (2010). Demasculinization and feminization of male gonads by atrazine: Consistent effects across vertebrate classes. *Journal of Steroid Biochemistry and Molecular Biology, 123*(2–3), 123–132 https://doi.org/10.1016/j.jsbmb.2011.03.015

Hekster, F. M., Laane, R. W. P. M., & de Voogt, P. (2003). Environmental and toxicity effects of perfluoroalkylated substances. *Reviews of Environmental Contamination and Toxicology, 179*, 99–121. https://doi.org/10.1007/0-387-21731-2_4

Higgins, C. P., & Luthy, R. G. (2006). Sorption of perfluorinated surfactants on sediments. *Environmental Science & Technology*, *40*(23), 7251–7256. https://doi.org/10.1021/es061000n

Jassal, P.S., Kaur, D., Kaur, M. & Sharma, D. (2023). Level of antibiotic contamination in the major river systems: A review on South Asian countries perspective. *Journal of Applied Pharmaceutical Science*, *13*(6), 010–017. https://doi.org/10.7324/JAPS.2023.56748

Karásková, P., Venier, M., Melymuk, L., Bečanová, J., Vojta, Š., Prokeš, R., Diamond, M. L., & Klánová, J. 2016. Perfluorinated alkyl substances (PFASs) in household dust in Central Europe and North America. *Environment International*, *94*, 315–324 https://doi.org/10.1016/j.envint.2016.05.031

Katz, B. G., Griffin, D. W., & Davis, J. H. (2011). Groundwater quality impacts from the land application of treated municipal wastewater in a karst region, Florida. *Hydrogeology Journal*, *19*(3), 529–542. https://doi.org/10.1016/j.scitotenv.2009.01.022

Kolpin, D. W., Furlong, E. T., Meyer, M. T., Thurman, E. M., Zaugg, S. D., Barber, L. B., & Buxton, H. T. (2002). Pharmaceuticals, hormones, and other organic wastewater contaminants in U.S. streams, 1999–2000: A national reconnaissance. *Environmental Science & Technology*, *36*(6), 1202–1211. https://doi.org/10.1021/es011055j

Kotthoff, M., Müller, J., Jürling, H., Schlummer, M., & Fiedler, D. (2015). Perfluoroalkyl and polyfluoroalkyl substances in consumer products. *Environmental Science and Pollution Research*, *22*(19), 14546–14559. https://doi.org/10.1007/s11356-015-4202-7

Kumar, M., Ram, B., Honda, R., Poopipattana, C., Canh, V. D., Chaminda, T., & Furumai, H. (2019). Concurrence of antibiotic resistant bacteria (ARB), viruses, pharmaceuticals and personal care products (PPCPs) in ambient waters of Guwahati, India: Urban vulnerability and resilience perspective. *Science of The Total Environment*, *693*, 133640. https://doi.org/10.1016/j.scitotenv.2019.133640

Kuroda, K., & Kobayashi, J. (2021). Pharmaceuticals, personal care products, and artificial sweeteners in Asian groundwater: A review. In Manish Kumar, Daniel D. Snow, Ryo Honda & Santanu Mukherjee (eds), *Contaminants in Drinking and Wastewater Sources: Challenges and Reigning Technologies*, pp. 3–36. Springer. https://doi.org/10.1007/978-981-15-4599-3_1

Lapworth, D. J., Baran, N., Stuart, M. E., & Ward, R. S. (2012). Emerging organic contaminants in groundwater: A review of sources, fate and occurrence. *Environmental Pollution*, *163*, 287–303 https://doi.org/10.1016/j.envpol.2011.12.034

Liu, Q., Luo, X., Zheng, Z., Zheng, B., Zhang, J., Zhao, Y., Yang, X., Wang, J., & Wang, L. (2011). Factors that have an effect on degradation of diclofenac in aqueous solution by gamma ray irradiation. *Environmental Science and Pollution Research*, *18*, 1243–1252.

Lohmann, R., Booij, K., Smedes, F., & Vrana, B. (2012). Use of passive sampling devices for monitoring and compliance checking of POP concentrations in water. *Environmental Science and Pollution Research*, *19*, 1885–1895. https://doi.org/10.1007/s11356-012-0748-9

Loos, R., Gawlik, B. M., Locoro, G., Rimaviciute, E., Contini, S., & Bidoglio, G. (2010). EU-wide survey of polar organic persistent pollutants in European river waters. *Environmental Pollution*, *157*(2), 561–568. https://doi.org/10.1016/j.envpol.2008.09.020

Lopez-Espinosa, M. J., Mondal, D., Armstrong, B. G., Eskenazi, B., & Fletcher, T. (2016). Perfluoroalkyl substances, sex hormones, and insulin-like growth factor-1 at 6–9 years of age: a cross-sectional analysis within the C8 Health Project. *Environmental Health Perspectives*, *124*(8), 1269–1275. https://doi.org/10.1289/ehp.1509869

Mackay, D., Shiu, W. Y., Ma, K. C., & Lee, S. C. (2006). *Handbook of physical-chemical properties and environmental fate for organic chemicals* (2nd ed.). CRC Press. https://doi.org/10.1201/9781420044393

Marino, L., Gagliano, E., Santoro, D., & Roccaro, P. (2024). Online control of UV and UV/H2O2 processes targeted for the removal of contaminants of emerging concern (CEC) by a fluorescence sensor. *Journal of Hazardous Materials*, *480*, 136075. https://doi.org/10.1016/j.jhazmat.2024.136075

Marino, L., Gagliano, E., Santoro, D., & Roccaro, P. (2025). Fluorescence sensor enabled control of contaminants of emerging concern in reclaimed wastewater using ozone-based treatment processes. *Water Research*, *268*, 122616. https://doi.org/10.1016/j.watres.2024.122616

Mastrantonio, M., Bai, E., Uccelli, R., Cordiano, V., Screpanti, A., & Crosignani, P. (2018). Drinking water contamination from perfluoroalkyl substances (PFAS): An ecological mortality study in the Veneto Region, Italy. *European Journal of Public Health*, *28*(1), 180–185. https://doi.org/10.1093/eurpub/ckx066

Morrissey, C. A., Mineau, P., Sanchez-Bayo, F., Lindenmayer, D. B., Ortiz-Santaliestra, M. E., & Skelly, D. K. (2015). Neonicotinoid contamination of global surface waters and associated risk to aquatic invertebrates: A review. *Environmental International*, 74, 291–303. https://doi.org/10.1016/j.envint.2014.10.024

Muir, D., Bossi, R., Carlsson, P., Evans, M., De Silva, A., Halsall, C., ... & Wilson, S. (2019). Levels and trends of poly- and perfluoroalkyl substances in the Arctic environment—An update. *Emerging Contaminants, 5*, 240–250. https://doi.org/10.1016/j.emcon.2019.06.002

Mutiyar, P. K., & Mittal, A. K. (2014). Occurrences and fate of selected pharmaceuticals in the Yamuna River, India. *Environmental Monitoring and Assessment*, *186*(1), 541–553. https://doi.org/10.1007/s10661-013-3398-6

Nakada, N., Kiri, K., Shinohara, H., Harada, A., Kuroda, K., Takizawa, S., & Takada, H. (2008). Evaluation of pharmaceuticals and personal care products as water-soluble molecular markers of sewage. *Environmental Science & Technology*, *42*(17), 6347–6353. https://doi.org/10.1021/es7030856

Nguyen, H. T., Thai, P., Kaserzon, S., O'Brien, J., & Mueller, J. F. (2018). Nationwide Occurrence and Discharge Mass Load of Per- and Polyfluoroalkyl Substances in Effluent and Biosolids: A Snapshot from 75 Wastewater Treatment Plants Across Australia. Available at SSRN: https://ssrn.com/abstract=4668557 or http://dx.doi.org/10.2139/ssrn.4668557

Nolan, B. T., Malone, R. W., Gronberg, J. A., Thorp, K. R., & Ma, L. (2012). Verifiable metamodels for nitrate losses to drains and groundwater in the Corn Belt, USA. *Environmental Science & Technology*, *46*(2), 901–908. https://doi.org/10.1021/es202875e

Postigo, C., & Barceló, D. (2015). Synthetic organic compounds and their transformation products in groundwater: Occurrence, fate and mitigation. *Science of The Total Environment*, *503*, 32–47. https://doi.org/10.1016/j.scitotenv.2014.06.019

Ricolfi, L., Yang, Y., Pottier, P., Morrison, K., Williams, C., Pollo, P., Hesselson, D., Neely, G., Taylor, M. D., Nakagawa, S., & Lagisz, M. (2025). A meta-analysis reveals PFAS concentrations double with each trophic level across aquatic and terrestrial food webs. https://doi.org/10.32942/X2SP92

Schwaiger, J., Ferling, H., Mallow, U., Wintermayr, H., & Negele, R. D. (2004). Toxic effects of the non-steroidal anti-inflammatory drug diclofenac: Impact on fish. *Aquatic Toxicology*, *68*(2), 141–150. https://doi.org/10.1016/j.aquatox.2004.03.014

Sharma, B. M., Bharat, G. K., Tayal, S., Nizzetto, L., & Larssen, T. (2016). Perfluoroalkyl substances (PFAS) in river and groundwater systems of India: Sources and implications. *Environmental Pollution*, *214*, 153–161. https://doi.org/10.1016/j.envpol.2015.10.050

Singer, A. C., Shaw, H., Rhodes, V., & Hart, A. (2016). Review of antimicrobial resistance in the environment and its relevance to environmental regulators. *Frontiers in Microbiology*, 7, 1728. https://doi.org/10.3389/fmicb.2016.01728

Sorensen, J. P. R., Lapworth, D. J., Nkhuwa, D. C. W., Stuart, M. E., Gooddy, D. C., Bell, R. A., & Pedley, S. (2015). Emerging contaminants in urban groundwater sources in the UK. *Science of The Total Environment*, *536*, 987–997. https://doi.org/10.1016/j.watres.2014.08.002

Strynar, M., Dagnino, S., McMahen, R., Liang, S., Lindstrom, A., Andersen, E., ... & Ball, C. (2015). Identification of novel perfluoroalkyl ether carboxylic acids (PFECAs) and other PFAS in U.S. surface waters using high-resolution mass spectrometry. *Environmental Science & Technology*, *49*(19), 11622–11630. https://doi.org/10.1021/acs.est.5b01215

Sunderland, E. M., Hu, X. C., Dassuncao, C., Tokranov, A. K., Wagner, C. C., & Allen, J. G. (2019). A review of the pathways of human exposure to poly- and perfluoroalkyl substances (PFAS) and present understanding of health effects. *Journal of Exposure Science & Environmental Epidemiology*, *29*(2), 131–147. https://doi.org/10.1038/s41370-018-0094-1

Ternes, T. A., Joss, A., & Siegrist, H. (2004). Scrutinizing pharmaceuticals and personal care products in wastewater treatment. *Environmental Science & Technology*, *38*(20), 392A–399A. https://pubs.acs.org/doi/pdf/10.1021/es040639t

Tran, N. H., Reinhard, M., Khan, E., Chen, H., Nguyen, V. T., Li, Y., Goh, S. G., Nguyen, Q. B., Saeidi, N., & Gin, K. Y. H. (2019). Emerging contaminants in wastewater, stormwater runoff, and surface water: Application as chemical markers for diffuse sources. *Science of The Total Environment*, *676*, 252–267. https://doi.org/10.1016/j.scitotenv.2019.04.160

Verlicchi, P., Lacasa, E., & Grillini, V. (2023). Quantitative and qualitative approaches for CEC prioritization when reusing reclaimed water for irrigation needs–A critical review. *Science of The Total Environment*, *900*, 165735. https://doi.org/10.1016/j.scitotenv.2023.165735

Wang, T., Wang, P., Meng, J., Liu, S., Lu, Y., & Khim, J. S. (2017). A review of sources, multimedia distribution and health risks of perfluoroalkyl acids (PFAAs) in China. *Chemosphere*, *171*, 592–605 https://doi.org/10.1016/j.chemosphere.2014.09.021

World Health Organization. (2017). *One Health: A framework for tackling complex global health challenges*. WHO Press. www.who.int/news-room/fact-sheets/detail/one-health#:~:text=One%20Health%20is%20an%20approach,animals%2C%20both%20wild%20and%20domestic

Zhang, Y., Liu, X., Yu, L., Hua, Z., Zhao, L., Xue, H., & Tong, X. (2022). Perfluoroalkyl acids in representative edible aquatic species from the lower Yangtze river: Occurrence, distribution, sources, and health risk. *Journal of Environmental Management*, *317*, 115390. https://doi.org/10.1016/j.jenvman.2022.115390

Zushi, Y., & Masunaga, S. (2009). First-flush loads of perfluorinated compounds in stormwater runoff from Hayabuchi River basin, Japan. *Science of The Total Environment*, *408*(2), 426–434. https://doi.org/10.1016/j.chemosphere.2009.04.004

3 Existing Policies and Regulatory Measures for Management of Emerging Contaminants

G. Shyamala, Ravindran Gobinath, and Shirisha Balle

3.1 INTRODUCTION

3.1.1 Emerging Contaminants: Definition and Significance

Synthetic or naturally occurring chemicals and microorganisms along with contaminants are emerging contaminants (ECs). Emerging contaminants are novel or previously unrecognized substances, often anthropogenic, that are present in the environment and have the potential to pose risks to human health or ecosystems. These contaminants are neither frequently detected nor regulated in environmental systems, yet present unknown risks to human and ecological health. Scientists have identified pharmaceuticals, endocrine-disrupting compounds, perfluoroalkyl substances (PFAS), and microplastics together with nanomaterials under emerging contaminants alongside other categories. The observation and discovery of emerging contaminants is typically associated with increased industrial activity and rapid urban growth, combined with improved analytical methods for measuring trace contaminants (S. Khan et al., 2022; Hassan et al., 2022; Pena-Pereira et al., 2021). Ecological concerns about ECs derive from their persistence and accumulation potential, which results in long-term ecological consequences. Surface water and wastewater samples contain pharmaceuticals and personal care products that damage the reproductive behaviors and biological responses of aquatic organisms. PFAS, also known as "forever chemicals," demonstrate such outstanding persistence in the environment that they create ongoing anxiety about water contamination and health-related concerns (X. Li et al., 2024).

3.1.2 Challenges in Managing Emerging Contaminants

Figure 3.1 shows the complexity of managing ECs arises from their distinct properties together with fluctuations in emerging pollutants. Ultra-trace EC detection

DOI: 10.1201/9781032623801-3

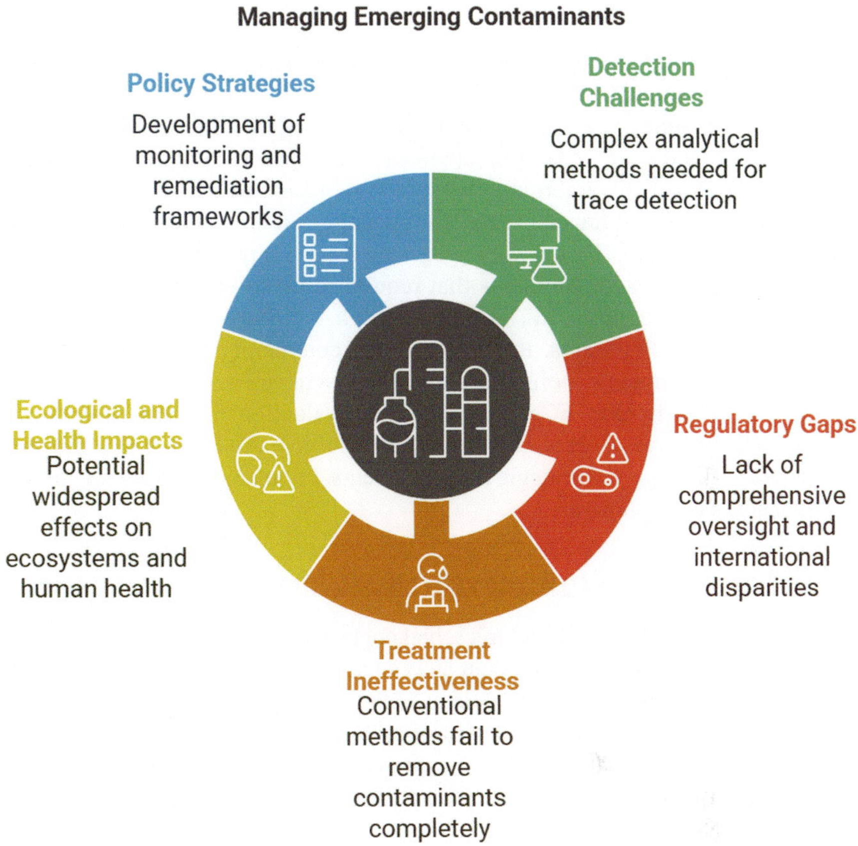

FIGURE 3.1 Key challenges and policy strategies for managing emerging contaminants.

requires sophisticated methods, including liquid chromatography-tandem mass spectrometry (LC-MS/MS) (Martín-Pozo et al., 2019). The expensive equipment coupled with the necessary specialized knowledge makes these technologies unreachable in many developing countries. Ecological research on numerous ECs remains incomplete because of insufficient toxicological data, which impede the development of protective exposure thresholds. Scientists are yet to fully understand the chronic impacts of pharmaceutical combination exposure. Classic environmental controls have established regulatory measures, mainly for pollutants that have already been discovered. Existing regulatory frameworks fail to include ECs, thus creating substantial holes in risk management planning). The combination of industrial procedures, worldwide trade patterns, and extensive chemical usage produces EC distribution throughout water spaces, as well as terrestrial and atmospheric environments. To handle these challenges properly, nations must work together with unified monitoring guidelines (Tran et al., 2019). Antibiotics found in aquatic ecosystems lead to the development of antimicrobial resistance, which creates concerning public health dangers.

3.1.3 Objective of the Chapter

This chapter presents a complete assessment of ecocentric policy frameworks, combined with international regulatory controls designed for the effective management of emerging contaminants. The European Union's (EU's) Water Framework Directive includes the Environmental Quality Standards Directive, and its Contaminants of Emerging Concern Watch List, which serve as dynamic water pollution surveillance tools for aquatic environments. Through its Contaminant Candidate List (CCL) program, the US Environmental Protection Agency (EPA) selects contaminants that require research or regulatory attention (Hendry, 2017; Sudarsan et al., 2024). Through the Stockholm Convention on Persistent Organic Pollutants, the international community acts to monitor and control persistent organic pollutants, and their standard-setting includes types such as PFAS and other ECs. A systematic evaluation of these frameworks exposes their positive aspects, together with insufficient elements and potential growth areas, leading to strategic improvements for managing ECs in modern interconnected environments.

3.2 DEFINING AND CLASSIFYING EMERGING CONTAMINANTS

Synthetic or natural chemicals and microbial contaminants, known as ECs, typically remain unmonitored in the environment and can negatively affect both environmental and human health based on their unique monitoring profiles. Various substances known as emerging contaminants consist of multiple chemical categories, including drugs, anti-pollution products (PPCPs), nanomaterials, assorted industrial elements, and contraceptives (Shanmuganathan et al., 2023; Rathi et al., 2021; Osuoha et al., 2023; Patel et al., 2019).

3.2.1 Comprehensive Definition of Emerging Contaminants

Throughout the environment, ECs are found everywhere because they originate from human activities in the water, soil, and atmosphere. Despite being detected at low concentrations, the potential of ECs to cause environmental harm, even in minute amounts, remains unregulated. Various technical definitions illustrate these compounds as recently produced substances that resist traditional waste processing systems and show potential for biological build-up. According to the EPA, ECs are untracked substances that demonstrate potential harm to both environmental and human health systems (Mohapatra & Kirpalani, 2019).

3.2.2 Classification Schemes for Emerging Contaminants

Given the diverse chemical properties and environmental behaviors of ECs, several classification schemes have been proposed to effectively categorize them, as shown in Figure 3.2. These schemes help in understanding their sources, mechanisms of toxicity, and regulatory implications.

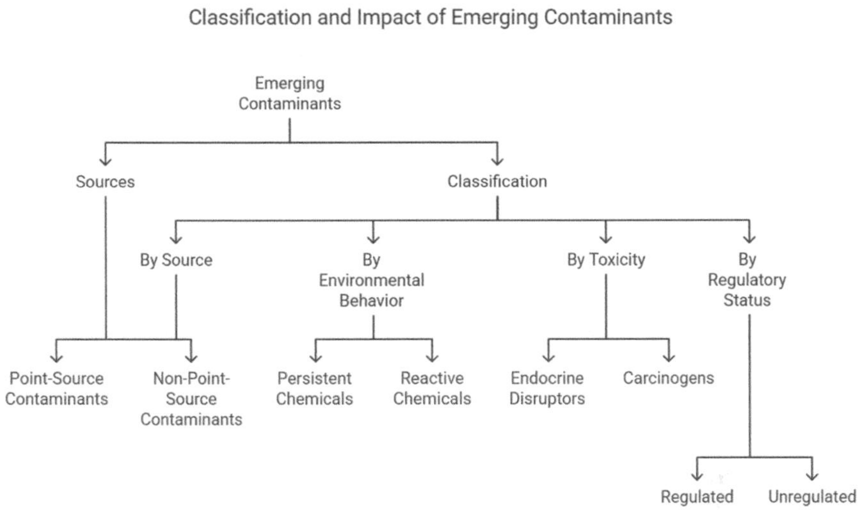

FIGURE 3.2 Classification and impact of emerging contaminants.

- **Based on Source**
  ECs originate from specific discharge points such as pharmaceutical industries, hospitals, and municipal wastewater treatment plants. ECs are released from diffuse sources, including agricultural run-off containing pesticides and fertilizers (Chen et al., 2024).
- **Based on Environmental Behavior**
  Compounds with high environmental persistence and bioaccumulative potential include polychlorinated biphenyls (PCBs) (Kümmerer, 2024). Substances that degrade into equally or more toxic metabolites, such as pharmaceuticals, undergo photolysis (Postigo, 2023).
- **Based on Toxicity**
  Chemicals that interfere with hormonal systems, such as bisphenol A (BPA) and phthalates can have significant health impacts. Substances that cause immediate or long-term toxic effects, such as nanomaterials, can negatively affect aquatic species (Ismanto et al., 2022; Kasonga et al., 2021).
- **Based on Regulatory Status**
  Chemicals are subject to existing legal frameworks, such as arsenic, under the Safe Drinking Water Act. Compounds without formal standards are monitored under programs such as the EPA's Unregulated Contaminant Monitoring Rule (UCMR) (Puri et al., 2023; Sousa et al., 2018).
- **Based on Physical and Chemical Properties**
  Pharmaceuticals with high water solubility can pose environmental risks due to their persistence in water systems. Persistent compounds, such as polybrominated diphenyl ethers (PBDEs), accumulate in sediments and biota (Ghattas et al., 2017).

The classification framework aims to provide a structured approach for studying the sources, fate, and impacts of ECs, enabling policymakers and scientists to prioritize mitigation strategies.

3.3 EXISTING REGULATORY FRAMEWORKS

Current regulatory guidelines for managing ECs work through international agreements along with domestic operational programs. Treaties, including the Stockholm Convention on POPs and the Rotterdam Convention on PIC, control dangerous chemicals at the worldwide level, while the World Health Organization (WHO) and United Nations Environment Programme (UNEP) work together with organizations to deliver research directions and policy recommendations. Governments at the national level apply legal frameworks for water quality, waste disposal, and chemical and air pollution management to deal with emerging contaminants explicitly or through related measures. The EPA and the ECHA work under regulatory supervision to assess risks through authorization and monitoring, which establishes exposure restrictions and implements pollution control requirements. These entities play a critical role in linking scientific analysis with policy development to protect environmental and public health through their activities.

3.3.1 Relevant International Treaties and Conventions

Through the Stockholm Convention, countries have worked to ban or tightly control the creation and application of persistent organic pollutants. The Stockholm Convention functions worldwide as an agreement intended to defend human well-being and environmental preservation from chemicals that survive long periods in natural settings while migrating across broad areas, then transferring to human and wild creature fatty tissue and demonstrating damaging effects on health and the environment (Liu et al., 2021). The Convention maintains strong connections with the Rotterdam Convention and the Basel Convention because these instruments provide integrated solutions to handle hazardous chemical and waste management issues (Templeton, 2020). Through the Rotterdam Convention, countries gain access to the vital information needed to make decisions about importing hazardous chemicals. Through the Convention, parties establish a system for united responsibility that leads to cooperative actions between exporters and importers in managing hazardous chemicals to safeguard environmental well-being (Templeton, 2020). The Convention specifies that exporters must achieve importer authorization before executing transactions involving these chemicals (Liu et al., 2021).

WHO functions as an essential leader in supervising research activities and generating directions for the management of environmental contaminants. The organization evaluates harmful chemical exposure and then develops protective guidelines for public wellness. Through partnerships with international organizations, the organization strives to embed health perspectives through environmental policies and treaty frameworks (Elsässer, 2024). Through UNEP policy development initiatives, countries receive technical assistance to establish waste and chemical management frameworks, as shown in Figure 3.3, while maintaining a sustainable development alignment (He et al., 2022).

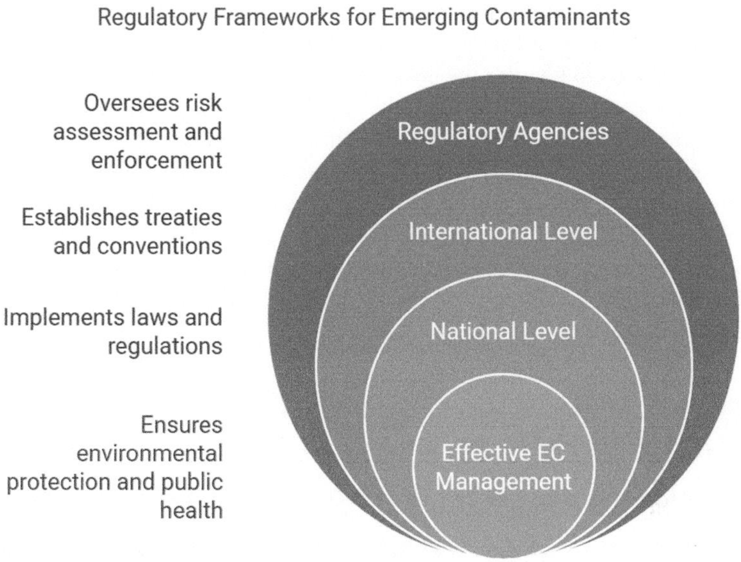

FIGURE 3.3 Regulatory frameworks for emerging contaminants.

3.3.2 National Level

Legislation and regulations regarding water quality, along with air quality, waste management, and chemical control, exhibit robust regional differences across countries. The Water Framework Directive (WFD) operates as the EU's primary law for protecting water resources against chemical pollutants (Hendry, 2017). The WFD encounters numerous obstacles to extensive chemical pollution control (Brack et al., 2017). Chinese environmental authorities prioritize wastewater treatment plant effluent regulations by focusing on pharmaceutical contaminants and endocrine-disrupting chemicals (Yang et al., 2024). The Latin American region has minimal regulatory laws regarding emerging pollutants, resulting in major contamination issues (Vargas-Berrones et al., 2020). Current legal frameworks deal with emerging contaminants, particularly through general provisions instead of detailed regulations, as shown in Figure 3.3. The EU's WFD promotes chemical status assessment procedures that indirectly target emerging contaminants through an effect-based methodology and prioritizes contaminant selection (Brack et al., 2017). The authorities in China are working on setting enduring water quality benchmarks for hazardous suspended compounds, although most pollutants do not yet satisfy regulatory thresholds (Yang et al., 2024). Specific regulations governing endocrine-disrupting chemicals such as nonylphenol remain absent throughout Latin America, thus requiring dedicated regulatory structures (Vargas-Berrones et al., 2020).

Regulatory agencies fulfill critical duties in the management of Environmental Control systems, as shown in Figure 3.3. The EPA works alongside the ECHA of the EU to assess and handle chemical risks. These agencies fulfill the mission to create guidelines alongside risk assessment duties and active monitoring responsibilities to protect both public health and environmental health (Parida et al., 2021;

Petrie et al., 2015). Oversight bodies use diverse methods to evaluate risks, authorize ECs, and monitor these substances. The EPA and ECHA rely on established risk assessment methods to identify the possible health and environmental consequences of ECs. Assessment frameworks require toxicological evaluation procedures to measure permissible drinking water value limits (Fischer et al., 2017). Efforts to monitor contaminants combine effect-based testing with passive sample collection methods to identify accumulating chemicals together with consolidated systems to establish prioritization. Business decision support systems operating in specific regions present stakeholders with resourceful techniques to direct EC management throughout neighborhood water routines beyond standard engineering solutions (Baken et al., 2018).

3.4 POLICY APPROACHES FOR MANAGING EMERGING CONTAMINANTS

3.4.1 Policies to Minimize the Release of Emerging Contaminants (ECs)

Source reduction serves as a method to reduce EC utilization while preventing the initial release. The reduction of EC-containing waste requires both selective and waste-generation reduction strategies. Covering an area with urine-diverting toilets constitutes a recommended strategic approach for reducing EC pollution in groundwater, specifically in South African territories that present unregulated waste as a significant contamination source. Wastewater treatment facilities can employ comprehensive water management systems through technological updates that reduce environmental ECs (Gaston et al., 2019; Xabadia et al., 2021).

Manufacturers maintain product stewardship duties by responsible environmental management of their products, from design through operational phases to end-of-life disposal. Manufacturers should develop environment-friendly products while building disposal systems to protect ambient conditions. The absence of official regulations for many exposed chemicals, including pharmaceuticals and personal care products, requires manufacturers to undertake proactive steps toward minimizing the environmental impact of their products (Sutherland & Ralph, 2019). Manufacturers should create products with decreased EC content as part of a strategy to reduce release amounts, and customers should be educated about how their products affect the environment (Pereira et al., 2015).

The core principle of green chemistry is to create chemical products and processes designed to suppress the use of hazardous substances, as well as their quantity in production. Using this method, scientists can develop alternative products that do not remain in the environment while supporting the protection of natural systems. Biodegradable flame retardants and pesticides created by developers decrease negative environmental consequences (Pereira et al., 2015). Furthermore, innovative treatment technologies, such as the use of microalgae for bioremediation, offer promising solutions for detoxifying ECs through bioadsorption, biouptake, and biodegradation, transforming them into less harmful compounds (Sutherland & Ralph, 2019).

3.4.2 Policies for Controlling the Release and Fate of Emerging Contaminants (ECs)

Emerging contaminants (ECs) pose significant challenges to conventional wastewater treatment plants (WWTPs), which are not typically designed to effectively remove these pollutants. Traditional methods, such as activated sludge and membrane bioreactors, often fail to eliminate ECs such as pharmaceuticals, personal care products, and microplastics (Corpus et al., 2024; Gogoi et al., 2018; Lv et al., 2019). Advanced treatment technologies, including advanced oxidation processes (AOPs), adsorption-based methods, and novel biological treatments, have been explored to enhance their removal efficiencies (Rizzo et al., 2019). Hybrid systems combining biological and physicochemical treatments, such as membrane bioreactors followed by reverse osmosis, have shown promise for effectively removing a range of ECs (Dhangar & Kumar, 2020). However, these advanced methods often result in increased energy consumption and cost, highlighting the need for sustainable and cost-effective solutions (Shah et al., 2020).

The core responsibility of the EC management approach is to manage water systems; however, preventive air control actions remain vitally important because they block the vaporization of treatment process contaminants. WWTPs benefit from air pollution control systems that reduce the emission of volatile organic compounds and other airborne ECs; however, the academic literature examines these solutions to a lesser extent than those focused on water treatment (Khan et al., 2021). Successful waste management techniques need to be developed to limit the environmental discharge of ECs. Sludge and biosolid disposal practices need to be properly managed because wastewater treatment processes allow EC accumulation in the generated solids. The reduction of microplastic fibers from laundry facilities, along with a plastic microbead ban, represents essential strategies to limit the entry of ECs into wastewater systems. Environmental protection requires the establishment of rules and recommendations for treating waste materials containing ECs (Tran et al., 2018).

3.4.3 Policies for Monitoring Emerging Contaminants (ECs)

The scientific community has identified ECs as an important environmental problem because of their destructive impact on both ecosystems and human health. Monitoring policies for ECs requires a complete framework that includes standardized sampling procedures, advanced analytical technology, and regulated national monitoring frameworks. Standardized methods for environmental sample collection strengthen the quality of representative data collected from different environments, such as water bodies, soil, and sludge deposits. Comprehensive data can help researchers identify spatial relationships and temporal fluctuations in environmentally released endocrine disruptors. The detection of multiple environmental chemicals in both liquid and solid matrices has become possible through the development of analytical approaches that use ultra-high-performance liquid chromatography-tandem mass spectrometry (Petrie et al., 2016).

National-level tracking systems support the evaluation of the environmental risks generated by ECs. The Indian population requires a detailed monitoring system to evaluate exposure risks, because area studies have documented excessive EC concentrations (Philip et al., 2018). The European NORMAN Association facilitates scientific data exchange about ECs through its collaborative network to advance the harmonization of measurement methods throughout the region.

Early warning systems are vital for predicting and mitigating the impacts of ECs on the environment and public health. Such systems rely on environmental fate models and innovative monitoring tools. Environmental fate models (EFMs) are used to predict the distribution and behavior of ECs in environmental matrices. These models help in understanding the life-cycle distribution of ECs and their potential ecological impacts (Suhendra et al., 2020). EFMs can be classified into material flow analysis, multimedia compartmental, and spatial river/watershed models, each offering different insights into the fate of ECs (Suhendra et al., 2020). The development of new assessment and modeling tools, such as effect-directed analysis (EDA) and toxicity identification and evaluation (TIE), helps establish cause-effect relationships between chemical contamination and ecological quality deterioration. Voltammetric techniques are emerging as cost-effective and sensitive methods for detecting ECs in water samples, thus providing valuable data for early warning systems (das Chagas et al., 2023).

The ongoing evaluation and tracking of ECs across different ecosystems requires environmental surveillance programs. These programs unify various methodologies to deliver complete assessment capabilities. Researchers need to implement integrated analysis techniques that incorporate both targeted screening and non-targeted screening with biological assays for a complete understanding of the ecological effects of ECs. This analytic strategy enables the identification of compound transformation products that generate toxic effects comparable to those of their original chemical agents. Evaluation through programmed thermal analysis and Raman spectroscopy revealed the necessity for advanced analytical approaches in surveillance programs that measure carbon nanotubes in complex matrices. Studies examining how microbial agents break down ECs (synthetic musks and pharmaceuticals) suggest that biological clean-up methods should become an integral part of environmental tracking initiatives. New bacterial strains that perceive the EC degradation ability strengthen natural ecosystem protection methods and apply them to remediation solutions (Azaroff et al., 2021).

3.4.4 Policies Promoting Research and Development

Enhancing analytical capabilities and standard detection approaches should be the focus of policies designed to identify and characterize new ECs. To identify ECs effectively and understand their environmental effects, researchers require analytical solutions that merge specific substance screening methods with biological detection systems. New policies must fund advancements in high-resolution mass spectrometry that enable researchers to identify and measure transformation products and enantiomers despite their limited detection rates and substantial harmful effects on organisms. Joint efforts between regulatory agencies and investigative research facilities will promote EC data sharing along with methodology exchange for an expanded understanding of their composition (Naidu et al., 2016).

Environmental assessment of ECs requires research policies that embrace investigations of their pathways in different environments while studying their longevity and ability to concentrate within ecosystems. Research investments are needed to evaluate EC physicochemical properties, specifically the octanol-water partitioning coefficient and solid-water distribution coefficient, because they help anticipate environmental movements (Rout et al., 2021). Public policies must adopt standardized risk assessment tools that blend ecotoxicological research to gauge how ECs affect ecosystems while protecting human health (Parida et al., 2021). A full understanding of contaminant behavior requires an assessment of suspended particulate matter in sludge samples within environmental monitoring frameworks.

Open policies must establish powerful incentives to accelerate the creation and deployment of advanced removal techniques that efficiently extract ECs from wastewater, together with other environmental zones. Research support for advanced oxidation processes and membrane bioreactors, alongside hybrid treatment methods, continues to receive attention because these methods demonstrate robust removal capabilities for various ECs (Yap et al., 2019; Shahid et al., 2021). The implementation of constructed wetlands and algae-based systems, together with sustained efforts to find economically friendly technological approaches, supports greater efficiency in EC remediation practices (Zhou et al., 2024). Public policy should promote research on enzyme-assisted biodegradation as a promising method for the biotransformation of contaminated sites (Bilal et al., 2019). Treatment technology innovation enables regulatory policies to reduce environmental pollution risks and health risks from ECs.

3.5 CHALLENGES AND LIMITATIONS OF CURRENT POLICIES

The provided Table 3.1 presents crucial issues and constraints in the existing policies and regulatory systems aimed at managing emerging ECs. The lack of necessary data along with difficult analysis and insufficient risk evaluation hinders regulators from creating successful regulations. The insufficient standardization of monitoring protocols alongside regulatory boundaries allows many ECs to remain free from any oversight even though their dangers are known. ECs present international collaborative obstacles because they cross borders and nations have different regulatory systems. The solution to environmental risks and health problems resulting from ECs requires worldwide cooperation together with enhanced analytical tools, also combined with enhanced regulatory practices.

3.6 FUTURE DIRECTIONS

Table 3.2 presents directions which will enhance the management strategies for ECs. The table shows that accomplished management needs learning capacities from new science findings, while innovation drives purity procedures development and public education plays a key role in fighting EC risks. A critical element for managing ECs consists of strengthening international partnerships, which enables data exchanges between entities while implementing harmonized rules and regulations. The measures established today will lead to the development of successful strategies that reduce environmental as well as health consequences from ECs.

TABLE 3.1
Challenges and limitations of current policies in managing emerging contaminants

Challenge	Explanation	Citations
Data Gaps	There is a significant lack of comprehensive data on the occurrence, fate, and long-term effects of many ECs. This hinders the ability to assess their risks and develop appropriate regulations. Many ECs remain unmonitored due to inadequate environmental surveillance programs.	Parida et al., 2021; Offiong et al., 2019
Analytical Challenges	Detecting and quantifying ECs in complex environmental matrices is difficult due to their low concentrations and varied chemical properties. Current analytical methods require improvements in sensitivity and accuracy to detect a broader range of ECs effectively.	Rasheed et al., 2019; Sophia A. & Lima, 2018
Risk Assessment Challenges	Conducting robust risk assessments for ECs is difficult due to their complex chemical mixtures, potential long-term effects, and uncertainties in exposure pathways. Traditional toxicology models may not account for the combined effects of multiple ECs.	Bilal et al., 2019; Gogoi et al., 2018
Regulatory Gaps	Existing legislation often fails to address ECs due to the lack of standardized limits, monitoring frameworks, and enforcement mechanisms. Many ECs remain unregulated despite increasing evidence of their environmental and health risks.	Parida et al., 2021; Ouda et al., 2021
International Cooperation	Addressing ECs requires global collaboration due to their transboundary nature. Differences in regulatory frameworks across countries create challenges in developing unified strategies for EC management. Strengthened international agreements and shared research initiatives are needed.	Johnson & Bell, 2023; Gwenzi, 2023

TABLE 3.2
Future directions for emerging contaminants (EC) management

Future Direction	Explanation	Citations
Adaptive Management	Policies must incorporate adaptive management approaches that allow for continuous learning and policy adjustments based on new scientific discoveries and monitoring data. This ensures that EC management remains dynamic and responsive to evolving environmental conditions.	Saidulu et al., 2021; Parida et al., 2021

TABLE 3.2 (Continued)
Future directions for emerging contaminants (EC) management

Future Direction	Explanation	Citations
Innovation and Technology	The development of new remediation technologies, such as advanced oxidation processes, membrane filtration, and bioremediation, is critical for the effective control and removal of ECs. Emerging methods, including AI-driven environmental monitoring, hold promise for future EC mitigation.	Zhou et al., 2024; Rout et al., 2021
Public Awareness and Engagement	Educating the public on EC risks and encouraging behavioral changes, such as reducing pharmaceutical waste and adopting sustainable practices, can enhance pollution control. Policies should incorporate community engagement initiatives to foster public participation in EC management.	Ouda et al., 2021; Johnson & Bell, 2023
Strengthening International Collaboration	Given the transboundary nature of EC pollution, international cooperation is necessary for data sharing, harmonized regulations, and joint research initiatives. Strengthened global frameworks can enhance EC risk management and policy enforcement.	Gwenzi, 2023; Gogoi et al., 2018

3.7 CONCLUSION

Ecological and human health considerations, diverse nature, and persistent qualities create major obstacles to ECs management. This chapter analyses existing international and national regulatory frameworks while demonstrating monitoring limitations and assessment and enforcement deficiencies. The Stockholm Convention on Persistent Organic Pollutants (POPs) and the Rotterdam Convention on Prior Informed Consent (PIC) provide regulatory governance, but many emerging contaminants receive insufficient oversight. The EU Water Framework Directive, alongside the EPA Contaminant Candidate List, uses structured policies across Europe and the United States, but these regulations do not sufficiently address all new pollutants.

Multiple critical barriers prevent effective handling of ECs. Current data deficiencies lead to constrained risk assessments, and regulatory choices while analyzing the presence of ECs at low concentrations remain troublesome because of analytical issues. Regional regulatory differences lead to unequal efforts in pollution reduction, and global economic and technological disparities worsen the situation. EC pollution persists as an environmental and public health threat because states lack established pollutant limits and mandatory enforcement practices. Various policy strategies must be implemented to address these difficulties. The combination of source reduction practices with product stewardship measures, along with green chemistry techniques, ensures reduced EC emissions. Wastewater treatment improvement, air pollution control systems, and proper disposal practices must be pursued to minimize environmental contamination. Early warning detection systems in combination with national monitoring frameworks and standardized analytical protocols will increase both risk surveillance and mitigation capabilities. Investment in AI-driven monitoring along with innovative remediation technologies must receive funding to create stronger long-term procedures for managing EC.

A successful EC governance framework requires an anticipatory combined system of actions. The development of adaptive policies requires scientific breakthroughs together with regulatory unification through active public participation in the policy-making process. International data exchange, along with common regulatory standards and a joint approach to management strategies, becomes possible through enhanced cross-border cooperation. EC management will advance through stronger multinational intergovernmental relationships coupled with public education efforts, while enabling technological developments to reduce environmental threats. Consistent adaptive policies are needed to develop scientific discoveries alongside contemporary technologies that support sustainable treatment systems through collaborative community-based programs. Transitioning EC management into anticipatory instead of reactive methods through a forward-focused approach will stop environmental and health damage, which will lead to a safer future, along with improved environmental quality.

REFERENCES

Azaroff, A., Monperrus, M., Miossec, C., Gassie, C., & Guyoneaud, R. (2021). Microbial degradation of hydrophobic emerging contaminants from marine sediment slurries (Capbreton Canyon) to pure bacterial strain. *Journal of Hazardous Materials*, *402*(123477), 123477. https://doi.org/10.1016/j.jhazmat.2020.123477

Baken, K. A., Sjerps, R. M. A., Schriks, M., & van Wezel, A. P. (2018). Toxicological risk assessment and prioritization of drinking water relevant contaminants of emerging concern. *Environment International, 118*, 293–303. https://doi.org/10.1016/j.envint.2018.05.006

Bilal, M., Adeel, M., Rasheed, T., Zhao, Y., & Iqbal, H. M. N. (2019). Emerging contaminants of high concern and their enzyme-assisted biodegradation – A review. *Environment International*, *124*, 336–353. https://doi.org/10.1016/j.envint.2019.01.011

Brack, W., Dulio, V., Ågerstrand, M., Allan, I., Altenburger, R., Brinkmann, M., Bunke, D., Burgess, R. M., Cousins, I., Escher, B. I., Hernández, F. J., Hewitt, L. M., Hilscherová, K., Hollender, J., Hollert, H., Kase, R., Klauer, B., Lindim, C., Herráez, D. L., … Vrana, B. (2017). Towards the review of the European Union Water Framework Directive: Recommendations for more efficient assessment and management of chemical contamination in European surface water resources. *Science of The Total Environment*, *576*, 720–737. https://doi.org/10.1016/j.scitotenv.2016.10.104

Chen, P., Wang, J., Xue, Y., Wang, C., Sun, W., Yu, J., & Guo, H. (2024). From challenge to opportunity: Revolutionizing the monitoring of emerging contaminants in water with advanced sensors. *Water Research*, *265*(122297), 122297. https://doi.org/10.1016/j.watres.2024.12229730.

Corpus, R. M. B., Bayani, M. S., Aguilar, J. L., Aguilar, J. C. B., & Aguilar, H. B. (2024). Emerging pollutants in waste water: Challenges and advancements in treatment technology. *IOP Conference Series. Earth and Environmental Science*, *1372*(1), 012037. https://doi.org/10.1088/1755-1315/1372/1/012037

Dhangar, K., & Kumar, M. (2020). Tricks and tracks in removal of emerging contaminants from the wastewater through hybrid treatment systems: A review. *Science of The Total Environment*, *738*(140320), 140320. https://doi.org/10.1016/j.scitotenv.2020.140320

Elsässer, J. P. (2024). Managers of complex change? How United Nations treaty secretariats jointly govern institutional interplay in global environmental governance. *Environmental Policy and Governance*, *34*(6), 610–622. https://doi.org/10.1002/eet.2105

Fischer, A., ter Laak, T., Bronders, J., Desmet, N., Christoffels, E., van Wezel, A., & van der Hoek, J. P. (2017). Decision support for water quality management of contaminants of emerging concern. *Journal of Environmental Management*, *193*, 360–372. https://doi.org/10.1016/j.jenvman.2017.02.002

Gaston, L., Lapworth, D. J., Stuart, M., & Arnscheidt, J. (2019). Prioritization approaches for substances of emerging concern in groundwater: A critical review. *Environmental Science & Technology*, *53*(11), 6107–6122. https://doi.org/10.1021/acs.est.8b04490

Ghattas, A.-K., Fischer, F., Wick, A., & Ternes, T. A. (2017). Anaerobic biodegradation of (emerging) organic contaminants in the aquatic environment. *Water Research*, *116*, 268–295. https://doi.org/10.1016/j.watres.2017.02.001

Gogoi, A., Mazumder, P., Tyagi, V. K., Tushara Chaminda, G. G., An, A. K., & Kumar, M. (2018). Occurrence and fate of emerging contaminants in water environment: A review. *Groundwater for Sustainable Development*, *6*, 169–180. https://doi.org/10.1016/j.gsd.2017.12.009

Gwenzi, W. (2023). Closing blank spots and illuminating blind spots in research on emerging contaminants: The Source–Pathway–Receptor–Impact–Mitigation (SPRIM) continuum as an organizing framework. *Water*, *15*(3), 526. https://doi.org/10.3390/w15030526

Hassan, M. H., Khan, R., & Andreescu, S. (2022). Advances in electrochemical detection methods for measuring contaminants of emerging concerns. *Electrochemical Science Advances*, *2*(6). https://doi.org/10.1002/elsa.202100184

He, Y., Liu, X., & Wang, X. (2022). How can environment get better? A research review of pollution governance. *Management of Environmental Quality*, *33*(2), 406–418. https://doi.org/10.1108/meq-08-2021-0187

Hendry, S. (2017). The eu Water Framework Directive – challenges, gaps and potential for the future. *Journal for European Environmental & Planning Law*, *14*(3–4), 249–268. https://doi.org/10.1163/18760104-01403001

Ismanto, A., Hadibarata, T., Kristanti, R. A., Maslukah, L., Safinatunnajah, N., & Kusumastuti, W. (2022). Endocrine disrupting chemicals (EDCs) in environmental matrices: Occurrence, fate, health impact, physio-chemical and bioremediation technology. *Environmental Pollution (Barking, Essex: 1987)*, *302*(119061), 119061. https://doi.org/10.1016/j.envpol.2022.119061

Johnson, C., & Bell, S. J. (2023). Linking emerging contaminants to production and consumption practices. *WIREs. Water*, *10*(1). https://doi.org/10.1002/wat2.1615

Kasonga, T. K., Coetzee, M. A. A., Kamika, I., Ngole-Jeme, V. M., & Benteke Momba, M. N. (2021). Endocrine-disruptive chemicals as contaminants of emerging concern in wastewater and surface water: A review. *Journal of Environmental Management*, *277*(111485), 111485. https://doi.org/10.1016/j.jenvman.2020.111485

Khan, M. T., Shah, I. A., Ihsanullah, I., Naushad, M., Ali, S., Shah, S. H. A., & Mohammad, A. W. (2021). Hospital wastewater as a source of environmental contamination: An overview of management practices, environmental risks, and treatment processes. *Journal of Water Process Engineering*, *41*(101990), 101990. https://doi.org/10.1016/j.jwpe.2021.101990

Khan, S., Naushad, M., Govarthanan, M., Iqbal, J., & Alfadul, S. M. (2022). Emerging contaminants of high concern for the environment: Current trends and future research. *Environmental Research*, *207*(112609), 112609. https://doi.org/10.1016/j.envres.2021.112609.

Li, X., Shen, X., Jiang, W., Xi, Y., & Li, S. (2024). Comprehensive review of emerging contaminants: Detection technologies, environmental impact, and management strategies. *Ecotoxicology and Environmental Safety*, *278*(116420), 116420. https://doi.org/10.1016/j.ecoenv.2024.116420

Liu, L., Wu, G., Ren, Y., Chen, Y., & Feng, Q. (2021). China's synergy in the implementation of international conventions on chemicals and waste. *Emerging Contaminants*, *7*, 132–138. https://doi.org/10.1016/j.emcon.2021.03.002

Lv, X., Dong, Q., Zuo, Z., Liu, Y., Huang, X., & Wu, W.-M. (2019). Microplastics in a municipal wastewater treatment plant: Fate, dynamic distribution, removal efficiencies, and control strategies. *Journal of Cleaner Production*, *225*, 579–586. https://doi.org/10.1016/j.jclepro.2019.03.321

Martín-Pozo, L., de Alarcón-Gómez, B., Rodríguez-Gómez, R., García-Córcoles, M. T., Çipa, M., & Zafra-Gómez, A. (2019). Analytical methods for the determination of emerging contaminants in sewage sludge samples. A review. *Talanta*, *192*, 508–533. https://doi.org/10.1016/j.talanta.2018.09.056

Mohapatra, D. P., & Kirpalani, D. M. (2019). Advancement in treatment of wastewater: Fate of emerging contaminants. *Canadian Journal of Chemical Engineering*, *97*(10), 2621–2631. https://doi.org/10.1002/cjce.23533

Naidu, R., Arias Espana, V. A., Liu, Y., & Jit, J. (2016). Emerging contaminants in the environment: Risk-based analysis for better management. *Chemosphere*, *154*, 350–357. https://doi.org/10.1016/j.chemosphere.2016.03.068

Offiong, N.-A. O., Inam, E. J., & Edet, J. B. (2019). Preliminary review of sources, fate, analytical challenges and regulatory status of emerging organic contaminants in aquatic environments in selected African countries. *Chemistry Africa*, *2*(4), 573–585. https://doi.org/10.1007/s42250-019-00079-6

Osuoha, J. O., Anyanwu, B. O., & Ejileugha, C. (2023). Pharmaceuticals and personal care products as emerging contaminants: Need for combined treatment strategy. *Journal of Hazardous Materials Advances*, *9*(100206), 100206. https://doi.org/10.1016/j.hazadv.2022.100206

Ouda, M., Kadadou, D., Swaidan, B., Al-Othman, A., Al-Asheh, S., Banat, F., & Hasan, S. W. (2021). Emerging contaminants in the water bodies of the Middle East and North Africa (MENA): A critical review. *Science of The Total Environment*, *754*(142177), 142177. https://doi.org/10.1016/j.scitotenv.2020.142177

Parida, V. K., Saidulu, D., Majumder, A., Srivastava, A., Gupta, B., & Gupta, A. K. (2021). Emerging contaminants in wastewater: A critical review on occurrence, existing legislations, risk assessment, and sustainable treatment alternatives. *Journal of Environmental Chemical Engineering*, *9*(5), 105966. https://doi.org/10.1016/j.jece.2021.105966

Patel, M., Kumar, R., Kishor, K., Mlsna, T., Pittman, C. U., Jr, & Mohan, D. (2019). Pharmaceuticals of emerging concern in aquatic systems: Chemistry, occurrence, effects, and removal methods. *Chemical Reviews*, *119*(6), 3510–3673. https://doi.org/10.1021/acs.chemrev.8b00299

Pena-Pereira, F., Bendicho, C., Pavlović, D. M., Martín-Esteban, A., Díaz-Álvarez, M., Pan, Y., Cooper, J., Yang, Z., Safarik, I., Pospiskova, K., Segundo, M. A., & Psillakis, E. (2021). Miniaturized analytical methods for determination of environmental contaminants of emerging concern – A review. *Analytica Chimica Acta*, *1158*(238108), 238108. https://doi.org/10.1016/j.aca.2020.11.040

Pereira, L. C., de Souza, A. O., Franco Bernardes, M. F., Pazin, M., Tasso, M. J., Pereira, P. H., & Dorta, D. J. (2015). A perspective on the potential risks of emerging contaminants to human and environmental health. *Environmental Science and Pollution Research International*, *22*(18), 13800–13823. https://doi.org/10.1007/s11356-015-4896-6

Petrie, B., Barden, R., & Kasprzyk-Hordern, B. (2015). A review on emerging contaminants in wastewaters and the environment: Current knowledge, understudied areas and recommendations for future monitoring. *Water Research*, *72*, 3–27. https://doi.org/10.1016/j.watres.2014.08.053

Petrie, B., Youdan, J., Barden, R., & Kasprzyk-Hordern, B. (2016). Multi-residue analysis of 90 emerging contaminants in liquid and solid environmental matrices by ultra-high-performance liquid chromatography tandem mass spectrometry. *Journal of Chromatography A*, *1431*, 64–78. https://doi.org/10.1016/j.chroma.2015.12.036

Philip, J. M., Aravind, U. K., & Aravindakumar, C. T. (2018). Emerging contaminants in Indian environmental matrices – A review. *Chemosphere*, *190*, 307–326. https://doi.org/10.1016/j.chemosphere.2017.09.120

Puri, M., Gandhi, K., & Kumar, M. S. (2023). Emerging environmental contaminants: A global perspective on policies and regulations. *Journal of Environmental Management*, *332*(117344), 117344. https://doi.org/10.1016/j.jenvman.2023.117344.

Rasheed, T., Bilal, M., Nabeel, F., Adeel, M., & Iqbal, H. M. N. (2019). Environmentally-related contaminants of high concern: Potential sources and analytical modalities for detection, quantification, and treatment. *Environment International*, *122*, 52–66. https://doi.org/10.1016/j.envint.2018.11.038

Rathi, B. S., Kumar, P. S., & Show, P.-L. (2021). A review on effective removal of emerging contaminants from aquatic systems: Current trends and scope for further research. *Journal of Hazardous Materials*, *409*(124413), 124413. https://doi.org/10.1016/j.jhazmat.2020.124413.

Rizzo, L., Malato, S., Antakyali, D., Beretsou, V. G., Đolić, M. B., Gernjak, W., Heath, E., Ivancev-Tumbas, I., Karaolia, P., Lado Ribeiro, A. R., Mascolo, G., McArdell, C. S., Schaar, H., Silva, A. M. T., & Fatta-Kassinos, D. (2019). Consolidated vs new advanced treatment methods for the removal of contaminants of emerging concern from urban wastewater. *Science of The Total Environment*, *655*, 986–1008. https://doi.org/10.1016/j.scitotenv.2018.11.265

Rout, P. R., Zhang, T. C., Bhunia, P., & Surampalli, R. Y. (2021). Treatment technologies for emerging contaminants in wastewater treatment plants: A review. *Science of The Total Environment*, *753*(141990), 141990. https://doi.org/10.1016/j.scitotenv.2020.141990

Saidulu, D., Gupta, B., Gupta, A. K., & Ghosal, P. S. (2021). A review on occurrences, eco-toxic effects, and remediation of emerging contaminants from wastewater: Special emphasis on biological treatment based hybrid systems. *Journal of Environmental Chemical Engineering*, *9*(4), 105282. https://doi.org/10.1016/j.jece.2021.105282

Shah, A. I., Din Dar, M. U., Bhat, R. A., Singh, J. P., Singh, K., & Bhat, S. A. (2020). Prospectives and challenges of wastewater treatment technologies to combat contaminants of emerging concerns. *Ecological Engineering*, *152*(105882), 105882. https://doi.org/10.1016/j.ecoleng.2020.105882

Shahid, M. K., Kashif, A., Fuwad, A., & Choi, Y. (2021). Current advances in treatment technologies for removal of emerging contaminants from water – A critical review. *Coordination Chemistry Reviews*, *442*(213993), 213993. https://doi.org/10.1016/j.ccr.2021.213993

Shanmuganathan, R., Sibtain Kadri, M., Mathimani, T., Hoang Le, Q., & Pugazhendhi, A. (2023). Recent innovations and challenges in the eradication of emerging contaminants from aquatic systems. *Chemosphere*, *332*(138812), 138812. https://doi.org/10.1016/j.chemosphere.2023.138812

Singh, S., Sivaram, N., Dhanjal, D. S., Assefa, H., Singh, J., & Ramamurthy, P. C. (2024). Navigating the complexity of emerging contaminants: Sources, impacts, and remediation strategies. *Journal of the Indian Institute of Science*, *104*(2), 519–553. https://doi.org/10.1007/s41745-024-00446-4

Sophia, A. C., & Lima, E. C. (2018). Removal of emerging contaminants from the environment by adsorption. *Ecotoxicology and Environmental Safety*, *150*, 1–17. https://doi.org/10.1016/j.ecoenv.2017.12.026

Sousa, J. C. G., Ribeiro, A. R., Barbosa, M. O., Pereira, M. F. R., & Silva, A. M. T. (2018). A review on environmental monitoring of water organic pollutants identified by EU guidelines. *Journal of Hazardous Materials*, *344*, 146–162. https://doi.org/10.1016/j.jhazmat.2017.09.058

Sudarsan, J. S., Dogra, K., Kumar, R., Raval, N. P., Leifels, M., Mukherjee, S., Trivedi, M. H., Jain, M. S., Zang, J., Barceló, D., Mahlknecht, J., & Kumar, M. (2024). Tricks and tracks of prevalence, occurrences, treatment technologies, and challenges of mixtures of emerging contaminants in the environment: With special emphasis on microplastic. *Journal of Contaminant Hydrology*, *265*(104389), 104389. https://doi.org/10.1016/j.jconhyd.2024.104389

Suhendra, E., Chang, C.-H., Hou, W.-C., & Hsieh, Y.-C. (2020). A review on the environmental fate models for predicting the distribution of engineered nanomaterials in surface waters. *International Journal of Molecular Sciences*, *21*(12), 4554. https://doi.org/10.3390/ijms21124554

Sutherland, D. L., & Ralph, P. J. (2019). Microalgal bioremediation of emerging contaminants – Opportunities and challenges. *Water Research*, *164*(114921), 114921. https://doi.org/10.1016/j.watres.2019.114921

Templeton, J. (2020). Stockholm convention on persistent organic pollutants. In Jean-Frederic Morin & Amandine Orsini (eds), *Essential Concepts of Global Environmental Governance* (pp. 242–243). Routledge.

Tran, N. H., Reinhard, M., & Gin, K. Y.-H. (2018). Occurrence and fate of emerging contaminants in municipal wastewater treatment plants from different geographical regions-a review. *Water Research*, *133*, 182–207. https://doi.org/10.1016/j.watres.2017.12.029

Tran, N. H., Reinhard, M., Khan, E., Chen, H., Nguyen, V. T., Li, Y., Goh, S. G., Nguyen, Q. B., Saeidi, N., & Gin, K. Y.-H. (2019). Emerging contaminants in wastewater, stormwater runoff, and surface water: Application as chemical markers for diffuse sources. *Science of the Total Environment*, *676*, 252–267. https://doi.org/10.1016/j.scitotenv.2019.04.160.

Vargas-Berrones, K., Bernal-Jácome, L., Díaz de León-Martínez, L., & Flores-Ramírez, R. (2020). Emerging pollutants (EPs) in Latin América: A critical review of under-studied EPs, case of study -Nonylphenol-. *Science of The Total Environment*, *726*(138493), 138493. https://doi.org/10.1016/j.scitotenv.2020.138493

W. M. das Chagas, F., B. P. Mangas, M., & G. da Silva, J. (2023). Determination of emerging contaminants in water samples in Brazil by voltammetric techniques: A literature review. *Revista Virtual de Química*, *15*(1), 92–121. https://doi.org/10.21577/1984-6835.20220100.

Xabadia, A., Esteban, E., Martinez, Y., & Ortuzar, I. (2021). Contaminants of emerging concern: A review of biological and economic principles to guide water management policies. *International Review of Environmental and Resource Economics*, *15*(4), 387–430. https://doi.org/10.1561/101.00000138

Yang, W., Bu, Q., Shi, Q., Zhao, R., Huang, H., Yang, L., Tang, J., & Ma, Y. (2024). Emerging contaminants in the effluent of wastewater should be regulated: Which and to what extent? *Toxics*, *12*(5), 309. https://doi.org/10.3390/toxics12050309

Yap, H. C., Pang, Y. L., Lim, S., Abdullah, A. Z., Ong, H. C., & Wu, C.-H. (2019). A comprehensive review on state-of-the-art photo-, sono-, and sonophotocatalytic treatments to degrade emerging contaminants. *International Journal of Environmental Science and Technology: IJEST, 16*(1), 601–628. https://doi.org/10.1007/s13762-018-1961-y

Zhou, Q., Chen, H., Liu, G., & Wang, X. (2024). Occurrence, sustainable treatment technologies, potential sources, and future prospects of emerging pollutants in aquatic environments: A review. *Frontiers in Environmental Science*, 12. https://doi.org/10.3389/fenvs.2024.1455377

4 Comprehensive Overview of Polycyclic Aromatic Hydrocarbons (PAHs) with Remediation Techniques in Association with One Health

Shagufta Tanveer, Meghna Arya, and Preeti Chaturvedi Bhargava

4.1 INTRODUCTION

Polycyclic aromatic hydrocarbons (PAHs) are chemical-based compounds with multiple aromatic rings. On the basis of their ring structure, PAHs can be classified as alternant and non-alternant PAHs. Alternant PAHs are composed of six-carbon ring structures, whereas in non-alternant PAHs like fluorene, the six-carbon rings are associated with a supplementary ring containing fewer carbon atoms (Dhokale et al., 2023). The presence of dense electrons on these carbon rings contributes to the biochemical stability of PAHs, rendering them resistant to nucleophilic attacks. This environmental persistence results from the PAHs' chemical structure, which hampers their degradation by biological or chemical processes. PAHs have garnered significant attention because of their widespread distribution, environmental existence and stability, and potential side effects on the health of human beings and ecological communities. PAHs originate from natural and anthropogenic sources, with combustion processes being the primary contributor to their release into the environment. Once released, PAHs can undergo global atmospheric dispersion, leading to their widespread distribution across regional and global scales. While natural sources like volcanic eruptions and wildfires contribute minimally to PAH levels, anthropogenic activities are the principal drivers of PAH contamination. These activities can be categorized into industrial, mobile, domestic, and agricultural. Industrial waste production, metal processing, cement manufacturing, and various chemical production processes causing waste incineration significantly contribute to PAH emissions.

 DOI: 10.1201/9781032623801-4

These industries produce PAHs through incomplete combustion, particularly from processes like coal-tar pitch production and asphalt manufacturing (Abdel et al., 2016). In addition, diesel and gasoline engines used in machinery and vehicles emit PAHs, making mobile sources another important contributor. Domestic sources of PAHs include household activities like garbage fire, coal, and timber or cooking on oil/gas burners and stoves. Agricultural activities, such as open biomass burning and waste disposal, also release PAHs into the environment, especially when combustion occurs under suboptimal conditions. The distribution of PAH pollution varies across regions and seasons. Rural areas typically experience higher levels of PAHs from domestic and agricultural sources, whereas urban areas face additional pollution from industrial and mobile sources. PAH concentrations tend to be highest in winter and spring because of increased fossil fuel combustion for heating and lower photodegradation rates, exacerbated by geographical determinants like tranquil gusts and low temperatures. PAH concentrations can also be categorized based on production sources such as pyrogenic, petrogenic, and biogenic sources. Pyrogenic PAHs result from unintentional incomplete combustion at high temperatures, such as from industrial processes and certain pyrolytic activities. Petrogenic PAHs originate from petrol and its subsidiary products, often due to repository, transportation, and outflow (Patel et al., 2020).

Biogenic PAHs are produced by life forms during slow organic material conversion, including microorganisms, phytoplankton, algae, and plants. PAHs exhibit a spectrum of toxicological effects, with certain compounds classified as carcinogenic, mutagenic, and teratogenic (Chauhan et al., 2024). Several PAHs, including benzo[a]pyrene (BaP), have been identified as Group 1 carcinogens by the International Agency for Research on Cancer (IARC), highlighting their potential to cause malignant in humans. The US Environmental Protection Agency (USEPA) has categorized 16 different PAHs in the primary pollutants list (USEPA, 2016c). Table 4.1 depicts these 16 PAHs with their physical properties.

Furthermore, many health side effects have been connected with respiratory illnesses, cardiovascular diseases, reproductive disorders, and developmental abnormalities (Lee et al., 2023). The climate suitability transport of PAHs is persuaded by numerous elements, involving their physical and chemical properties, partitioning behaviour, and interactions with environmental matrices. Once released into the environment, PAHs can go through evaporation, photodegradation, sorption, and biodegradation, ultimately affecting their distribution, persistence, and bioavailability. Given their complex nature and adverse impacts on environmental and human health, addressing PAH contamination requires a multidisciplinary approach to integrate scientific expertise, technological innovation, regulatory frameworks, and public awareness (Li et al., 2024). Moreover, recognizing the association between environmental pollution, human health, and ecosystem integrity highlights the importance of adopting a holistic perspective, such as the One Health approach, in managing PAH-related challenges.

In the following chapter, we investigate the toxicological effects of PAHs, their environmental fate and transport mechanisms, remediation techniques, and the implications of adopting a One Health perspective in addressing PAH contamination

TABLE 4.1
PAHs with their physical properties

S.No.	PAHs	Molecular Formula	Melting Point (°C)	Boiling Point (°C)
1.	Naphthalene	$C_{10}H_8$	80.2	218
2.	Acetanapthene	$C_{12}H_{10}$	168	420
3.	Acetanapthylene	$C_{12}H_8$	91.8	280
4.	Fluorene	$C_{13}H_{10}$	117	295
5.	Phenanthrene	$C_{14}H_{10}$	100.5	340
6.	Anthracene	$C_{14}H_{10}$	216.4	342
7.	Fluoranthene	$C_{16}H_{10}$	108.8	375
8.	Pyrene	$C_{16}H_{10}$	393	150.4
9.	Benzo (a) anthracene	$C_{18}H_{12}$	158	438
10.	Chrysene	$C_{18}H_{12}$	255	440.7
11.	Benzo(b)fluoranthene	$C_{20}H_{12}$	168.3	481
12.	Benzo(k)fluoranthene	$C_{20}H_{12}$	215.7	480
13.	Benzo(a)pyrene	$C_{20}H_{12}$	178.7	493
14.	Benzo(g,h,i)perylene	$C_{22}H_{12}$	277	500
15.	Indeno[1,2,3,-cd]pyrene	$C_{22}H_{12}$	164	536
16.	Dibenzon(a,h)anthracene	$C_{22}H_{14}$	267	524

and strategies for mitigating their adverse effects and promoting sustainable environmental governance.

4.2 SOME IMPORTANT PAHS IN THE ENVIRONMENT

Some widely observed PAHs in the atmosphere come from natural and anthropogenic sources including benzo[a]pyrene (BaP), pyrene, anthracene, maphthalene, and fluoranthene, each with distinct sources and impacts.

Benzo[a]pyrene (BaP) largely results from inadequate combustion of petroleum and coal, smoke from tobacco, fried and grilled meals, and some industrial operations like aluminium manufacturing and coke ovens. The IARC has categorized it as a Group 1 carcinogen (IARC), BaP is mutagenic and can cause DNA mutations, leading to genetic damage (Jameson, 2021).

Naphthalene is found in mothballs, air fresheners, fossil fuel combustion, and industrial processes like phthalic anhydride production, and is classified as a possible human carcinogen (Group 2B by IARC). Irritation to the respiratory system, eyes, and skin can be caused by short-term contact, while long-term exposure is associated with haemolytic anaemia and liver damage (Yost et al., 2021).

Anthracene, derived from coal tar, petroleum, wood burning, and industrial processes such as dye, plastic, and pesticide production, is relatively low in toxicity compared to other PAHs. It can cause skin irritation (Brzezinski et al., 2024) with prolonged exposure and can bioaccumulate, impacting aquatic life.

Fluoranthene, which is produced by the partial burning of organic matter (Abdel et al., 2016) (e.g. forest fires, vehicle exhaust) and industrial operations such as steel and iron manufacturing, is a possible carcinogen recognized by the IARC as probably harmful to humans. It can cause mutations and reproductive issues in aquatic species, and is persistent in the environment, leading to long-term ecological effects (Vijayanand et al., 2023).

Pyrene, originating from fossil fuel combustion, industrial emissions, asphalt production, and charbroiled foods (Mallah et al., 2022), is not classified as carcinogenic to humans by the IARC. However, it can induce cancer when combined with other PAHs. Pyrene exposure can cause skin irritation and respiratory issues, and is persistent in soil and sediment, affecting terrestrial and aquatic ecosystems (Hussain et al., 2018).

4.3 TOXICITY AND HEALTH EFFECTS OF PAHS ON ONE HEALTH

4.3.1 Effects of PAHs on Humans

Individuals can be exposed to PAH threats for varying durations depending on their living or working environments. PAHs have been related to impaired lung function in asthma sufferers, potentially exacerbating the illness due to short-term exposure.

Additionally, a higher incidence of thrombosis has been observed among individuals with coronary artery disease. High levels of PAHs, particularly from occupational exposure, may result in symptoms like vomiting, bloating, anxiety, and inflammation of the skin and eyes. (Venkataraman et al., 2024). Even low-dose exposure to PAHs can lead to significant health issues, including deterioration in pulmonary function and various cancers, particularly affecting the lungs, skin, and digestive system (Figure 4.1) (Mallah et al., 2022).

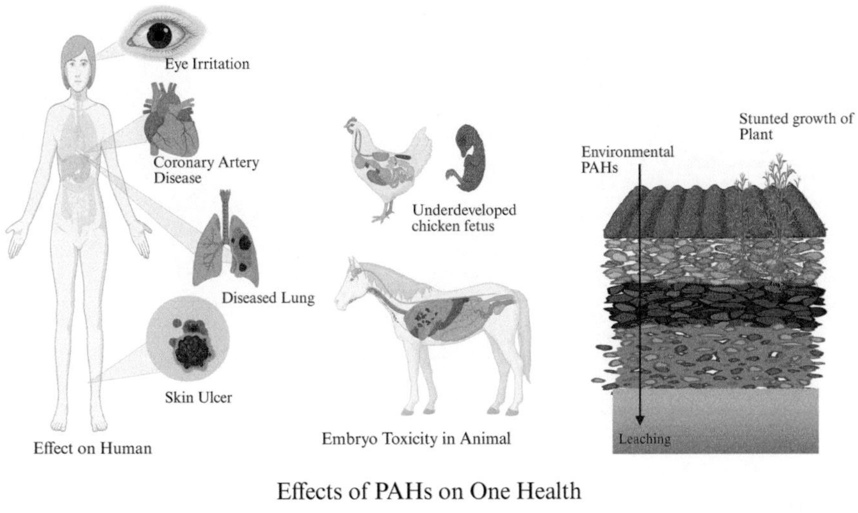

FIGURE 4.1 Effect of PAHs on One Health.

4.3.2 Effects of PAHs on Animals

PAHs pose significant toxicity and health risks to various animals, including mammals, birds, fish, and invertebrates, due to their persistence, bioaccumulative nature, and potential to cause adverse effects through multiple exposure routes. Animals can be exposed to PAHs by inhaling polluted air, consuming contaminated food or water, and coming into contact with polluted soil or water. Once inside an animal's body, PAHs undergo absorption, distribution, metabolism, and excretion. Exposure to PAHs can disrupt reproductive and developmental processes in animals, leading to reduced fertility, embryotoxicity (Figure 4.1), teratogenicity, and developmental abnormalities, which affect the health and viability of offspring (Ramesh et al., 2022).

The harmful effects of PAHs vary greatly among animal species, influenced by factors such as metabolic processes, detoxification capabilities, tissue distribution, and organ sensitivity (Abdel et al., 2016). PAHs can bioaccumulate in organisms through dietary intake, dermal contact, and respiratory exposure. Key factors influencing bioaccumulation and biomagnification of PAHs include lipid solubility, trophic level, and feeding habits. PAHs with higher lipid solubility accumulate more readily in lipid-rich organisms like fish and marine mammals (Schwaab, 2020) due to their ability to dissolve and persist in fat tissues. Predators at the highest food chain levels, including eagles or huge predatory fish, have higher PAH concentrations than their prey (Sakshi et al., 2019) (Figure 4.2).

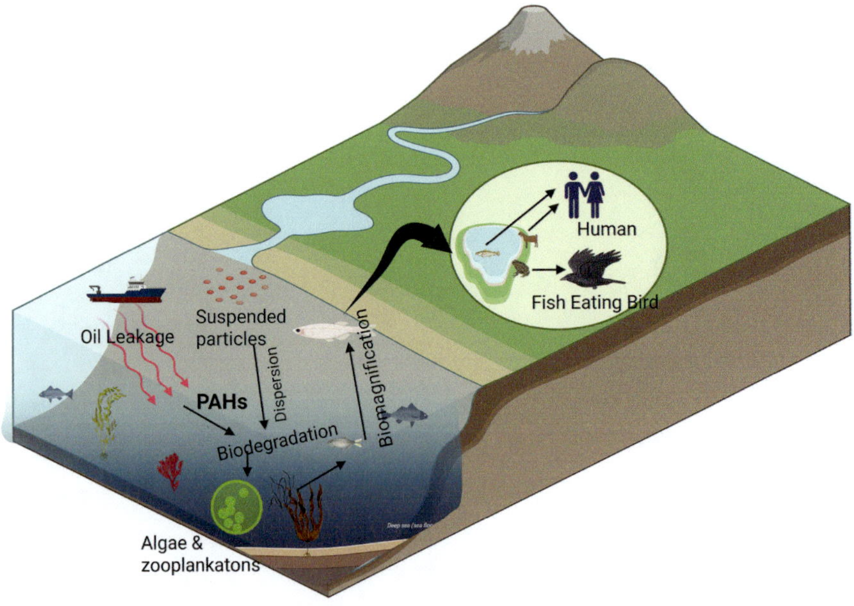

FIGURE 4.2 Biomagnification in the food chain.

4.3.3 Effects of PAHs on the Environment

PAHs are widely distributed environmental pollutants that accumulate in soil, water, sediments, and air, posing significant risks to various organisms, including aquatic species, soil microbes, plants, and animals. These compounds can disrupt biological processes, interfere with reproduction, and cause developmental defects such as reduced growth, immune suppression, and increased susceptibility to disease in exposed organisms like fish, amphibians, and other aquatic species (Sakshi et al., 2019). The ecotoxicity of PAHs poses substantial threats across different environmental compartments. In marine environments, PAHs can disrupt essential biological processes in fish and invertebrates, leading to altered metabolism, impaired immune responses, and behavioural changes (Onyena et al., 2024). Contaminated soils can potentially cause secondary contamination by leaching PAHs into groundwater or transporting them to surface water bodies during rainstorms. Soil microbes are crucial for nutrient cycling and soil health, and are highly sensitive to PAH contamination, which can inhibit microbial growth and activity, reduce decomposition rates, and alter soil fertility. High levels of PAHs in soil can impair plant growth and physiology, preventing seed germination, root growth, and nutrient uptake, which results in poor crop yields. Alkio et al. (2005) studied the PAH reaction to stress in *Arabidopsis* (*Arabidopsis thaliana*) upon exposure to the three-ring aromatic molecule phenanthrene. PAH stress morphological indicators included root and shoot growth loss, twisted trichomes, reduced root hairs, chlorosis, delayed flowering, and the appearance of white spots that eventually turned necrotic. Plants experienced stress from oxidation at both the tissue and the intracellular phase, as shown by localized H_2O_2 production and cell death. The findings revealed that *Arabidopsis* absorbs phenanthrene, implying that plant degradation and a PAH response in plants and animals may share stress mechanisms, as PAH detoxification causes oxidative stress in animal cells. The cumulative effects of PAHs on germination, root development, and nutrient uptake result in diminished crop yields with significant economic and ecological implications.

PAHs enter the atmosphere primarily through combustion, including vehicle emissions, industrial activities, residential heating, and natural events like wildfires. PAHs can exist in both gaseous and particulate forms, depending on their molecular weight and atmospheric conditions. Atmospheric deposition, including dry deposition (settling of particles and gases) and wet deposition (rainout and washout by precipitation), is the primary mechanism through which PAHs return to the earth's surface (Hussain et al., 2018).

4.4 FACTORS INFLUENCING FATE AND TRANSPORT PROCESSES OF PAHS

The fate and transport processes of PAHs in the environment are influenced by a multitude of factors that determine their persistence, distribution, and eventual impact on ecological and human health (Zhou et al., 2020). The complex interplay of physicochemical properties, environmental conditions, organic content, and biological processes governs the fate and transport of PAH. Understanding these factors is crucial for developing effective strategies to manage PAH contamination and mitigate its environmental and human health impacts. Several factors influence the fate and transport of PAHs in the environment as follows:

4.4.1 Physicochemical Properties

The physicochemical characteristics of PAHs, especially their molecular weight, water solubility, and vapor pressure, determine how they behave in the environment.

Low-molecular-weight PAHs (e.g. naphthalene, phenanthrene) are more volatile and water soluble, whereas high-molecular-weight PAHs (e.g. benzo[a]pyrene, dibenzo [a, h] anthracene) are more hydrophobic and tend to partition into organic phases (Lakhani, 2018). Solubility can influence the mobility of PAHs in aqueous environments. Vapor pressure affects the volatility and atmospheric transport of PAHs. Compounds with higher vapor pressures are more likely to volatilize from contaminated surfaces and undergo long-range atmospheric transport (Belis et al., 2010)

4.4.2 Environmental Conditions

Temperature, pH, and redox conditions are various environmental factors that significantly influence the fate and transport of PAHs in different environmental matrices (Hussain et al., 2018). Temperature significantly impacts the deposition rate from the atmosphere. At higher temperatures, a larger proportion of PAH exists in the vapor phase, while at lower temperatures it enhances the sorption of PAH.

pH affects the speciation and sorption behaviour of PAHs in soil and water environments; changes in pH can alter the solubility, dissociation, and partitioning of PAHs between aqueous and solid phases. Redox conditions determine the transformation pathways of PAHs in anaerobic environments, such as sediments and groundwater. Under anaerobic conditions, PAHs can undergo reductive transformations, forming metabolites and degradation products (Sakshi et al., 2019).

4.4.3 Organic Matter Content

Organic carbon content in soil and sediment plays a crucial role in PAH sorption and sequestration processes. High organic matter content enhances the sorption capacity of soils and sediments for PAHs, reducing their bioavailability and potential for leaching into groundwater or entering aquatic ecosystems (Ahangar, 2010). Sediment organic carbon content affects the sorption capacity of sediments for PAHs, controlling their partitioning between the water column and the sediment bed. Organic-rich sediments tend to accumulate higher concentrations of PAHs than mineral-dominated sediments (Pi et al., 2021). The fate and transport of PAHs can be influenced by the presence of other contaminants, including heavy metals, pesticides, and polychlorinated biphenyls (PCBs), through interactions, such as competition for sorption sites or co-metabolism by microbial communities (Merhaby et al., 2019).

4.5 REMEDIATION TECHNIQUES FOR PAH CONTAMINATION

PAHs are organic pollutants that persist and endanger human well-being and the environment. Addressing PAH contamination requires implementing effective remediation techniques to mitigate risks, restore environmental quality, and protect the ecosystem. Following are some commonly used remediation techniques for PAH contamination (Figure 4.3).

4.5.1 Physical Remediation

Physical remediation involves application of techniques for removal of contaminants from soil, sediment, and groundwater through physical means. It mainly includes soil washing, soil vapour extraction (SVE), and dredging for addressing contamination caused by PAHs in soil and sediment.

4.5.1.1 Soil Extraction

Due to their hydrophobic nature, PAHs bind to naturally occurring substances in soil. It is an excellent method for eliminating PAHs from polluted soils because of their solubility.

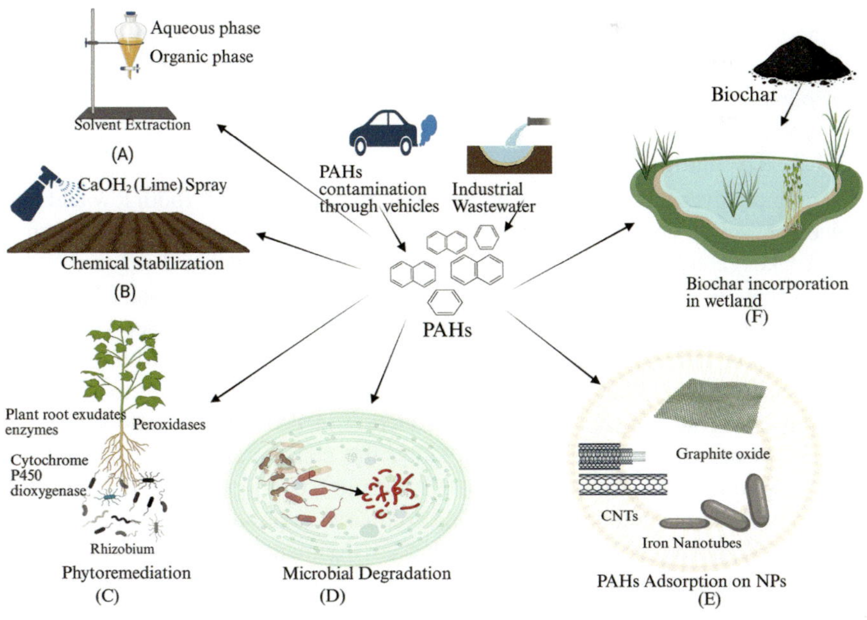

FIGURE 4.3 Traditional and emerging remediation processes for PAH contamination. A, B, C, and D are traditional techniques, and E and F are emerging remediation processes.

Various cyclodextrin solvents such as β-cyclodextrin, methyl-β-cyclodextrin, and hydroxypropyl-β-cyclodextrin, along with cooking oils, can be potentially used. The process involves two steps: desorption of PAHs from soil and leaching into the extraction fluid (Von Lau et al., 2014). The efficiency depends on the solvent type and the soil-to-solvent ratio. Ninety-five per cent of PAHs (naphthalene, phenanthrene, and benzo[a]pyrene) were removed from highly contaminated soils by a mixture of ethanol, 1-pentanol, and water. Similarly, a solvent mixture of cyclohexane and ethanol (3:1) achieves a 93% extraction efficiency for fluoranthene. The primary concern is the toxicity of solvents, necessitating careful selection and further treatment of the solvent containing extracted PAHs for complete degradation (Sakshi et al., 2019).

4.5.1.2 Soil Vapor Extraction (SVE)

Soil vapor extraction (SVE) is a remedial process that removes volatile PAH pollutants from the soil matrix by vacuuming the subsurface. Vacuum pumps are used to create a negative pressure zone in the soil, which causes volatile contaminants to volatilize and migrate towards extraction wells (Stewart et al., 2018). The extracted vapours are then treated using methods such as activated carbon uptake or thermal oxidation to remove or destroy the contaminants. Alternatively, the vapours can be collected for disposal or treatment off-site. SVE can be combined with other treatment methods such as air sparging or bioremediation, to enhance remediation efficiency. SVE is most effective for the removal of lower molecular weight PAHs with two to three rings, such as naphthalene, acenaphthylene, acenaphthene, fluorene, phenanthrene, and anthracene (Darajeh et al., 2020). However, larger PAHs have low volatility and are more likely to adsorb strongly to soil particles, making them less amenable. For these higher molecular weight PAHs, alternative or supplementary remediation methods may be required, such as bioremediation, chemical oxidation, or thermal desorption.

4.5.1.3 Dredging

Dredging involves the mechanical or hydraulic removal of PAH- (both low and high molecular weight compounds) contaminated sediments from water bodies such as rivers, lakes, and harbours. Dredging equipment such as suction dredges or clamshell buckets is used to remove contaminated sediments from the water body. The dredged sediments are dewatered to remove excess water and reduce volume for transportation and disposal. The dewatered sediments are transported to a disposal facility or treatment plant for proper disposal or treatment according to regulatory requirements. Data shows a significant reduction in PAH concentrations at a few sites following dredging activities. Notably, the concentrations of phenanthrene (0.81 μg/g to 0.069 μg/g) and pyrene (0.98 μg/g to 0.071 μg/g) were reduced by at least an order of magnitude post-dredging compared to pre-dredging levels (Walker et al., 2013). Dredging can have significant environmental impacts, including disturbance of aquatic habitats, sediment resuspension, and release of contaminants into the water column. Despite its environmental impacts, dredging is effective in addressing sediment contamination in aquatic environments and improving water quality (Wenger et al., 2017).

4.5.2 Chemical Remediation

Chemical remediation techniques, such as chemical oxidation and chemical stabilization, are effective strategies for addressing PAH contamination in the environment.

4.5.2.1 Chemical Oxidation

The process of chemical oxidation is an application of powerful oxidizing chemicals to convert PAHs into less hazardous compounds. In this procedure, hydrogen peroxide (H_2O_2), ozone (O_3), and potassium permanganate ($KMnO_4$) are common oxidizing agents. These agents react with PAH molecules, generating chemical reactions that degrade the chemicals into simpler, less harmful metabolites (Patel et al., 2020). There are two ways of chemical oxidation: in situ and ex situ chemical oxidation:

a) **In situ Chemical Oxidation:** In situ chemical oxidation involves injection of oxidizing agents directly into contaminated soil or groundwater. Once introduced, the oxidizing agent reacts with the PAHs, breaking them into harmless substances (Kuppusamy et al., 2017).
   This method is often used to treat subsurface contamination without the need for excavation. Pyrene degrades quickly through chemical oxidation. Peluffo et al. (2018) conducted a study in which pyrene was degraded at a faster pace (96% of 2,800 $mg.kg^{-1}$) than phenanthrene (91% of 1,200 $mg.kg^{-1}$) in a series of reactors of aged contaminated soil at ambient temperature in seven days by heat stimulated persulfate oxidation (Patel et al., 2020).
b) **Ex situ Chemical Oxidation:** Ex situ chemical oxidation involves the removal of contaminated soil, sediment, or groundwater from the site to be treated in a controlled environment. The oxidizing agent is applied to the removed material, either by mixing or spraying, to facilitate the degradation of PAHs. This method is suitable for treating highly contaminated materials or when in situ treatment is not feasible (Kuppusamy et al., 2017). In a study, PAH content is reduced to 100 mg/kg from 1200 mg/kg. Contaminated sites are oxidized by potassium permanganate, Fenton's reagent, and hydrogen peroxide.

4.5.2.2 Chemical Stabilization

Chemical stabilization immobilizes PAHs in contaminated soil or sediment, thereby reducing their bioavailability and potential for migration. This technique involves the addition of chemical amendments, such as lime, cement, or activated carbon, to the contaminated material. The chemical amendments used in stabilization react with PAH molecules, forming insoluble complexes or compounds that bind the contaminants and prevent them from leaching into groundwater or being taken up by plants. Lime ($CaOH_2$), for example, raises the pH of the soil, promoting the formation of insoluble calcium salts with PAHs. Cement can physically encapsulate PAHs, preventing their release into the environment. Activated carbon is another amendment used in chemical stabilization. It binds effectively with PAHs due to extensive surface area and strong adsorption capacity. When activated carbon is added to contaminated soil or sediment, it lowers the accessibility for adsorption by plants and microorganisms. This adsorption is particularly useful in waterlogged environments where traditional stabilization methods may be less effective (Dewangan et al., 2021).

4.5.3 Biological Remediation

4.5.3.1 Bioremediation

Bioremediation utilizes microorganisms, such as bacteria, fungi, or plants, to degrade PAHs into harmless compounds through biodegradation processes. It can be conducted in situ or ex-situ and may involve stimulating indigenous microbial populations or introducing specialized microbial consortia. Microbial communities that thrive in swamp soils and sediments play an important role in the degradation and transformation of PAHs via microbial-mediated processes such as biodegradation, biomineralization, and biofilm formation. Bioaugmentation (the introduction of specialized PAH-degrading microbes), biostimulation (the addition of nutrients or electron acceptors), and biofilm engineering (the immobilization of microbial consortia on natural or synthetic substrates) are three innovative approaches for increasing the activity and diversity of indigenous microbial populations (Balseiro et al., 2024). This cost-effective method improves aeration, moisture, and nutrient levels in polluted soil, stimulating microbial activity for PAH degradation. PAH degrading microorganisms include *Pseudomonas*, *Alcaligenes*, *Sphingomonas*, *Mycobacterium, Streptomyces*, *Burkholderia*, *Polaromonas*, *Ralstonia*, *Rhodococcus*, and *Stenotrophomonas Flavobacterium* (Srivastava and Kumar, 2019). It is particularly effective for low-molecular-weight PAHs (LMWPAHs) but less so for high-molecular-weight PAHs (HMWPAHs). Nonetheless, it is known that the main sources of energy and carbon for *Sphingomonads*, *Actinobacteria*, and *Mycobacterium* are the degradation of four or more ring HMWPAHs. The final electrophile used in aerobic degradation is oxygen. It serves as a co-substrate for hydroxylation and oxygenolytic breakage of carbon rings. While requiring minimal maintenance, it is limited by slow degradation rates and susceptibility to uncontrollable factors like heavy rainfall. Soil characteristics, including nutrient content, porosity, temperature, organic matter content, and pH, also play crucial roles in the bioremediation process (Dewangan et al., 2021). Microbial ecology factors such as surfactant production, substrate range, species abundance, metabolic pathways, substrate affinity, and microbial diversity are essential for effective PAH biodegradation.

4.5.4 Phytoremediation

Phytoremediation is the deliberate utilization of vegetative matter to remove, degrade, or sequester pollutants from soil, water, or air. PAHs may be processed or stored in plant biomass, after plants absorb them through their roots and transport them to above-ground tissues. Plants employ mechanisms like phytoextraction, phytovolatilization, and phytodegradation to mitigate pollutant levels. Careful plant selection based on growth characteristics and pollutant tolerance is crucial for successful phytoremediation. For example, *Echinacea purpurea* (purple coneflower) showed 92.41% removal of PAHs like fluoranthene, pyrene, benz[a]anthracene, chrysene, benzo[b]fluoranthene, benzo[k]fluoranthene, dibenzo[a,h]anthracene and a mixture of *Festuca arundinacea*, *Festuca elata*, and *Festuca gigantea* showed 64.57% removal of pyrene, chrysene, benzo[b]fluoranthene, benzo[k]fluoranthene (Dai et al., 2020). The efficiency of the process was enhanced using intercropping. It removed 31% of PAHs from contaminated soil, significantly outperforming monoculture and unplanted soils, which removed 20% and 0% of PAHs, respectively. Once pollutants are taken up by plants, enzymatic degradation breaks down the PAHs into simple molecules such as CO_2 and H_2O (Turkovskaya and Muratova, 2019).

Genetically modified plants with improved root design and metabolic pathways can influence PAH breakdown. Modifying root architecture can increase the surface area of roots exposed to contaminated soil, resulting in increased PAH absorption rates. This can be achieved by introducing genetic modifications that promote the growth of longer, thicker, or more branched roots. Root exudates that contain organic acids or surfactants can help mobilize and absorb PAHs from the earth's matrix. The metabolic pathways involved in PAH degradation can be altered to increase their efficacy in flora. Plants can be modified to produce specific binding proteins or peptides that have a high affinity for PAHs, increasing their ability to sequester pollutant compounds and lowering their environmental toxicities (Kuppusamy et al., 2020). Gene editing methods like CRISPR-Cas9 can be used to target specific genes that produce cytochrome P450 enzymes or peroxidases that govern PAH metabolism, resulting in higher enzyme levels and greater PAH breakdown (Das et al., 2023).

4.5.5 Rhizoremediation

Rhizoremediation represents a specific approach within phytoremediation, focusing on utilizing plant-associated rhizosphere microorganisms to treat polluted soils. This method offers intensified degradation of PAHs, with successful outcomes relying on the symbiotic partnership between plants and microbes possessing degradation capabilities (Praveen et al., 2019). In rhizoremediation, plants play a pivotal role by providing a vast root surface area for microbial growth, facilitating the remediation of pollutants to depths of approximately 10–15 metres within the soil. Plant roots serve as nutrient sources for PAH-degrading microbes, supplying amino acids, carbohydrates, organic acids, and flavonoids essential for their growth and activity. Simultaneously, microbes support plants by aiding them in overcoming stress induced by pollutants, thereby reducing phytotoxicity and enhancing overall remediation efficiency (Panchenko et al., 2022). Rhizoremediation occurs spontaneously when flavonoids and other root chemicals are released, stimulating bacterial growth and activity degrading the PAHs. Root growth improves soil aeration, which aids in the oxidative decomposition of PAHs (Patel et al., 2020).

4.5.6 Thermal Remediation

Applying heat to the contaminated area is a technique called thermal remediation that is used to treat soil or sediment contaminated with PAHs. This method utilizes high temperatures to volatilize or thermally decompose PAHs, effectively removing them from the contaminated material. As the temperature increases, PAH molecules break down into simpler compounds, such as water and carbon dioxide, or volatilize into the air, thereby reducing their concentration in the soil or sediment. Thermal remediation is capable of being carried out in situ or ex-situ, relying on the specific site conditions and remediation goals (Eldos et al., 2022). In-Situ Thermal Treatment (ISTT), a type of thermal remediation eliminates PAHs from soil and groundwater. Common ISTT techniques for removing PAHs from aquifers include radiofrequency heating, steam injection, and electrical resistance heating. This technique involves increasing the temperature of the soil or water until pollutants vaporize and get removed. ISTT has many benefits like being least disruptive to sites, removing pollutants faster than other methods, and reducing future problems with the environment (Sun et al., 2024).

4.6 EMERGING REMEDIATION TECHNOLOGIES

Wetland sequestration is an emerging remediation strategy encompassing a range of innovative novel approaches aimed at mitigating PAH contamination, offering promising solutions. It plays a crucial role due to wetland sequestration's unique hydrological and biogeochemical characteristics. Following are some novel approaches for wetland sequestration of PAHs:

4.6.1 Constructed Wetlands

Engineered systems known as constructed wetlands are created to replicate the natural processes of wetlands in terms of pollutant removal and water treatment. These wetlands utilize a combination of physical, chemical, and biological processes, including adsorption, sedimentation, microbial degradation, and plant uptake, to sequester and degrade PAHs. Innovative designs incorporating various substrates, flow patterns, and vegetation types can optimize PAH removal efficiency in constructed wetlands. Leung et al. (2016) report that *Phragmites australis* and other wetland plant species helped eliminate over 99% of PAHs from a subsurface flow created wetland. In order to treat PAHs like polychlorinated dibenzo-p-dioxins/dibenzofurans, wetlands with free surface flow types and subsurface flow systems were built.

Constructed wetlands systems utilize vegetation, soil, and microbial communities to treat contaminated water or soil, making them ideal for mitigating PAH pollution in wetland ecosystems (Figure 4.4). By leveraging ecological functions and processes, such as plant uptake, microbial degradation, and sedimentation, constructed wetlands and bioretention systems can effectively sequester PAHs and reduce their availability in the environment (Tsatsau et al., 2023).

4.6.2 Biochar Amendment

Biochar is an abundant carbon source created by biomass pyrolysis, and it is becoming more recognized for its capacity to trap organic contaminants like PAHs in marsh soil. Because of its special combination of structural, chemical, and physical properties,

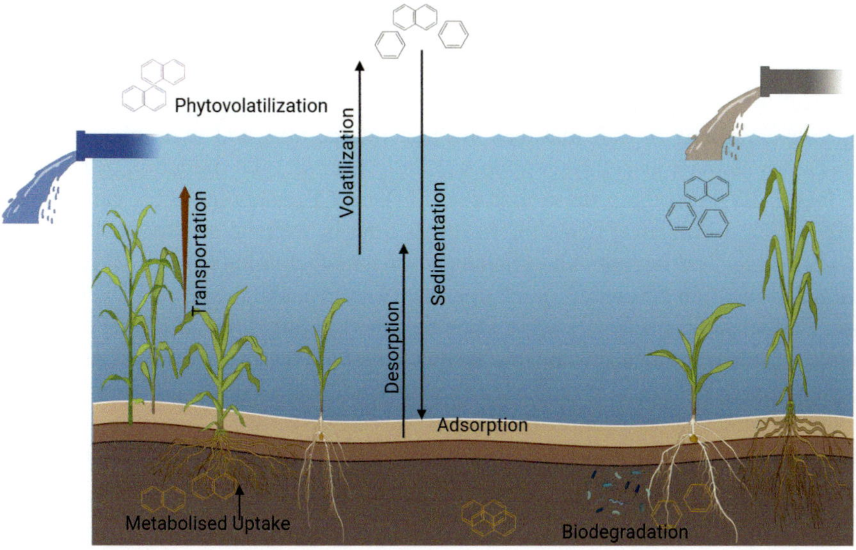

FIGURE 4.4 Wetland sequestration of PAHs.

biochar is perfect for removing PAHs. Its porous structure includes both macropores and micropores, which allow for the adsorption and retention of PAH molecules of various sizes. Biochar surfaces have functional groups like carboxyl, hydroxyl, and phenolic, which can form chemical connections with PAHs via π–π interactions, hydrogen bonding, and hydrophobic interactions. Biochar is resistant to microbial breakdown and can remain in the soil for a longer period of time, giving long-term benefits for PAH sequestration. The study demonstrated notable improvements in phenanthrene (Phe) removal efficiency with the modification of biochar. Specifically, adding Mn (IV) and Fe (III) in biochar resulted in 24.2% and 31.4%, respectively, higher removal of Phe from the water phase than unmodified biochar. Additionally, the Phe contents in the sediment were found to be 8.7% and 17.4% lower after Mn (IV) and Fe (III) modification, respectively, compared to unmodified biochar (Guo et al., 2023). Biochar effectively restricts PAHs within the soil matrix by adsorbing them to its surfaces, decreasing their mobility and bioavailability to plants, microorganisms, and aquatic creatures. Biochar can provide labile organic carbon and electron donors for microbial metabolism, allowing microbial consortia to degrade sorbed PAHs more efficiently.

4.6.3 Nanotechnology/Nanoparticle-Based Techniques

This technique examines the utilization of nanomaterials such as nanoscale zerovalent iron (nZVI) or carbon nanotubes (CNTs) for targeted adsorption or catalytic degradation of PAHs in wetland matrices. nZVI particles are helpful for the adsorption and degradation of PAHs in wetland soils and sediments due to their large surface area and reactivity. When applied to contaminated areas, nZVI can adsorb PAH molecules to its surface and catalyse their destruction via reductive dechlorination or Fenton-like reactions. Furthermore, the small size of nZVI allows it to penetrate soil pores and reach deeply buried contaminants, hence increasing removal efficiency. Functionalized CNTs can selectively adsorb PAHs from aqueous solutions or soil matrices via π–π interactions and hydrophobic forces. CNTs can also act as catalyst supports, aiding the catalytic oxidation or reduction of PAHs. By leveraging these features, CNT-based nanocomposites can improve PAH removal efficiency and shorten repair time in wetland environments (Chung et al., 2022). The benefits and challenges of nanotechnology-based PAH remediation include high efficiency, quick treatment kinetics, and the possibility for in-situ use. Furthermore, nanomaterials can be designed to target specific PAH chemicals while minimizing negative impacts on non-target organisms. However, issues such as nanomaterial durability, long-term destiny and travel in the environment, and potential dangers connected with nanoparticle release must be thoroughly evaluated to assure the safety and sustainability of nanotechnology-based techniques (Rajput et al., 2023).

4.7 ONE HEALTH APPROACH IN PAH REMEDIATION

In recent years, the One Health approach has gained recognition as a comprehensive framework for addressing complex health challenges that transcend disciplinary boundaries. One Health is fundamentally an interdisciplinary, cooperative approach that acknowledges the connection between environmental, animal, and human health. (Danasekaran et al., 2024). The approach brings together experts from diverse fields to develop innovative strategies for monitoring, mitigating, and remediating PAH contamination. One Health promotes integrated surveillance systems to detect PAH levels in the environment, in wildlife, and in human populations, enabling targeted interventions to minimize exposure and prevent health risks (Singh et al., 2024). Management approaches such as habitat restoration and biodiversity conservation enhance ecosystem resilience, which in turn supports essential services like water purification and food provisioning. Aligning with sustainable development principles, One Health ensures that PAH management strategies address environmental protection, social equity, and economic prosperity, fostering resilient, equitable, and inclusive communities that coexist harmoniously with their natural environment (WHO, 2022).

4.8 CARBON FOOTPRINT CONSIDERATIONS IN PAH REMEDIATION

Incorporating carbon footprint considerations into the planning and implementing of PAH remediation efforts holds significant potential to minimize environmental impacts, enhance sustainability, and contribute to global climate change mitigation strategies (Calvin et al., 2021). This section explores various strategies and approaches to integrate carbon footprint considerations into PAH remediation practices, ensuring a comprehensive and environmentally responsible approach.

4.8.1 Conducting Life Cycle Assessments (LCAs)

Life Cycle Assessments (LCAs) cover all remediation stages: raw material extraction, production, transportation, operation, and maintenance & disposal, and ensure a comprehensive environmental impact evaluation. An LCA uses functional units to standardize and compare remediation techniques. Data collection on energy use, resource consumption, transportation, and waste are vital for accurate assessments. LCAs measure impacts like global warming potential (GWP), resource depletion, and ecosystem damage. Interpreting LCA results identifies environmental hotspots, enabling stakeholders to optimize processes, adopt sustainable practices, and implement mitigation measures (Testa et al., 2022). Continuous improvement driven by LCAs ensures enhancement of environmental performance in PAH remediation.

4.8.2 Analysing Energy Requirements

Optimizing energy usage can minimize carbon emissions, enhance sustainability, and incorporate renewable energy sources like solar, wind, or biomass systems. Energy-efficient strategies, passive remediation, and integrated approaches can further reduce energy consumption. Carbon emissions reduction measures, such as carbon offsetting, carbon capture and storage (CCS), and sustainable land use practices, support broader climate goals. Integrating these measures enhances the sustainability of PAH remediation, mitigating climate impacts and promoting a low-carbon future (Guidi et al., 2023).

4.8.3 Implementing Emissions Reduction Strategies

Minimizing fuel consumption during transportation of equipment and personnel can significantly lower carbon emissions through efficient route planning, optimized vehicle loads, and use of fuel-efficient or electric vehicles. Transitioning to low-carbon alternatives, like electric-powered machinery, using eco-friendly chemical treatments, recycling by-products, optimizing procedures, and implementing energy management systems also help reduce emissions. Innovative technologies, such as biochar amendments and anaerobic digestion, present further opportunities (Ghosh et al., 2003).

4.9 CONCLUSION AND FUTURE DIRECTION

PAHs constitute prevalent environmental toxins that are damaging to the liver, kidneys, and lungs while also increasing the risk of cancer. To remediate PAH pollution, conventional remediation methods like excavation, soil vapor extraction (SVE), dredging, and chemical oxidation can be effective but often have high costs, environmental impacts, and incomplete results. Sustainable approaches such as constructed wetlands, biochar amendments, and nanotechnology-based techniques offer promising alternatives. Constructed wetlands utilize physical, chemical, and biological processes to remove PAHs, enhancing ecosystem resilience and biodiversity, and providing water filtration, flood control, and carbon sequestration. Biochar amendments increase soil PAH sequestration and microbial activity, providing a cost-effective solution. Nanotechnology, employing materials like nanoscale zerovalent iron (nZVI) and carbon nanotubes (CNTs), significantly improves PAH adsorption and degradation, though it requires careful evaluation of environmental risks and long-term impacts. By restoring contaminated environments, these techniques support agricultural productivity and food safety, directly benefiting human health and animal welfare. Addressing PAH pollution through a One Health perspective aligns with the principles of sustainable development, including environmental protection, social equity, and economic prosperity. By fostering resilience, equity, and inclusivity, One Health contributes to building healthier and more sustainable communities that thrive in harmony with their natural environment.

CONFLICTS OF INTEREST

The authors declare that they have no conflict of interest.

ACKNOWLEDGEMENT

This book chapter did not receive any specific grant from funding agencies in the public, commercial, or not-for-profit sectors. The authors would like to express their gratitude to the Director, CSIR-Indian Institute of Toxicology Research (CSIR-IITR), Lucknow (India), for his invaluable encouragement and assistance in completing this chapter. The CSIR-IITR manuscript number is IITR/SEC/MS/2024/43.

REFERENCES

Abdel-Shafy, H. I., & Mansour, M. S. (2016). A review on polycyclic aromatic hydrocarbons: Source, environmental impact, effect on human health and remediation. *Egyptian Journal of Petroleum*, 25(1), 107–123.

Ahangar, A. G. (2010). Sorption of PAHs in the soil environment with emphasis on the role of soil organic matter: A review. *World Applied Sciences Journal*, 11(7), 759–765.

Alkio, M., Tabuchi, T. M., Wang, X., & Colon-Carmona, A. (2005). Stress responses to polycyclic aromatic hydrocarbons in Arabidopsis include growth inhibition and hypersensitive response-like symptoms. *Journal of Experimental Botany*, 56(421), 2983–2994.

Balseiro-Romero, M., Wick, L. Y., Vila, J., Grifoll, M., & Ortega-Calvo, J. J. (2024). Drivers for efficient bioaugmentation and clean-up of contaminated soil. In Jose Julio Ortega-Calvo & Frederic Coulon (eds), *Soil Remediation Science and Technology* (pp. 261–291). Cham: Springer Nature Switzerland.

Belis, C. A., Offenthaler, I., & Weiss, P. (2010). Semivolatiles in the forest environment: The case of PAHs. In Peter Schröder and Christopher D. Collins (eds), *Organic xenobiotics and plants: From mode of action to ecophysiology* (pp. 47–73). Springer Netherlands.

Brzezinski, M., Martin, L., Simpson, K., Lu, K., Gan, N., Huang, C., ... & Xu, W. (2024). Photodegradation enhances the toxic effect of anthracene on skin. *Journal of Hazardous Materials,* 471, 134386.

Calvin, K., Cowie, A., Berndes, G., Arneth, A., Cherubini, F., Portugal-Pereira, J., … & Smith, P. (2021). Bioenergy for climate change mitigation: Scale and sustainability. *GCB Bioenergy*, 13(9), 1346–1371. https://doi.org/10.1111/gcbb.12863

Chauhan, P., Imam, A., Kanaujia, P. K., & Suman, S. K. (2024). Bioremediation of high-molecular-weight polycyclic aromatic hydrocarbons: An insight into state of art and cutting-edge approaches. In Ajay Kumar, Joginder Singh Panwar, & Lucas Carvalho Basilio de Azevedo (eds), *Microbiome-based decontamination of environmental pollutants* (pp. 179–212). Elsevier Science.

Chung, J. H., Hasyimah, N., & Hussein, N. (2022). Application of carbon nanotubes (CNTs) for remediation of emerging pollutants-a review. *Tropical Aquatic and Soil Pollution,* 2(1), 13–26.

Dai, Y., Liu, R., Zhou, Y., Li, N., Hou, L., Ma, Q., & Gao, B. (2020). Fire Phoenix facilitates phytoremediation of PAH-Cd co-contaminated soil through promotion of beneficial rhizosphere bacterial communities. *Environment International,* 136, 105421.

Danasekaran, R. (2024). One health: A Holistic approach to tackling global health issues. *Indian Journal of Community Medicine*, 49(2), 260–263.

Darajeh, N., Alizadeh, H., Farraji, H., Park, J., Barghi, A., & Rezania, S. (2020). Removal of polycyclic aromatic hydrocarbons (PAHs) by different physicochemical methods: A mini-review. *Journal of Energy and Environmental Pollution*, 1(2), 44–50.

Das, N., Das, A., Das, S., Bhatawadekar, V., Pandey, P., Choure, K., ... & Pandey, P. (2023). Petroleum hydrocarbon catabolic pathways as targets for metabolic engineering strategies for enhanced bioremediation of crude-oil-contaminated environments. *Fermentation, 9*(2), 196.

Dewangan, S., Bhatia, A. K., Singh, A. K., & Carabineiro, S. A. (2021). Removal of hydrophobic contaminants from the soil by adsorption onto carbon materials and microbial degradation. *C*, 7(4), 83.

Dhokale, B., Saeed, Z. M., Ali, W. A., & Mohamed, S. (2023). Recent advances in quantification and remediation technologies for toxic PAH mitigation from the environment. In Morajkar, P., & Naik, M. (eds), *Progress in Biochemistry and Biotechnology, Advances in Nano and Biochemistry* (vol 15, pp. 233–258). Academic Press. ISBN 9780323952538.

Eldos, H. I., Zouari, N., Saeed, S., & Al-Ghouti, M. A. (2022). Recent advances in the treatment of PAHs in the environment: Application of nanomaterial-based technologies. *Arabian Journal of Chemistry*, 15(7), 103918.

Guidi Nissim, W., Castiglione, S., Guarino, F., Pastore, M. C., & Labra, M. (2023). Beyond cleansing: Ecosystem services related to phytoremediation. *Plants*, 12(5), 1031.

Guo, Z., Kang, Y., Wu, H., Li, M., Hu, Z., & Zhang, J. (2023). Enhanced removal of phenanthrene and nutrients in wetland sediment with metallic biochar: Performance and mechanisms. *Chemosphere,* 327, 138523.

Ghosh, U., Zimmerman, J. R., & Luthy, R. G. (2003). PCB and PAH speciation among particle types in contaminated harbor sediments and effects on PAH bioavailability. *Environmental Science & Technology*, 37(10), 2209–2217. https://doi.org/10.1021/es020965t

Hussain, K., Hoque, R. R., Balachandran, S., Medhi, S., Idris, M. G., Rahman, M., & Hussain, F. L. (2018). Monitoring and risk analysis of PAHs in the environment. *Handbook of Environmental Materials Management,* 10, 978–3.

Jameson, C. W. (2019). Polycyclic aromatic hydrocarbons and associated occupational exposures. In: *Tumour Site Concordance and Mechanisms of Carcinogenesis*. International Agency for Research on Cancer, Lyon (FR). PMID: 33979079.

Kuppusamy, S., Maddela, N. R., Megharaj, M., Venkateswarlu, K., Kuppusamy, S., Maddela, N. R., ... & Venkateswarlu, K. (2020). Approaches for remediation of sites contaminated with total petroleum hydrocarbons. In *Total petroleum hydrocarbons: Environmental fate, toxicity, and remediation* (pp. 167–205). Springer International Publishing. ISBN 978-3-030-24034-9 ISBN 978-3-030-24035-6 (eBook).

Kuppusamy, S., Thavamani, P., Venkateswarlu, K., Lee, Y. B., Naidu, R., & Megharaj, M. (2017). Remediation approaches for polycyclic aromatic hydrocarbons (PAHs) contaminated soils: Technological constraints, emerging trends and future directions. *Chemosphere*, 168, 944–968.

Lakhani, A. (2018). Polycyclic aromatic hydrocarbons: Sources, importance and fate in the atmospheric environment. *Current Organic Chemistry*, 22(11), 1050–1069.

Lee, K. J., & Choi, K. (2023). Non-carcinogenic health outcomes associated with polycyclic aromatic hydrocarbons (PAHs) exposure in humans: An umbrella review. *Exposure and Health*, 15(1), 95–111.

Leung, J. Y., Cai, Q., & Tam, N. F. (2016). Comparing subsurface flow constructed wetlands with mangrove plants and freshwater wetland plants for removing nutrients and toxic pollutants. *Ecological Engineering*, 95, 129–137.

Li, X., Shen, X., Jiang, W., Xi, Y., & Li, S. (2024). Comprehensive review of emerging contaminants: Detection technologies, environmental impact, and management strategies. *Ecotoxicology and Environmental Safety*, 278, 116420.

Mallah, M. A., Changxing, L., Mallah, M. A., Noreen, S., Liu, Y., Saeed, M., ... & Zhang, Q. (2022). Polycyclic aromatic hydrocarbon and its effects on human health: An overeview. *Chemosphere*, 296, 133948.

Merhaby, D., Rabodonirina, S., Net, S., Ouddane, B., & Halwani, J. (2019). Overview of sediments pollution by PAHs and PCBs in mediterranean basin: Transport, fate, occurrence, and distribution. *Marine Pollution Bulletin*, 149, 110646.

Onyena, A. P., Manohar, C. S., Nkwoji, J. A., & Chukwu, L. O. (2024). Characterization of the molecular differential responses in marine benthic macroinvertebrates exposed to polycyclic aromatic hydrocarbons. *Aquatic Ecology*, 58(2), 263–288.

Panchenko, L. V., Kuzyanov, D. A., Pleshakova, Y. V., Pozdnyakova, N. N., Muratova, A. Y., & Turkovskaya, O. V. (2022). Effect of plant root exudate constituents on the degradation of Phenanthrene by the Rhizobacterium Mycolicibacteriumgilvum (Mycobacteriaceae, Actinobacteria). *Biology Bulletin*, 49(10), 1958–1964.

Patel, A. B., Shaikh, S., Jain, K. R., Desai, C., & Madamwar, D. (2020). Polycyclic aromatic hydrocarbons: Sources, toxicity, and remediation approaches. *Frontiers in Microbiology*, 11, 562813.

Peluffo, M., Rosso, J. A., Morelli, I. S., & Mora, V. C. (2018). Strategies for oxidation of PAHs in aged contaminated soil by batch reactors. *Ecotoxicology and Environmental Safety*, *151*, 76–82.

Pi, X., Sun, Y., Fu, G., Wu, Z., & Cheng, J. (2021). Effect of processing on soybean allergens and their allergenicity. *Trends in Food Science & Technology*, *118*, 316–327.

Praveen, A., Pandey, V. C., Marwa, N., & Singh, D. P. (2019). Rhizoremediation of polluted sites: Harnessing plant–microbe interactions. In Pandey, V. C. & Bauddh, K. (eds), *Phytomanagement of polluted sites* (pp. 389–407). Elsevier.

Rajput, V. D., Kumari, S., Minkina, T., Sushkova, S., & Mandzhieva, S. (2023). Nano-enhanced microbial remediation of PAHs contaminated soil. *Air, Soil and Water Research*, *16*, 11786221231170099.

Ramesh, A., Harris, K. J., & Archibong, A. E. (2022). Reproductive toxicity of polycyclic aromatic hydrocarbons. In Gupta, R. C. (ed), *Reproductive and developmental toxicology* (3rd ed., pp. 759–778). Academic Press. ISBN 9780323897730

Sakshi, Singh, S. K., & Haritash, A. K. (2019). Polycyclic aromatic hydrocarbons: Soil pollution and remediation. *International Journal of Environmental Science and Technology*, 16, 6489–6512.

Schwaab, M. R. (2020). *Hepatobiliary polycyclic aromatic hydrocarbons in pelagic fishes of the Gulf of Mexico*. University of South Florida.

Singh, S., Sharma, P., Pal, N., Sarma, D. K., Tiwari, R., & Kumar, M. (2024). Holistic one health surveillance framework: Synergizing environmental, animal, and human determinants for enhanced infectious disease management. *ACS Infectious Diseases*, 10, 808–826.

Srivastava, S., & Kumar, M. (2019). Biodegradation of Polycyclic Aromatic Hydrocarbons (PAHs): A Sustainable Approach. In Shah, S., Venkatramanan, V., & Prasad, R. (eds). *Sustainable Green Technologies for Environmental Management*. Springer, Singapore. https://doi.org/10.1007/978-981-13-2772-8_6

Stewart, L., Truesdale, R., Redmon, J., Barth, E. D., Northeim, C., & McKernan, J. (2018). *Engineering issue: Soil vapor extraction (SVE) technology*. EPA/600/R-18/053.

Sun, X., Zhao, L., Huang, M., Hai, J., Liang, X., Chen, D., & Liu, J. (2024). In-situ thermal conductive heating (TCH) for soil remediation: A review. *Journal of Environmental Management,* 351, 119602.

Testa, F., Tessitore, S., Buttol, P., Iraldo, F., & Cortesi, S. (2022). How to overcome barriers limiting LCA adoption? The role of a collaborative and multi-stakeholder approach. *International Journal of Life Cycle Assessment*, 27(7), 944–958.

Tsatsou, A., Frantzeskaki, N., & Malamis, S. (2023). Nature-based solutions for circular urban water systems: A scoping literature review and a proposal for urban design and planning. *Journal of Cleaner Production,* 394, 136325.

Turkovskaya, O., & Muratova, A. (2019). Plant–bacterial degradation of polyaromatic hydrocarbons in the rhizosphere. *Trends in Biotechnology*, 37(9), 926–930.

USEPA. (2016c). *Summary of Results for the 2011 National-scale Assessment.* U.S. Environmental Protection Agency, Washington, D.C.

Venkatraman, G., Giribabu, N., Mohan, P. S., Muttiah, B., Govindarajan, V., Alagiri, M., ... & Karsani, S. A. (2024, March). Environmental impact and human health effects of polycyclic aromatic hydrocarbons and remedial strategies: A detailed review. *Chemosphere*, 351, 141227.

Vijayanand, M., Ramakrishnan, A., Subramanian, R., Issac, P. K., Nasr, M., Khoo, K. S., ... & Ravindran, B. (2023). Polyaromatic hydrocarbons (PAHs) in the water environment: A review on toxicity, microbial biodegradation, systematic biological advancements, and environmental fate. *Environmental Research, 227*, 115716.

Von Lau, E., Gan, S., Ng, H. K., & Poh, P. E. (2014). Extraction agents for the removal of polycyclic aromatic hydrocarbons (PAHs) from soil in soil washing technologies. *Environmental Pollution*, 184, 640–649.

Walker, T. R., MacAskill, D., Rushton, T., Thalheimer, A., & Weaver, P. (2013). Monitoring effects of remediation on natural sediment recovery in Sydney Harbour, Nova Scotia. *Environmental Monitoring and Assessment*, 185(10), 8089–8107.

Wenger, A. S., Harvey, E., Wilson, S., Rawson, C., Newman, S. J., Clarke, D., ... & Evans, R. D. (2017). A critical analysis of the direct effects of dredging on fish. *Fish and Fisheries*, 18(5), 967–985.

World Health Organization. (2022). *A health perspective on the role of the environment in One Health* (No. WHO/EURO: 2022-5290-45054-64214). World Health Organization. Regional Office for Europe.

Yost, E. E., Galizia, A., Kapraun, D. F., Persad, A. S., Vulimiri, S. V., Angrish, M., ... & Druwe, I. L. (2021). Health effects of naphthalene exposure: A systematic evidence map and analysis of potential considerations for dose–response evaluation. *Environmental Health Perspectives*, 129(7), 076002.

Zhou, Y., Wang, J., Zou, M., Jia, Z., Zhou, S., & Li, Y. (2020). Microplastics in soils: A review of methods, occurrence, fate, transport, ecological and environmental risks. *Science of the Total Environment,* 748, 141368.

5 Implementation of Cradle-to-Cradle Concept in PFA Management and Potential Impact on One Health Concept

Alaa El Din Mahmoud and Eduardo Alberto Lopez Maldonado

5.1 INTRODUCTION

The term "cradle-to-cradle" (C2C) was introduced in 2002 by Michael Braungart and William McDonough in their book "*Cradle-to-Cradle: Remaking the Way We Make Things*", as a conceptual framework aimed at fostering both economic and environmental benefits. It emphasizes eco-effectiveness as a design alternative to generate positive environmental and economic impacts (Rodríguez. 2019).

The core objective of this concept is to close the material cycle within production by minimizing waste generation and reintegrating residues into the production process through biological or technical cycles, ultimately making the process sustainable (Ismayilova & Silvius. 2021). Therefore C2C and eco-effectiveness design provides an alternative strategy by recycling waste materials into new forms and uses to serve both the economy and the environment. Recent literature explores waste valorization as a means to implement circular economy principles in wastewater treatment, including studies by Nagy et al. (2023), Mahmoud et al. (2024), Mahmoud et al. (2023), Mahmoud et al. (2022b,c) and Mahmoud (2020). This approach represents a significant shift in service design by aligning with nature's principle that waste serves as a resource for new processes (Mahmoud & Fawzy 2023). Waste generated from one process becomes a nutrient input for another. There are two types of nutrients, biological and technical. Technical nutrients, which are inorganic materials, should be non-toxic, non-harmful, and environmentally benign synthetic substances. Biological nutrients are organic materials that decompose naturally without causing environmental harm. Technical nutrients are considered for industrial recycling, while biological nutrients return to the soil. An increase in waste production results in a greater

DOI: 10.1201/9781032623801-5

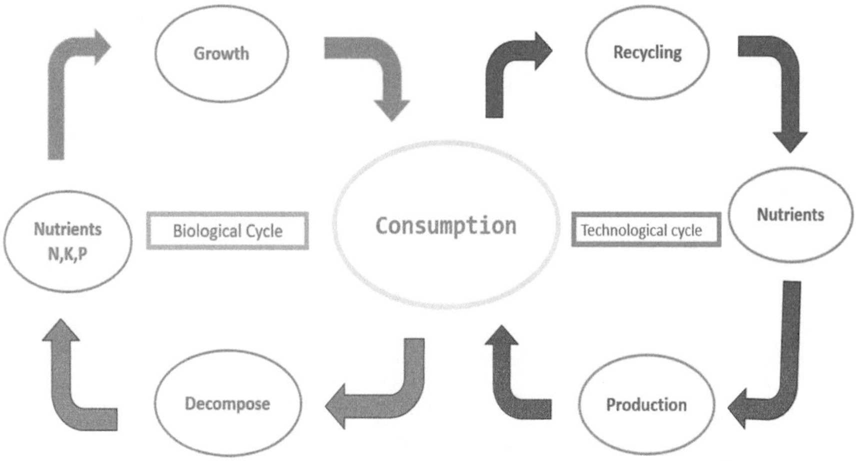

FIGURE 5.1 The technical and biological cycle of nutrients.

availability of nutrients (Mohajan, 2021). The concept of both nutrient cycles is illustrated in Figure 5.1.

Before discussing per- and polyfluoroalkyl substances (PFAs) management through the C2C concept, it is essential to highlight the issue and its impact. PFAs comprise a class of chemicals used extensively in industrial and commercial applications (Kwiatkowski et al., 2020). These compounds consist of fluorinated carbon chains, either long or short, and are found in a wide range of PFAS-containing products currently and newly introduced to the global market (Anderson et al., 2022; Grandjean et al., 2017). The unique physicochemical properties of PFAs contribute to their diverse functional applications. PFAs are prominently used in applications such as bicycle lubricants, carpets, cleaning products, food packaging, textiles, leather, firefighting foams, chrome plating, chemically enhanced oil production, fluoropolymer synthesis, and the semiconductor industry (see Figure 5.2) (Barry et al., 2013; Gewurtz et al., 2013; Glüge et al., 2021a). The functionality of PFAs includes water and grease repellency, chemical degradation resistance, and flame retardance. These properties render PFAs highly persistent in the environment and pose significant health risks (Cousins et al., 2020; Fenton et al., 2021; Langenbach & Wilson. 2021). Excessive PFA usage has been linked to global public health concerns (Glüge et al., 2021b). The primary adverse health effects reported are related to immune function, thyroid, liver disease and cancer, lipid and insulin dysregulation, kidney disease, uric acid and kidney cancer, and reproductive and developmental outcomes (see Figure 5.3) (Barry et al., 2013; Cousins et al., 2020; Mousazadeh et al., 2021). Many other health effects have not been scientifically proven but require timely and reliable data on the relationship between PFAs and health effects (Grandjean et al., 2017).

The high oxidation resistance of PFAs and low solubility in water are physicochemical properties that make them a class of highly dangerous chemicals and generate a high bioaccumulation in the environment (Anderson et al., 2022; Glüge

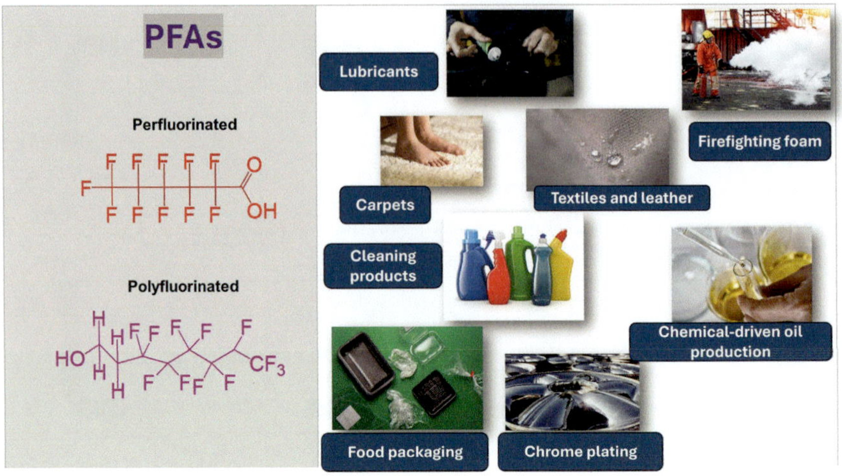

FIGURE 5.2 Functions and uses of PFAs.

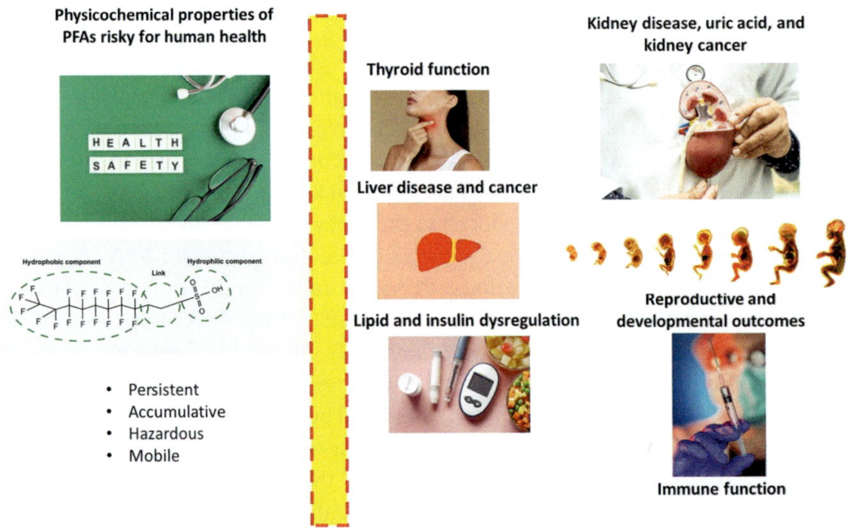

FIGURE 5.3 Toxic effects on the health of people and structures of PFAs.

et al., 2021a; Kwiatkowski et al., 2020). For all this, many efforts have been made to create and implement stricter regulations on the use, handling, and commercialization of PFAs. However, current regulations are not sufficient to ensure minimal impact from PFA exposure. The US Environmental Protection Agency (EPA) through the Toxic Substances Control Act (TSCA) has attempted to regulate PFAs to some degree (Carmody et al., 2021). Many challenges need to be overcome since there is not complete information on the toxicity risks for all the PFAs that are on the market. It is known that thanks to the strength of the C-F bond of PFAs, these compounds are hardly degraded by biological systems, leading to low-dose exposure of PFAs that have long-term toxic effects (Fenton et al., 2021; Glüge et al., 2021).

Unlike the problems of contamination of bodies of water by other contaminants (Hossein et al., 2023; Mahmoud et al., 2022), the case of PFAs is of great concern because conventional water treatment systems are not effective and sufficient for their destruction or removal from the water (Carmody et al., 2021). Separation is typically achieved through filtration or ion exchange processes. However, these technologies generate residues from PFAs that must be disposed of in sanitary landfills. The oxidation of PFAs requires a high level of energy and produces other substances with the same or a higher level of toxicity. These practices for the control of PFAs are not eco-friendly and are very expensive, and they are not equally effective for all types of PFAs, hence the need to implement management systems on PFAs in an integral way, considering the need for the function of PFAs, environmental impacts, health risks, and cost analysis. Other materials cover these functions without the need to use PFAs, being biodegradable and with different production strategies (Anderson et al., 2022; Kwiatkowski et al., 2020).

This chapter will highlight the C2C concept and principles as well as its merits over cradle-to-grave (C2G) due to its alignment with the One Health concept. It will discuss the potential and the challenges of implementing the C2C concept in PFA management.

5.2 CRADLE-TO-CRADLE CONCEPT AND PRINCIPLES

The circular economy is an economic system designed to minimize raw material use and value loss by maximizing reuse, thereby reducing operational costs and emphasizing resource restoration (Mahmoud & Mahmoud, 2023; Mahmoud & Desokey, 2022). Unlike simply minimizing negative impacts, C2C evaluates the substances used in processes to ensure they can be reused as nutrients, allowing them to safely re-enter the environment through component disassembly (Mannan & Al-Ghamdi, 2022). In contrast, cradle-to-grave (C2G) considers a life cycle assessment from a linear economic perspective as it assesses the environmental and social impacts of specific services through its whole life cycle from production to disposal. This approach systematically captures energy flow from raw material extraction to final disposal, helping to identify key areas for optimizing resource utilization (Rodríguez, 2019). A comparison between the circular and the linear economy is shown in Figure 5.4.

The implementation of C2C is based on three core principles: (1) eliminating waste by ensuring materials are designed for continual reuse through upcycling, (2) utilizing renewable energy to prevent pollution, and (3) fostering technological and social diversity to encourage creative, locally adaptive solutions rather than a one-size-fits-all approach (Ismayilova & Silvius, 2021). As a vital resource, water in C2C processes should be used responsibly and returned cleaner than before, ensuring water quality is maintained and local ecosystems are preserved (Awotoye et al., 2024; Kathi & Mahmoud, 2024).

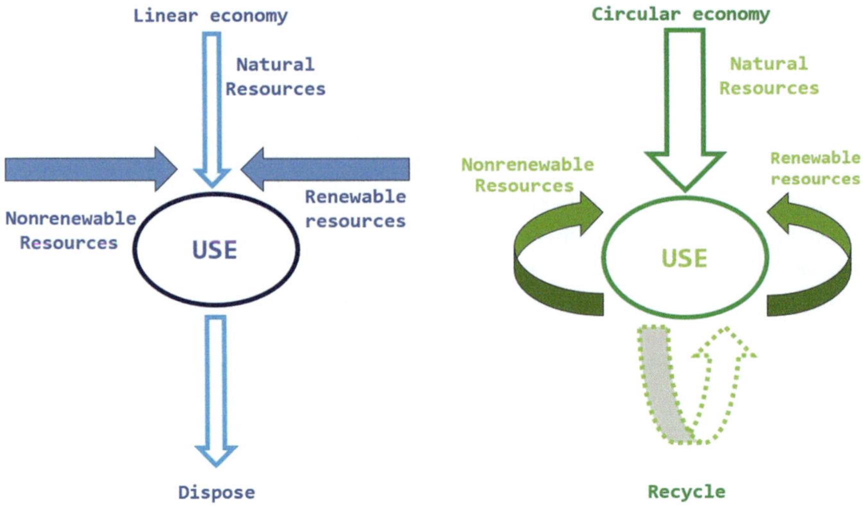

FIGURE 5.4 Differentiation between linear and circular economy in a simple manner.

The C2C concept not only reduces ecological footprints but also generates a positive environmental impact by enhancing the upcycling of by-products and waste materials into valuable resources, leading to waste and cost reductions. Integrating C2C into business strategies enhances resource efficiency, boosts economic competitiveness, generates employment opportunities, and promotes waste reduction (Mohajan, 2021). The five pillars of C2C framework define the standards for designing products and services to be upcyclable. These pillars ensure sustainability throughout the product or service life cycle and include: i) material health, All materials used must be assessed for their impact on human and environmental health. Components are classified as either biological (safe to return to the environment) or technological (suitable for continuous reuse in industrial cycles), and must be safe during and after use, ii) material recyclability to make sure that the components remain in the life cycle of the products or the service, iii) water management or stewardship as the service should not affect the drinking water quality, iv) renewable energy accreditation during the manufacturing process which contribute to minimize the carbon footprint of the service, v) social fairness, compliance with social responsibility standards is essential, including fair labor practices, human rights protection, and contributions to the well-being of local communities through job creation and equitable treatment (Rodríguez, 2019; Kathi et al., 2023; Mahmoud et al., 2021).

5.3 CRADLE-TO-CRADLE AS A TOOL TO ACHIEVE SUSTAINABLE DEVELOPMENT GOALS

The C2C contributes to sustainable development goals by improving water quality, implementing integrated water management systems, increasing water use efficiency, reducing the number of people experiencing water stress, and restoring ecosystems dependent on water bodies (Mahmoud et al., 2023). It will be crucial to achieve the following Goals: Goal 1 (no poverty), paying above poverty-level wages; Goal 3 (good health and well-being) identifying, assessing, and enhancing the safety of chemical materials that are used in the production cycle; Goal 6 (clean water and sanitation) reducing water consumption and avoiding water stress; Goal 7 (affordable and clean energy) implementing renewable energy strategies; Goal 8 (decent work and economic growth) improving awareness of input materials with a high risk and preventing waste and decreasing raw material consumption; Goal 9 (industry, innovation and infrastructure) providing an approach to sustainable manufacture; Goal 11 (sustainable cities and communities) making a transition to renewable resources to optimize the effectiveness of the tool; Goal 12 (responsible consumption and production) selecting sustainable processes and re-entering materials into the cycle of production to lead to a reduction in overall consumption; Goal 13 (climate action) ensuring awareness of climate action, requires changes in procedures and services to reduce these impacts' Goal 14 (life below water) identifying negative impacts on marine life, and instead of releasing waste into water, integrating it into the service cycle; and Goal 15 (life on land) identifying negative impacts on sensitive ecosystems and implementing conservation strategies and preventing pollutants (Khattab et al., 2023; Fawzy et al., 2022). Figure 5.5 shows the interconnection between C2C and related Sustainable Development Goals. The benefits of C2C can be categorized as: (i) Environmental: Reduces waste and pollution, encourages renewable resource use, and preserves ecosystems through product design for reuse and safe decomposition; (ii) Economic: Reduces waste management and resource extraction costs, while fostering innovation in renewable energy and sustainable design; (iii) Social: Creates safer environments for consumers and workers, supports ethical labor practices, and strengthens local economies through sustainable practices.

5.4 IMPLEMENTATION OF C2C ON WASTEWATER TREATMENT PLANTS (WWTPS)

Water reuse for agricultural and irrigation practices offers a replacement for agricultural demand for freshwater, where secondary treatment wastewater is suitable for non-food crops, and tertiary treatment wastewater can be safely utilized for irrigating food crops (Dotaniya et al., 2023). Energy recovery involves biogas production and heat pumps from anaerobic digestion, which can be utilized, or for combined heat and power applications. Finally, sewage sludge can be utilized in construction practices by recovering ash from burned sludge for the production of cement, concrete, and mortar (Neczaj & Grosser, 2018). From an economic perspective, implementing C2C in wastewater treatment helps reduce management costs, exposure to market fluctuations, and dependence on fossil fuels, leading to lower waste and emissions, increasing employment creation, raising environmental quality, and improving collaboration innovation capability (European Union, 2020).

The Tenorio wastewater treatment plant in San Luis Potosi, Mexico, sells treated wastewater to a nearby power plant for cooling towers, replacing the use of freshwater. Forty-five per cent of the wastewater generated by the city is treated by the

FIGURE 5.5 C2C and Sustainable Development Goals interconnections.

plant, and it is 33% cheaper than freshwater. This resulted in savings of $18 million for the power utility in a six-year period, and the revenue for the WWTP covers most of the operational costs. The treated wastewater is utilized for different purposes. For cooling purposes, water irrigation and environmental enhancement, it reduces groundwater extraction and improves environmental conditions, enhances local living standards, and supports better public health outcomes. . In San Francisco, the implementation of co-digestion for methane-generated electricity resulted in a $31 million project with a 15 MW capacity, saving $2.5 million annually in energy costs, generating $500,000 from electricity sales to the power grid, and earning $8 million annually from waste disposal fees, while reducing greenhouse gas emissions and landfill usage (Hernández-Chover et al., 2023). In Rio de Janeiro, Brazil's capital, biosolids were utilized in agriculture by reclaiming degraded lands in railway operational areas, demonstrating higher efficiency compared to conventional fertilizers. Results have shown the efficiency of this implementation compared to normal fertilizers, with an equivalent amount of nutrients such that all yields were higher than average. Biosolids were 21% more effective than normal fertilizer. Another study on soybeans shows that biosolids were 18% more effective than normal fertilizers. This leads to better crop yields and savings in landfill costs and production of normal fertilizers (RodríGuez et al., 2020). The next sections will focus on PFA management.

5.5 IMPLEMENTATION OF C2C IN PFA MANAGEMENT

Building upon the foundational understanding of PFAs' persistence discussed in Section 5.1, it is essential to evaluate current management strategies and their inherent limitations. Conventional methods predominantly target containment, removal, and destruction, but their long-term efficacy remains uncertain, particularly given PFAs' resistance to degradation. Remediation techniques such as adsorption, filtration, and advanced oxidation have been widely applied. However, as previously noted, the chemical robustness of PFAs significantly complicates complete elimination. Moreover, many existing treatment methods exhibit high energy consumption and generate secondary wastes, raising concerns about environmental sustainability.

Given scientific reports and environmental studies on the persistence, accumulation, and toxicity of PFAs, a structured system for PFAs management—spanning design, implementation, and monitoring—is essential. All these impacts have for decades been the product of the use of a linear economic or production system with PFAs (Glüge et al., 2021). Some systems that have tried to minimize the impacts of PFAs are the concept of green chemistry for example; however, it is not enough due to the structural complexity of PFAs. Another alternative is to search for substances that can carry out the same function with equal or better performance than PFAs. Today the C2C concept is an innovative strategy that involves considering all aspects and stages of the use of PFAs, comprehensively analyzing environmental, economic, social, and human health aspects.

To carry out the implementation of a robust management system for PFAs, actions are required in each stage of this PFA life cycle (Carmody et al., 2021). The first stage involves documenting and analyzing PFAs' primary functions across industrial and commercial applications to identify potential alternatives. This will help PFA manufacturing industries access other substances that they can use which are free of PFAs. In this first stage, it is extremely important that in-depth medical studies be carried out to provide evidence of the danger of existing PFAs and that research and development of more PFA compounds be limited. All these stages must be aligned and supervised by organizations and government regulations, to establish permits for manufacturing, uses, handling, and quantities allowed in any industrial activity (Figure 5.6). The second stage, the handling and transport of PFAs, must be carried out safely, with well-documented records on the quantities and destinations of use. In the third stage, PFA consumption should be strictly limited to essential applications to minimize environmental and health risks. Users must have complete information on the characteristics and risk of the use of per- and polyfluorinated substances. Here it would be very appropriate for users to have a catalog of other substances that could also perform the same function but with eco-friendly, biodegradable, and sustainable substances. In the fourth stage, all the PFAs already in use must be characterized and evaluated as potentially usable materials for other purposes before considering them as useless materials. Here many actions or alternative uses can be carried out, in such a way that these PFA compounds must be reused-recycled as many times as are sufficient for a given function and that generate the minimum amount of hazardous waste. In the final stage, materials containing PFAs should be classified based on their potential for reuse in the same or alternative applications, minimizing waste generation. The objective of this sustainable life cycle of PFAs is to increasingly reduce the amount of PFAs used in the world (see Figure 5.7).

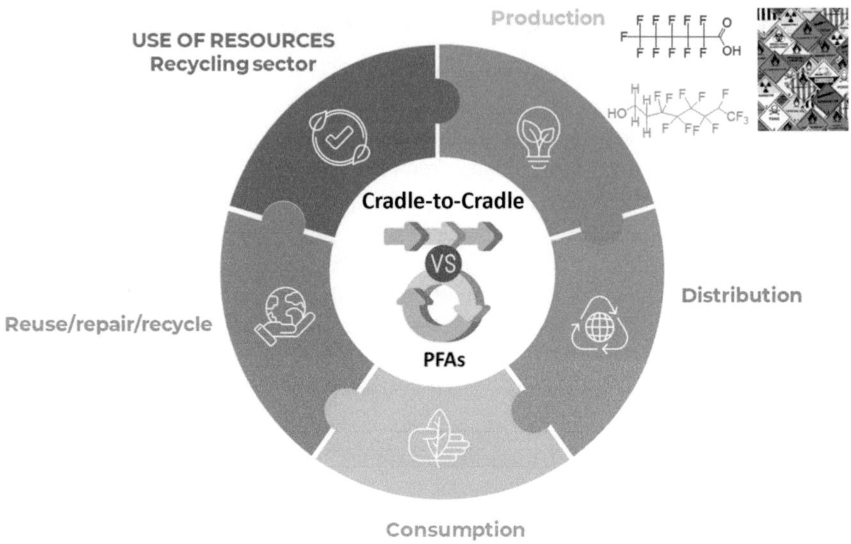

FIGURE 5.6 C2C concept in PFAS management.

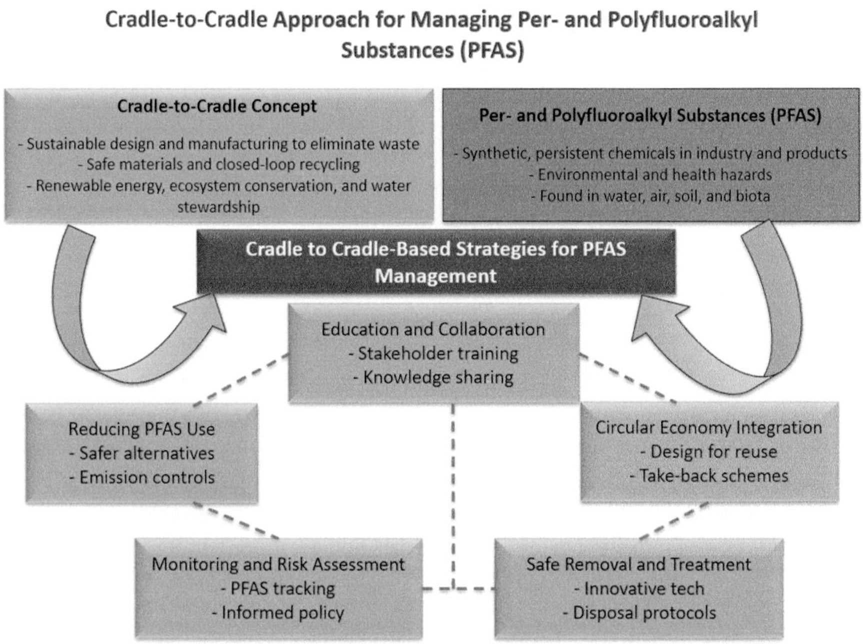

FIGURE 5.7 Main attributes of the C2C concept in PFA management.

In summary, we designed the following scheme (Figure 5.7) to overview the application approach of C2C for the management of PFAs.

5.6 THE POTENTIAL BENEFITS OF USING THE CRADLE-TO-CRADLE CONCEPT IN PFA MANAGEMENT

Transitioning from the evaluation of conventional PFA treatment methods, it becomes evident that innovative, sustainable frameworks are urgently needed. The C2C philosophy offers a promising model by emphasizing material health, continuous resource cycling, and positive environmental impacts. The C2C concept focuses on designing products in a manner that, after their useful life, enables full reintegration into either biological or technological cycles. In terms of material health, substances are classified as either biological or technological nutrients, depending on their capacity for safe assimilation or recovery. This paradigm encourages industries to adopt non-toxic, renewable, and endlessly recyclable materials, aligning perfectly with One Health priorities. As such, integrating C2C into PFA management could radically shift current practices from reactive remediation toward proactive, sustainable design.

The main advantages represented by the use of the C2C concept aligned with the concept of health for the management of PFAs is that throughout their life cycle, there are opportunities for improvement and alternatives that will help minimize the generation, use, disposition, and accumulation of these highly toxic substances in the environment (Ellis et al., 2023). Another important aspect is of an economic nature: the manufacturing and synthesis processes of PFAs are expensive and the search for new eco-friendly substances are business competitiveness strategies that could encourage entrepreneurs in these markets.

5.7 EXAMPLES OF SUCCESSFUL IMPLEMENTATION OF THE CRADLE-TO-CRADLE CONCEPT IN PFA MANAGEMENT

Some of the documented cases on the use of the C2C concept in PFA management involve the development of separation systems for PFAs from water using filtration operations based on granular carbon or anion exchange membranes. These reports show that it is technically and economically feasible to decontaminate water from PFAs, with the reuse of the filter media and reaching minimum PFA concentrations in the effluents (Ellis et al., 2023; Mu et al., 2022). We know that these pollution control systems have a lifetime (reusability cycles), and they must be evaluated economically along with the environmental impacts they represent. At the end of the use of these exhausted filter media, it is necessary to have management and disposal strategies, as current incineration actions are not friendly to the environment (Mahmoud & Mostafa, 2023). Other cases consist of the analysis of the essential use of PFAs in each application, looking for other compounds with the same physicochemical properties but with a more ecological or biodegradable character. Some PFA-free lubricants are already on the market and are extracted from plants. The manufacture of carpets without impregnation of PFAs is also viable. Here it is necessary for consumers to identify particular seals that these products have, such as the Blue Angel certificate based on the C2C concept. Regarding the use of industrial and household cleaning products, there are a variety of substances of natural origin that can also fulfill this function. It is reported that PFAs are not technically necessary in cleaning product formulations. Another of the actions that involves the use of PFAs is in the electroplating processes based on Cr^{6+}, since these substances prevent the evaporation of chromium, which is highly toxic and carcinogenic. An alternative consists in the use of a closed and pressurized technology in the coating baths (Gewurtz et al., 2013; Glüge et al., 2021).

5.8 CHALLENGES IN IMPLEMENTING THE CRADLE-TO-CRADLE CONCEPT IN PFA MANAGEMENT

PFAs are often dubbed "forever chemicals" due to their high chemical stability and resistance to degradation. This persistent nature hinders C2C principles, as these substances do not naturally reintegrate into the biosphere. Cousins et al. (2020) highlight that the fundamental persistence of PFAs in environmental systems makes it challenging to close the life cycle loop in a safe, environmentally-friendly manner. One major challenge in implementing the C2C concept for PFA management is the lack of publicly accessible information. There is a lack of publicly accessible databases containing health outcome data on the direct effects of PFA exposure across various applications. Effective coordination between government agencies, industries, the healthcare sector, academia, and environmental regulatory bodies is crucial for successful PFA- management (Mu et al., 2022). A lack of alignment among these key stakeholders disrupts the ability to manage PFAs within a sustainable life cycle framework. Figure 5.8 summarizes the key challenges in implementing the C2C concept in PFA management.

5.8.1 Technical Challenges

- **Persistence and Toxicity of PFAs**

  A major technical challenge in implementing a C2C approach for PFA management is their exceptional persistence in the environment. Due to their resistance to degradation, PFAs accumulate over time in ecosystems and living organisms, posing long-term environmental risks. Their persistent nature complicates efforts to integrate PFAs into a sustainable closed-loop system for safe reuse and recycling.

  PFAs contain exceptionally strong carbon-fluorine bonds, which render them highly resistant to conventional environmental degradation processes. Consequently, conventional waste management techniques like bioremediation and composting are largely ineffective in breaking down PFAs (Wee & Aris, 2023). Rather than degrading, PFAs persist in soils, groundwater, and aquatic organisms, further complicating efforts to recycle and reuse these substances.

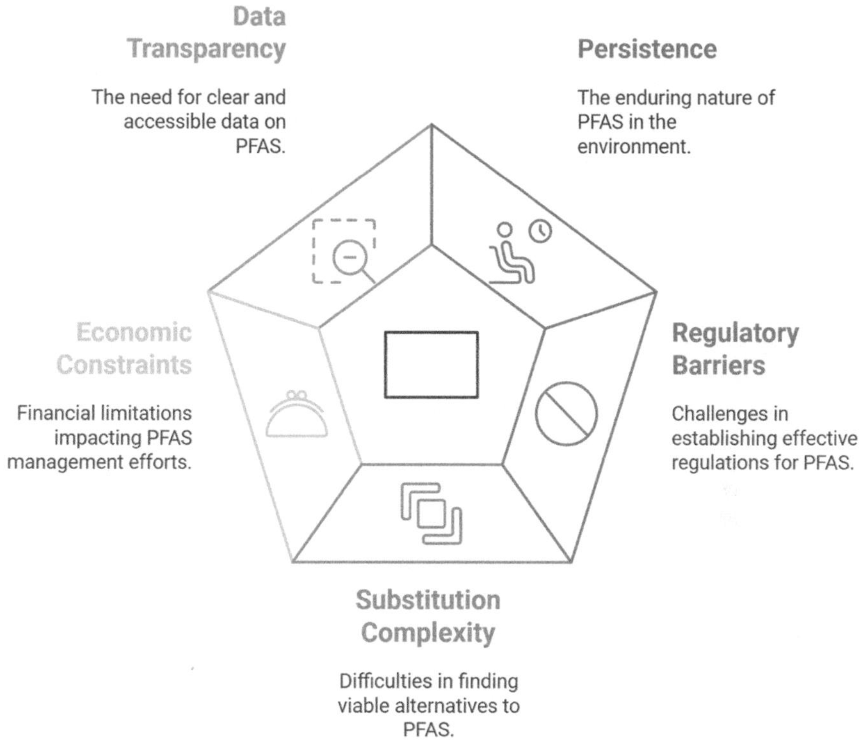

FIGURE 5.8 Key challenges in implementing the C2C concept in PFA management.

- **Limited Recycling Technologies**
  Existing PFA elimination technologies, including advanced oxidation and plasma treatment, are expensive and often inefficient at large-scale implementation (Bayode et al., 2024). Furthermore, these treatments can produce by-products that retain PFAs or other toxic substances, making reintegration into production cycles unfeasible.
  High-temperature incineration has been suggested for PFA destruction, but concerns remain about the potential release of toxic emissions during combustion. Additionally, isolating PFAs from materials such as plastics or textiles is difficult, as cross-contamination can compromise the quality of recycled products.
- **Environmental and Energy Costs of PFA Management**
  The management of PFAS, especially through high-energy destruction techniques, poses significant environmental and energy challenges. These methods require substantial energy input, which conflicts with the principles of the C2C model that emphasize resource efficiency and low-energy processes in production and recycling. The high energy demands of current PFAS disposal practices raise sustainability concerns and hinder progress toward a truly circular material system. (Wee & Aris, 2023).

5.8.2 Regulatory Challenges

- **Fragmented Regulatory Frameworks**
  One of the most significant obstacles to the implementation of C2C strategies for PFAs is the fragmentation of regulatory frameworks. Some countries, like the United States, have introduced stricter PFA regulations, including limits on their presence in consumer products and drinking water, whereas others have minimal or no regulations governing PFA management. The absence of consistent international regulatory standards creates uncertainty for manufacturers and recyclers in managing PFA-containing products. Many recycling systems operate on the assumption that materials can be safely reused or repurposed, an assumption that does not apply to PFAs due to their persistent and hazardous nature (Bayode et al., 2024).
- **Disposal and Recycling Regulations for PFAs**
  Regulations governing PFA disposal and recycling are frequently inadequate or lack proper enforcement. Numerous recycling policies overlook the presence of PFAs, increasing the risk of ongoing contamination within recycling streams. Advancing toward a C2C system requires updating regulatory frameworks to ensure PFA-containing products are managed in a manner that reduces risks to human health and the environment.

5.8.3 Social and Economic Challenges

- **Public Awareness and Acceptance of New Technologies**
  Limited public awareness of PFA-related risks and potential solutions represents a significant social barrier to implementing C2C strategies. Despite increasing media coverage of PFA contamination and health risks, public understanding of their presence in everyday products remains limited. The adoption of novel PFA management technologies, such as advanced destruction and recycling methods, may encounter resistance due to public skepticism regarding their safety, effectiveness, and cost implications. The successful implementation of large-scale C2C systems for PFAs hinges on securing public trust and widespread support (Lukić Bilela et al., 2023).
- **Implementation Costs**
  The transition to a C2C model for PFA management requires significant investment in research and development of new technologies, as well as infrastructure to support more sustainable practices. Such a transition could impose a considerable financial burden on industries dependent on PFA-containing products for their manufacturing processes. Balancing the economic costs of developing and scaling new PFA management technologies with environmental goals remains a major challenge to widespread C2C adoption (Lukić Bilela et al., 2023).

5.9 CONCLUSION AND RECOMMENDATIONS

The toxic effects of PFAs on human health and the environment are well documented and widely acknowledged. Addressing these harmful consequences requires a fundamental shift from the traditional linear economic model, which continues to prioritize the synthesis of new PFA compounds over the pursuit of sustainable alternatives. The pervasive presence of PFAs in water bodies, soil, and air underscores the critical need for comprehensive monitoring systems and effective separation strategies to mitigate their impact and prevent their bioaccumulation through the food chain. Integrating circular economy principles into PFA governance not only aligns with One Health priorities—recognizing the interconnectedness of human, animal, and environmental health—but also establishes a clear and actionable pathway toward sustainable chemical management. To effectively combat PFA contamination, it is essential to foster a coordinated approach that encompasses governmental, social, environmental, and economic actions. By doing so, we can ensure a meaningful impact on public health and the environment, paving the way for a healthier future for all.

REFERENCES

Anderson, J.K., Brecher, R.W., Cousins, I. T., DeWitt, J., Fiedler, H., Kannan, K., Kirman, C. R., et al. 2022. Grouping of PFAS for human health risk assessment: Findings from an independent panel of experts. *Regulatory Toxicology and Pharmacology*, *134*, p.105226. https://doi.org/10.1016/J.YRTPH.2022.105226.

Awotoye, D., Omoarukhe, F.O., Mahmoud, A.E.D., Akande, O., Adeleke, A.A., Ikubanni, P. and Ogbaga, C.C., 2024. Exploring the future of nanomaterials: Insights into synthesis, characterization, and potential applications. In J.A. Okolie, E.I. Epelle, A. Mukherjee, and A.E.D. Mahmoud (eds), *Nanomaterials for Sustainable Hydrogen Production and Storage* (pp. 10–35). CRC press.

Barry, V., Winquist, A. and Steenland, K., 2013. Perfluorooctanoic acid (PFOA) exposures and incident cancers among adults living near a chemical plant. *Environmental Health Perspectives*, *121*, pp.1313–1318. https://doi.org/10.1289/EHP.1306615.

Bayode, A.A., Emmanuel, S.S., Akinyemi, A.O., Ore, O.T., Akpotu, S.O., Koko, D.T., Momodu, D.E. and López-Maldonado, E.A., 2024. Innovative techniques for combating a common enemy forever chemicals: A comprehensive approach to mitigating per- and polyfluoroalkyl substances (PFAS) contamination. *Environmental Research*, *261*, p. 119719.

Carmody, J.E., Page, K., Petersen, P.D. and Weinberg, M., 2021. *Beyond Cradle to Grave: EPA Proposes to Regulate PFAS "Forever Chemicals" Under RCRA*. https://perkinscoie.com/insights/update/beyond-cradle-grave-epa-proposes-regulate-PFAs-forever-chemicals-under-rcra

Cousins, I.T., Dewitt, J.C., Glüge, J., Goldenman, G., Herzke, D., Lohmann, R., Ng, C.A. Scheringer, M. and Wang, Z., 2020. The high persistence of PFAS is sufficient for their management as a chemical class. *Environmental Science. Processes & Impacts, 22*, pp. 2307–2312. https://doi.org/10.1039/D0EM00355G.

Dotaniya, M.L., Meena, V.D., Saha, J.K., Dotaniya, C.K., Mahmoud, A.E.D., Meena, B.L., Meena, M.D., Sanwal, R.C., Meena, R.S., Doutaniya, R.K. and Solanki, P., 2023. Reuse of poor-quality water for sustainable crop production in the changing scenario of climate. *Environment, Development and Sustainability*, *25*(8), pp. 7345–7376.

Ellis, A. C., Boyer, T.H., Fang, Y., Liu, C.J. and Strathmann, T.J., 2023. Life cycle assessment and life cycle cost analysis of anion exchange and granular activated carbon systems for remediation of groundwater contaminated by per- and polyfluoroalkyl substances (PFASs). *Water Research, 243*, p.120324. https://doi.org/10.1016/J.WATRES.2023.120324.

European Union. 2020. *Circular Economy of Water – Waste Water Treatment*. China Europe Water Platform. www.cewp.eu/sites/default/files/2021-05/CEWP_Whitepaper_Circular%20Economy%20of%20Water%20in%20Waste%20Water%20Treatment_final.pdf

Fawzy, M., Mahmoud, A.E.D. and Abdelfatah, A.M., 2022. Environmental and industrial applications using Internet of Things (IoT). In Jyotir Moy Chatterjee, Harish Garg, & R. N. Thakur (eds), *A Roadmap for Enabling Industry 4.0 by Artificial Intelligence* (pp. 141–167). Wiley.

Fenton, S.E., Ducatman, A., Boobis, A., DeWitt, J.C., Lau, C., Ng, C., Smith, J.S. and Roberts, S.M., 2021. Per- and Polyfluoroalkyl substance toxicity and human health review: Current state of knowledge and strategies for informing future research. *Environmental Toxicology and Chemistry, 40*, p. 606. https://doi.org/10.1002/ETC.4890.

Gewurtz, S. B., Backus, S.M., De Silva, A.O., Ahrens, L., Armellin, A., Evans, M., Fraser, S., et al. 2013. Perfluoroalkyl acids in the Canadian environment: Multi-media assessment of current status and trends. *Environment International, 59*, pp. 183–200. https://doi.org/10.1016/j.envint.2013.05.008.

Glüge, J., London, R., Cousins, I.T., Dewitt, J., Goldenman, G., Herzke, D., Lohmann, R., et al. 2021a. Information requirements under the essential-use concept: PFAS case studies. *Environmental Science and Technology, 2022,* p. 6242.

Grandjean, P., Heilmann, C., Weihe, P., Nielsen, F., Mogensen, U.B., Timmermann, A. and Budtz-Jørgensen, E., 2017. Estimated exposures to perfluorinated compounds in infancy predict attenuated vaccine antibody concentrations at age 5-years. *Journal of Immunotoxicology, 14*, pp. 188–195. https://doi.org/10.1080/1547691X.2017.1360968.

Hernández-Chover, V., Castellet-Viciano, L., Fuentes, R. and Hernández-Sancho, F., 2023. Circular economy and effciency to ensure the sustainability in the wastewater treatment plants. *Journal of Cleaner Production*, *384*, p. 135563. https://doi.org/10.1016/j.jclepro.2022.135563.

Hossein, M., Ripanda, A., Eunice, M., Chande, O., Ngassapa, F. and Mahmoud, A.E.D., 2023. Monitoring of contaminants in aquatic ecosystems using Big Data. In A.E.D. Mahmoud, M. Fawzy, and N. Khan (eds), *Artificial Intelligence and Modeling for Water Sustainability* (pp. 129–157). CRC Press.

Ismayilova, A. and Silvius, G., 2021. Cradle-to-Cradle in Project Management: A case study. *International Journal of Circular Economy and Waste Management*, *1*(1), January-June. DOI: 10.4018/IJCEWM.20210101.

Kathi, S., Singh, S., Yadav, R., Singh, A.N. and Mahmoud, A.E.D., 2023. Wastewater and sludge valorisation: A novel approach for treatment and resource recovery to achieve circular economy concept. *Frontiers in Chemical Engineering*, *5*, p. 1129783.

Kathi, S. and Mahmoud, A.E.D., 2024. Trends in effective removal of emerging contaminants from wastewater: A comprehensive review. *Desalination and Water Treatment*, *317*, p.100258.

Khattab, M., El Haggar, S. and El Gendy, A., 2023. Cradle-to-Cradle novel approach for wastewater management in sustainable urban communities. *Journal of Environmental Protection, 14*, pp. 163–171. https://doi.org/10.4236/jep.2023.143011.

Kwiatkowski, C.F., Andrews, D.Q., Birnbaum, L.S., Bruton, T.A., Dewitt, J.C., Knappe, D.R.U., Maffini, M.V., et al. 2020. Scientific basis for managing PFAS as a chemical class. *Environmental Science and Technology Letters*, *7*, pp. 532–543.

Langenbach, B., and Wilson, M., 2021. Per- and Polyfluoroalkyl Substances (PFAS): Significance and considerations within the regulatory framework of the USA. *International Journal of Environmental Research and Public Health, 18*, p. 11142. https://doi.org/10.3390/IJERPH182111142.

Lukić Bilela, L., Matijošytė, I., Krutkevičius, J., Alexandrino, D.A.M., Safarik, I., Burlakovs, J., Gaudêncio, S.P. and Carvalho, M.F., 2023. Impact of per- and polyfluorinated alkyl substances (PFAS) on the marine environment: Raising awareness, challenges, legislation, and mitigation approaches under the One Health concept. *Marine Pollution Bulletin*, *194*, p. 115309. https://doi.org/10.1016/J.MARPOLBUL.2023.115309.

Mahmoud, A.E.D., 2020. Eco-friendly reduction of graphene oxide via agricultural byproducts or aquatic macrophytes. *Materials Chemistry and Physics*, *253*, p. 123336.

Mahmoud, A.E.D. and Desokey, O., 2022. 19 Artificial intelligence-based smart waste management for the circular economy. In *Environmental Management Technologies: Challenges and Opportunities*. CRC Press.

Mahmoud, A.E.D. and Fawzy, M., 2023. Decontamination of levofloxacin from water using a novel chitosan–walnut shells composite: Linear, nonlinear, and optimization modeling. *Applied Water Science*, *13*(12), p. 244.

Mahmoud, A.E.D. and Mahmoud, E., 2023. Circular economy models in water and wastewater sectors. In A.E.D. Mahmoud, M. Fawzy, and N. Khan (eds), *Artificial Intelligence and Modeling for Water Sustainability* (pp. 253–272). CRC Press.

Mahmoud, A.E.D. and Mostafa, E., 2023. Nanofiltration membranes for the removal of heavy metals from aqueous solutions: Preparations and applications. *Membranes*, *13*(9), p. 789.

Mahmoud, A.E.D., Ali, R. and Fawzy, M., 2024. Insights into levofloxacin adsorption with machine learning models using nano-composite hydrochars. *Chemosphere*, *355*, p. 141746.

Mahmoud, A.E.D., Fawzy, M. and Khan, N., 2023. *Artificial Intelligence and Modeling for Water Sustainability: Global Challenges*. 1st Edition ed. CRC Press.

Mahmoud, A.E.D., Franke, M. and Braeutigam, P., 2022. Experimental and modeling of fixed-bed column study for phenolic compounds removal by graphite oxide. *Journal of Water Process Engineering*, *49*, p. 103085.

Mahmoud, A.E.D., El-Maghrabi, N., Hosny, M. and Fawzy, M., 2022b. Biogenic synthesis of reduced graphene oxide from Ziziphus spina-christi (Christ's thorn jujube) extracts for catalytic, antimicrobial, and antioxidant potentialities. *Environmental Science and Pollution Research*, *29*(59), pp. 89772–89787.

Mahmoud, A.E.D., Hosny, M., El-Maghrabi, N. and Fawzy, M., 2022c. Facile synthesis of reduced graphene oxide by Tecoma stans extracts for efficient removal of Ni (II) from water: Batch experiments and response surface methodology. *Sustainable Environment Research*, *32*(1), pp. 1–16.

Mahmoud, A.E.D., Umachandran, K., Sawicka, B. and Mtewa, T.K., 2021. Water resources security and management for sustainable communities. In Andrew G. Mtewa and Chukwuebuka Egbuna (eds), *Phytochemistry, the Military and Health* (pp. 509–522). Elsevier.

Mannan, M. and Al-Ghamdi, S. G. 2022. Complementing circular economy with life cycle assessment: Deeper understanding of economic, social, and environmental sustainability. *Circular Economy and Sustainability*, *1*, pp. 145–160. https://doi.org/10.1016/B978-0-12-819817-9.00032-6.

Mohajan, H.K. 2021. Cradle to Cradle is a sustainable economic policy for the better future. *Annals of Spiru Haret University Economic Series*, *21*(4), pp. 559–572.

Mousazadeh, M., Naghdali, Z., Rahimian, N., Hashemi, M., Paital, B., Al-Qodah, Z., Mukhtar, A., Karri, R.R., Mahmoud, A.E.D., Sillanpää, M. and Dehghani, M.H., 2021. Management of environmental health to prevent an outbreak of COVID-19: A review. In M.H. Dehghani, R.R. Karri and S. Roy (eds), *Environmental and Health Management of Novel Coronavirus Disease (COVID-19)* (pp. 235–267).

Mu, H., Li, J., Chen, L., Hu, H., Wang, J., Gu, C., Zhang, X.X., Ren, H.Q. and Wu, B., 2022. Distribution, source and ecological risk of per- and polyfluoroalkyl substances in Chinese municipal wastewater treatment plants. *Environment International, 167,* 107447. https://doi.org/10.1016/J.ENVINT.2022.107447.

Nagy, H., Fawzy, M., Hafez, E. and Mahmoud, A.E.D. 2023. Potentials of mono-and multi-metal ion removal from water with cotton stalks and date palm stone residuals. *Environmental Science and Pollution Research, 31*, pp. 1–17.

Neczaj, E. and Grosser, A., 2018. Circular economy in wastewater treatment plant–challenges and barriers. *Insights on the Water-Energy-Food Nexus*, *2*(11). https://doi.org/10.3390/proceedings2110614.

Rodríguez, D. J., Serrano, H. A., Delgado, A., Nolasco, D. and Saltiel, G., 2020. From waste to resource. In *World Bank*, eBooks, Washington, DC. https://doi.org/10.1596/33436.

Rodríguez Lima, R.M., 2019. Campus greening and sustainable development. In W. Leal Filho (ed), *Encyclopedia of Sustainability in Higher Education*. Springer Nature, Switzerland AG. https://doi.org/10.1007/978-3-030-11352-0.

Wee, S.Y. and Aris, A.Z., 2023. Environmental impacts, exposure pathways, and health effects of PFOA and PFOS. *Ecotoxicology and Environmental Safety*, *267*, p. 115663. https://doi.org/10.1016/J.ECOENV.2023.115663.

6 Role of Microplastics in the Global Plastics Pollution Crisis with Respect to One Health in the Post-COVID-19 Era

Francis Aibuedefe Igiebor,
Orhue Osayamen Amengialue,
Uzoamaka Nneka Ngwoke,
Philip Idemudia Edogun,
Philip Oseikhuenmhen Ogboye,
Henrietta Omokhele Obueh,
Precious Ahosegbamen Egharevba,
Aishuehien Imas Iyoha, Haruna Ochoyama, and
Nosa Jude Uwuigiaren

6.1 INTRODUCTION

The global plastics pollution crisis has emerged as a critical environmental issue over the past several decades. Since the mid-20th century, the production and consumption of plastics have skyrocketed, with global plastic production reaching approximately 368 million metric tons in 2019 (PlasticsEurope, 2020). The unique properties of plastics such as durability, lightweight, and versatility have made them indispensable in various sectors, including packaging, construction, healthcare, and consumer goods (Igiebor *et al.*, 2024). However, these same properties have also contributed to significant environmental pollution, as plastics persist in the environment for centuries, due to their resistance to biodegradation. The pervasive nature of plastics pollution stems from both the sheer volume of plastic produced and its extensive use in disposable products and packaging. Once discarded, plastics do not biodegrade but instead break down into smaller fragments known as microplastics, defined as particles less than 5mm in diameter (Thompson *et al.*, 2004). These microplastics have been detected in

DOI: 10.1201/9781032623801-6

virtually all ecosystems, from deep ocean sediments to Arctic ice, highlighting their widespread distribution (Eriksen *et al.*, 2014).

Plastics pollution affects both marine and terrestrial environments. In marine ecosystems, plastics are now ubiquitous, affecting over 700 marine species through ingestion, entanglement, and habitat disruption (Gall and Thompson, 2015). Marine organisms, including fish, seabirds, and marine mammals, often mistake plastic debris for food, leading to physical harm and exposure to toxic substances that adhere to plastic surfaces (Wright *et al.*, 2013). Terrestrial environments are similarly impacted, with plastic waste disrupting soil ecosystems and agricultural productivity (de Souza Machado *et al.*, 2018).

Human exposure to plastics occurs through various pathways, including the consumption of contaminated food and water, inhalation of airborne microplastics, and dermal contact. The potential health impacts include endocrine disruption, reproductive toxicity, and carcinogenic effects due to hazardous chemicals such as phthalates and bisphenol A (BPA) present in plastics (Rochman *et al.,* 2013). Furthermore, microplastics can act as vectors for pathogens and harmful chemicals, exacerbating their health risks (Galloway, 2015). The socioeconomic dimensions of the plastics pollution crisis are significant. The economic cost of marine plastic pollution alone is estimated to be billions of dollars annually, impacting sectors such as fisheries, tourism, and shipping (Beaumont *et al.,* 2019). Developing countries, in particular, face disproportionate challenges due to inadequate waste management infrastructure, leading to greater environmental and health burdens (Lebreton and Andrady, 2019).

The COVID-19 pandemic has significantly influenced plastics use and pollution, exacerbating the global plastics crisis. During the pandemic, the demand for single-use plastics surged, driven by the increased use of personal protective equipment (PPE) such as masks, gloves, and face shields, as well as heightened reliance on plastic packaging for online shopping and food delivery services (Silva *et al.,* 2021). The sudden rise in single-use plastics has led to a substantial increase in plastic waste. Many countries have struggled to manage this surge in waste, resulting in increased environmental leakage and pollution. Improper disposal of PPE has been particularly problematic, with masks and gloves frequently found littering streets and waterways (Ardusso*et al.,* 2021). This improper disposal not only contributes to environmental pollution but also poses public health risks due to the potential spread of the virus from contaminated waste. The pandemic has also disrupted recycling programs worldwide. Lockdowns and social distancing measures have led to the suspension or reduction of recycling services in many regions, exacerbating the accumulation of plastic waste in landfills and the environment (Vanapalli *et al.*, 2021). The decline in oil prices during the pandemic further undermined recycling efforts by making the production of virgin plastics cheaper than recycling.

Addressing the plastics pollution crisis requires comprehensive policy and management strategies. International initiatives, such as the European Union's Single-Use Plastics Directive, aim to reduce the production and consumption of the most problematic plastic products (European Commission, 2019). National and local governments have implemented bans on plastic bags and microbeads, and there is a growing emphasis on promoting a circular economy, which aims to keep resources in use for as long as possible through recycling and sustainable design (Ellen MacArthur Foundation, 2017). Public awareness and education campaigns are also crucial in driving behavioral changes and reducing plastic waste. The role of individual action, combined with systemic policy changes, is essential in mitigating the impacts of plastics pollution. The One Health concept is a holistic approach that recognizes the interconnectedness of human, animal, and environmental health. It emphasizes the need for collaborative efforts across various disciplines to achieve optimal health outcomes. This integrated approach is particularly relevant in the context of microplastics pollution, as these pollutants impact all three domains of One Health. Microplastics are pervasive in the environment, affecting terrestrial and aquatic ecosystems. They are ingested by wildlife, enter the food chain, and ultimately impact human health through the consumption of contaminated food and water (Wright *et al.,* 2020). The One Health approach underscores the importance of understanding these interconnected pathways and developing comprehensive strategies to mitigate the adverse effects of microplastics. Research has shown that microplastics can cause physical and chemical harm to wildlife, leading to reduced fitness, altered behavior, and increased mortality (Gall and Thompson, 2015). These impacts on animal health can have cascading effects on ecosystem stability and biodiversity. Furthermore, microplastics can absorb toxic chemicals from the environment, which may then be transferred to humans through the food web, posing potential risks such as endocrine disruption and carcinogenicity (Galloway, 2015).

6.2 MICROPLASTICS IN THE ENVIRONMENT

Microplastics are plastic particles less than 5mm in diameter. They are pervasive environmental pollutants that originate from the degradation of larger plastic debris (secondary microplastics) or are intentionally manufactured as small particles (primary microplastics), such as those used in personal care products and industrial abrasives (Thompson *et al.*, 2004). Their small size allows them to infiltrate various ecosystems, leading to widespread environmental contamination and posing significant risks to ecological and human health.

6.3 DISTRIBUTION AND PREVALENCE IN MARINE AND TERRESTRIAL ECOSYSTEMS

Globally, microplastics have been detected in diverse environments; from the remotest parts of the oceans to urban soils (Figure 6.1). In marine ecosystems, microplastics are found in surface waters, water columns, seabeds, and even Arctic ice (Eriksen *et al.*, 2014; Obbard *et al.*, 2014). Marine species, ranging from plankton to whales, ingest microplastics, mistaking them for food, which can lead to physical harm and exposure to toxic substances (Gall and Thompson, 2015). In terrestrial ecosystems, microplastics are prevalent in agricultural soils, urban environments, and freshwater systems. Agricultural soils receive microplastics through the application of sewage sludge as fertilizer and the use of plastic mulches (Nizzetto *et al.,* 2016). Urban

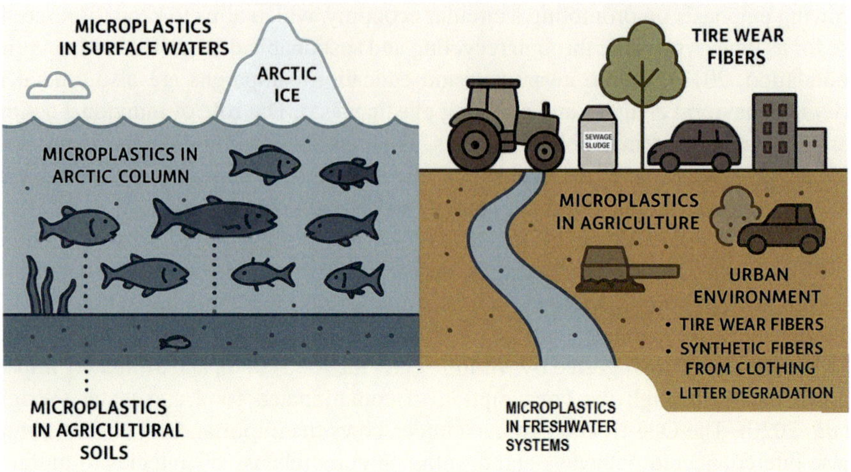

FIGURE 6.1 Global distribution and ecological impact of microplastics in marine and terrestrial ecosystems.

environments contribute to microplastic pollution through tire wear particles, synthetic fibers from clothing, and litter degradation (Kole *et al.*, 2017).

6.4 PATHWAYS AND TRANSPORT MECHANISMS

Microplastics enter the environment through various pathways and are transported by wind, water, and biotic vectors (Figure 6.2). Terrestrial sources include urban run-off, wastewater effluents, and atmospheric deposition. Urban run-off carries microplastics from streets and land surfaces into rivers and oceans, while wastewater treatment plants discharge microplastics into aquatic systems despite filtration processes (Carr *et al.*, 2016).In marine environments, ocean currents, waves, and wind-driven processes disperse microplastics over vast distances. Microplastics can be transported vertically through the water column, influenced by their buoyancy and biofouling, which causes them to sink (Reisser *et al.*, 2015). Atmospheric transport mechanisms can carry microplastic particles over long distances, leading to their deposition in remote areas such as the Arctic and high-altitude regions (Bergmann *et al.*, 2019).

6.5 PERSISTENCE AND DEGRADATION

Microplastics exhibit high persistence in the environment due to their resistance to biodegradation. Environmental factors such as ultraviolet radiation, mechanical abrasion, and chemical interactions contribute to the physical and chemical weathering of plastics, breaking them into smaller fragments but not fully degrading them (Andrady, 2011). This persistence allows microplastics to accumulate in ecosystems over time, posing long-term environmental risks. The degradation rates of microplastics vary depending on polymer type, environmental conditions, and

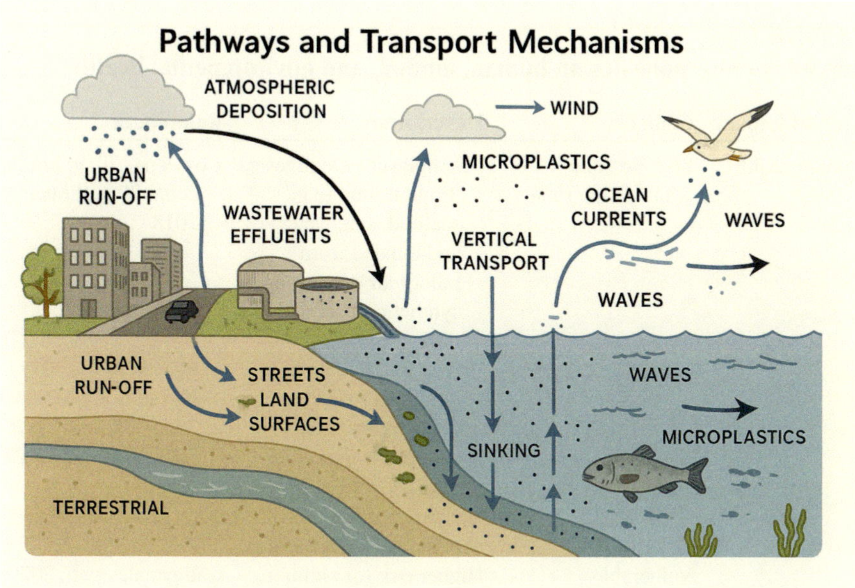

FIGURE 6.2 Pathways and transport mechanisms of microplastics in the environment.

the presence of additives. For instance, polyethylene and polypropylene, common polymers in consumer products, degrade more slowly compared to biodegradable plastics like polylactic acid (PLA) (Emadian *et al.*, 2017). However, even biodegradable plastics can persist under certain conditions, leading to continued environmental impacts.

The durability and persistence of microplastics highlight the need for effective management strategies to mitigate their release and accumulation. Understanding their distribution, pathways, and degradation processes is crucial for developing policies and technologies aimed at reducing microplastic pollution and protecting ecosystem and human health.

6.6 HEALTH IMPACTS OF MICROPLASTICS

6.6.1 Human Health

Ingestion and Inhalation Routes: Microplastics can enter the human body through various pathways, primarily ingestion and inhalation (Table 6.1). These tiny plastic particles are found in a wide range of food products, including seafood, table salt, and drinking water, posing a significant risk of ingestion (Cox *et al.*, 2019; Smith *et al.*, 2018). Inhalation of microplastics occurs through the air, especially in indoor environments where fibers from synthetic textiles and dust are prevalent (Prata, 2018). The ubiquitous presence of microplastics in daily life makes it difficult to avoid exposure.

TABLE 6.1
Impacts of microplastics on human, animal, and environmental health

Area of impact	Category	Description	References
Human health	Ingestion and inhalation routes	Microplastics are ingested through food (e.g., seafood, salt, water) and inhaled from indoor air containing synthetic fibers and dust.	Cox *et al.*, 2019; Smith *et al.*, 2018; Prata, 2018
	Toxicological effects	Induce inflammation, oxidative stress, cytotoxicity, and endocrine disruption due to associated chemicals like phthalates and BPA.	Wright & Kelly, 2017; Hirt & Body-Malapel, 2020; Galloway *et al.*, 2017; An *et al.*, 2021
	Vulnerable populations	Higher risk for children, individuals with respiratory diseases, and people in polluted or poorly managed environments	Campanale *et al.*, 2020; Wright *et al.*, 2020; Bouwmeester *et al.*, 2015
Animal health	Effects on wildlife and aquatic organisms	Ingestion causes blockages, false satiation, impaired reproduction and growth in marine species.	Wright *et al.*, 2013; Galloway *et al.*, 2017; Cole *et al.*, 2013; Critchell & Hoogenboom, 2018; Bessa *et al.*, 2018
	Bioaccumulation and biomagnification	Accumulation in tissues and magnification up the food chain, leading to increased exposure and toxic effects in predators.	Batista de Castro *et al.*, 2018; Carbery *et al.*, 2018; Chae & An, 2017; Rochman *et al.*, 2016
Environmental health	Impact on ecosystems and biodiversity	Microplastics alter habitats, affect soil and sediment quality, and harm species like mussels and crops.	Bergmann *et al.*, 2019; Lusher *et al.*, 2015; Sussarellu*et al.*, 2016; de Souza Machado *et al.*, 2018; Rillig*et al.*, 2019
	Interaction with other pollutants	Adsorb and transport POPs, heavy metals, enhancing toxicity and disrupting soil microbes and plant health.	Hartmann *et al.*, 2017; Bakir *et al.*, 2014; Rochman *et al.*, 2013; Chua *et al.*, 2014; Guo *et al.*, 2020

Toxicological Effects: Once inside the body, microplastics can induce several toxicological effects (Table 6.1). Studies have indicated that microplastics can cause inflammation, oxidative stress, and cytotoxicity, potentially leading to adverse health outcomes (Wright and Kelly, 2017; Hirt and Body-Malapel, 2020). One of the critical concerns is endocrine disruption, as microplastics often contain or adsorb endocrine-disrupting chemicals (EDCs) such as phthalates and bisphenol A (BPA). These substances can interfere with hormonal functions and have been linked to a range of health issues, including reproductive disorders, developmental problems, and metabolic diseases (Galloway *et al.*, 2017; An *et al.*, 2021).

Vulnerable Populations: Certain populations are more vulnerable to the health effects of microplastics. Children, for instance, are at higher risk due to their developing systems and behaviors that increase exposure, such as hand-to-mouth activities (Campanale *et al.*, 2020). Also, individuals with pre-existing health conditions, such as respiratory diseases, may experience exacerbated symptoms due to microplastic inhalation (Wright *et al.*, 2020). People living in areas with high levels of environmental pollution or inadequate waste management systems are also more likely to encounter higher concentrations of microplastics, further increasing their risk of adverse health effects (Bouwmeester *et al.*, 2015).

6.6.2 Animal Health

Effects on wildlife and aquatic organisms: Microplastics have been found to adversely affect a wide range of wildlife, particularly aquatic organisms (Table 6.1). Studies have demonstrated that ingestion of microplastics can lead to physical blockages, internal abrasions, and a false sense of satiation, which can reduce feeding and growth rates in marine animals (Wright *et al.*, 2013; Galloway *et al.*, 2017). For instance, zooplankton, a critical component of the marine food web, can ingest microplastics, which can impair their reproductive and growth processes (Cole *et al.*, 2013). Similarly, fish, birds, and other marine species are known to ingest microplastics, which can lead to toxicological effects such as liver toxicity, disruption of lipid metabolism, and changes in behavior (Critchell and Hoogenboom, 2018; Bessa *et al.*, 2018).

Bioaccumulation and Biomagnification: Microplastics pose a significant threat through the processes of bioaccumulation and biomagnification. Bioaccumulation occurs when organisms absorb microplastics at a rate faster than they can excrete them, leading to an accumulation of these particles in their tissues over time (Batista de Castro *et al.*, 2018). Biomagnification refers to the increasing concentration of microplastics and their associated toxic chemicals as they move up the food chain. Predatory species, including humans, are at risk of higher exposure levels due to consuming prey that haveaccumulated microplastics in their bodies (Carbery *et al.*, 2018; Chae and An, 2017). This magnification of microplastic concentrations can have severe health implications for top predators in the ecosystem, including potential endocrine disruption, reproductive harm, and increased mortality rates (Rochman*et al.*, 2016).

6.6.3 Environmental Health

Impact on Ecosystems and Biodiversity: Microplastics significantly impact ecosystems and biodiversity (Table 6.1). These particles have been found in diverse environments, from the deepest ocean trenches to remote mountain peaks, indicating their pervasive nature (Bergmann *et al.*, 2019). Microplastics can alter habitats, reduce the quality of sediments, and affect the health and survival of various organisms (Lusher *et al.*, 2015). For instance, filter-feeding organisms such as mussels and oysters can ingest microplastics, leading to physical harm and reduced feeding efficiency, which can cascade through the food web, impacting predator species and overall ecosystem health (Sussarellu *et al.*, 2016). Microplastics also pose a threat to terrestrial ecosystems. Soil health can be compromised by the presence of microplastics, which can alter soil structure and function, affecting water retention, nutrient cycling, and soil organisms' health (de Souza Machado *et al.*, 2018). The presence of microplastics in agricultural soils, primarily due to the application of sewage sludge and plastic mulching, can impact crop health and productivity (Rillig*et al.*, 2019).

Interaction with Other Pollutants: Microplastics interact with other environmental pollutants, exacerbating their impact. These interactions can occur through the adsorption of persistent organic pollutants (POPs), heavy metals, and other harmful substances onto the surfaces of microplastic particles (Hartmann *et al.*, 2017). This process can increase the persistence and toxicity of these pollutants in the environment. For example, microplastics can act as vectors for pollutants, facilitating their transport and uptake by organisms, which can lead to bioaccumulation and biomagnification within the food web (Bakir *et al.*, 2014; Rochman *et al.*, 2013).The interaction between microplastics and pollutants poses a significant risk to both aquatic and terrestrial ecosystems. Aquatic organisms exposed to microplastics contaminated with pollutants can experience enhanced toxicological effects, such as increased oxidative stress, impaired reproduction, and reduced survival rates (Chua *et al.*, 2014). In terrestrial environments, the combined effects of microplastics and pollutants can affect soil microbial communities and plant health, potentially disrupting ecosystem services and agricultural productivity (Guo *et al.*, 2020).

6.7 ONE HEALTH PERSPECTIVE

6.7.1 Integration of Human, Animal, and Environmental Health

The One Health perspective emphasizes the interconnectedness of human, animal, and environmental health. This approach is crucial in understanding and addressing the widespread implications of microplastic pollution. In humans, microplastics have been detected in various biological samples, raising concerns about their potential health risks, such as inflammatory responses, oxidative stress, and toxicological effects (Prata *et al.*, 2020). Research has shown that microplastics can carry harmful chemicals and pathogens, exacerbating their health impacts (Wright and Kelly, 2017). For example, microplastics have been found in the placenta of pregnant women, indicating their ability to cross biological barriers and potentially affect fetal development (Ragusa *et al.*, 2021). Similarly, animals, particularly marine organisms, ingest microplastics, leading to physical harm, such as blockages in the digestive tract and chemical contamination from adsorbed pollutants (Wright and Kelly, 2017). Studies have documented microplastics in a wide range of species, from plankton to fish to marine mammals, illustrating the pervasiveness of this issue (Karbalaei *et al.*, 2018). These pollutants can biomagnify up the food chain, ultimately affecting apex predators and humans who consume seafood (Wright and Kelly, 2017). The environmental health impact of microplastics is evident in the disruption of ecosystems and the bioaccumulation of toxic substances in food webs. Microplastics can alter soil structure and function, affecting plant growth and soil organisms (de Souza Machado *et al.*, 2018). In aquatic environments, microplastics can affect the behavior and reproduction of aquatic organisms, leading to ecological imbalances (Browne *et al.*, 2015). Addressing microplastic pollution requires an integrated approach that considers the health implications for humans, animals, and ecosystems collectively, as advocated by the One Health framework.

6.7.2 Interdisciplinary Approaches to Studying Microplastics

The complexity of the microplastic pollution crisis necessitates interdisciplinary approaches, combining expertise from environmental science, toxicology, public health, and engineering. Researchers from these diverse fields are working together to develop a comprehensive understanding of the sources, distribution, impacts, and mitigation of microplastics. Environmental scientists are crucial in identifying the sources and pathways of microplastics, which include both primary sources, such as microbeads in personal care products, and secondary sources, such as the breakdown of larger plastic debris (Galloway *et al.*, 2017). They also study the fate and transport of microplastics in different environments, providing critical data for risk assessment and management strategies (Andrady, 2011). Toxicologists focus on the health impacts of microplastics on living organisms, including their potential to carry harmful chemicals and pathogens (Campanale *et al.*, 2020). Studies have shown that microplastics can induce inflammation, oxidative stress, and other adverse effects in a range of organisms (Prata *et al.*, 2020). Understanding these mechanisms is essential for assessing the risks of microplastic exposure to humans and wildlife.Public health experts are involved in evaluating the potential health risks of microplastic exposure to humans, including vulnerable populations such as children and pregnant women (Prata *et al.*, 2020). They also develop guidelines and recommendations for minimizing exposure and promoting safer alternatives to plastic products. Engineers play a vital role in developing technologies to mitigate microplastic pollution, such as filtration systems for removing microplastics from wastewater and stormwater (Dey *et al.*, 2021). Innovative materials science approaches are also being explored to create biodegradable and sustainable alternatives to conventional plastics (Napper and Thompson, 2016). Collaborative efforts across these disciplines are essential for a comprehensive understanding and effective mitigation of microplastic pollution. Interdisciplinary research can inform evidence-based policies and interventions, ensuring a coordinated and holistic response to this global challenge.

6.7.3 Policy and Management Strategies

Effective management and policy strategies are critical to addressing the microplastic crisis from a One Health perspective. Policymakers need to implement comprehensive measures that address the entire lifecycle of plastics, from production to disposal, and promote sustainable alternatives.

a. **Policy Frameworks**: Policies need to be informed by scientific evidence and encompass measures such as banning single-use plastics, improving waste management systems, and promoting sustainable alternatives (European Commission, 2018). For instance, the European Union has implemented a directive to reduce the consumption of single-use plastics and promote the use of recycled materials (European Commission, 2018). Similar policies have been adopted by other countries and regions, demonstrating the effectiveness of regulatory approaches in reducing plastic pollution. In addition to banning or restricting certain plastic products, policymakers can incentivize the development and adoption of biodegradable and sustainable materials. This can be achieved through subsidies, tax incentives, and public procurement policies that prioritize eco-friendly products (UN Environment Programme, 2019).
b. **International Cooperation:** This is paramount to tackle the transboundary nature of plastic pollution. Plastic waste can travel long distances through ocean currents, affecting countries far from the original source of pollution (Lebreton *et al.*, 2017). International agreements and collaborations can facilitate the sharing of best practices, technologies, and resources, ensuring a coordinated global response to microplastic pollution (UN Environment Programme, 2019).For example, the United Nations Environment Assembly (UNEA) has adopted several resolutions to address marine plastic litter and microplastics, encouraging member states to develop national action plans and engage in regional and international initiatives (UN Environment Programme, 2019). Such frameworks can help harmonize policies and standards across countries, enhancing the effectiveness of global efforts to combat plastic pollution.
c. **Waste Management:** Enhancing waste management systems is crucial to mitigating microplastic pollution. This includes optimizing recycling processes, improving the collection and processing of plastic waste, and preventing plastic debris from entering waterways and marine environments (Pandey *et al.*, 2023). Advanced technologies for detecting and removing microplastics from wastewater and stormwater systems are also vital (Dey *et al.*, 2021). Investing in infrastructure for waste collection, sorting, and recycling can significantly reduce the amount of plastic waste that ends up in the environment. Extended producer responsibility (EPR) schemes, which hold producers accountable for the end-of-life management of their products, can incentivize more sustainable product design and waste management practices (European Commission, 2018). In addition to improving

waste management at the local and national levels, international cooperation is necessary to address the illegal trade and dumping of plastic waste. Strengthening regulations and enforcement mechanisms can help prevent the flow of plastic waste to countries with inadequate waste management infrastructure (UN Environment Programme, 2019).

6.8 MICROPLASTICS IN THE POST-COVID-19 ERA

6.8.1 Increased Plastic Use during the Pandemic

The COVID-19 pandemic led to a significant increase in plastic consumption due to the widespread use of personal protective equipment (PPE) such as masks, gloves, and faces shields, as well as increased demand for plastic packaging to prevent contamination. Studies have shown that the surge in plastic use was necessary to curb the virus spread, but it also exacerbated plastic pollution issues (Silva *et al.*, 2021; Parashar and Hait, 2021). For instance, it is estimated that globally, 129 billion face masks and 65 billion gloves were used monthly during the pandemic (Prata *et al.*, 2020). This unprecedented increase in plastic use has highlighted the interconnectedness of human health, environmental integrity, and wildlife exposure, central to the One Health framework by demonstrating how pandemic-driven health responses can inadvertently impact broader ecological systems.

6.8.2 Disposal Challenges and Environmental Leakage

The rapid increase in plastic waste generation posed significant challenges for waste management systems worldwide. Many countries struggled to keep up with the proper disposal and recycling of PPE and other plastic products, leading to environmental leakage. Inefficient waste management practices resulted in a considerable amount of plastic waste entering oceans and other ecosystems, further contributing to the microplastic pollution crisis (Pourebrahimi and Pirooz 2023; Benson *et al.*, 2021). Improper disposal, such as littering and inadequate landfill operations, increased the risk of PPE breakdown into microplastics, which can have detrimental effects on wildlife and human health (Fadare and Okoffo, 2020).

6.8.3 Long-Term Implications of Pandemic-Related Plastic Waste

The long-term environmental impacts of pandemic-related plastic waste are profound. Microplastics, resulting from the degradation of discarded PPE and other plastics, persist in the environment and pose risks to various aspects of the One Health framework, which encompasses human, animal, and environmental health. Studies indicate that microplastics can enter food chains, affect biodiversity, and contribute to chemical pollution, raising concerns about their long-lasting ecological and health consequences (Schnurr *et al.*, 2018; Wang *et al.*, 2021). The pandemic has thus underscored the urgent need to address plastic pollution comprehensively.

6.8.4 Innovations and Solutions in Plastic Management Post-Pandemic

In response to the challenges posed by increased plastic use and waste during the pandemic, several innovations and solutions have emerged. These include the development of biodegradable and reusable PPE, improved waste management systems, and enhanced recycling technologies. Researchers are also exploring methods to break down plastics more efficiently and reduce the environmental impact of plastic waste (Silva *et al.*, 2021; Torres and De-la-Torre, 2021). Additionally, policy measures aimed at reducing single-use plastics and promoting sustainable alternatives are being implemented globally to mitigate future plastic pollution and its effects on One Health.

6.9 MITIGATION STRATEGIES AND POLICY RECOMMENDATIONS

6.9.1 Reduction and Prevention of Microplastic Pollution

Reducing and preventing microplastic pollution involves addressing the sources of microplastics and implementing measures to minimize their release into the environment. This can be achieved through a combination of regulatory actions, technological innovations, and behavioral changes.

a. **Regulatory Actions:** Governments can implement regulations to ban or restrict the use of products that contribute to microplastic pollution. For example, banning single-use plastics, such as straws, cutlery, and plastic bags, can significantly reduce the amount of plastic waste generated (Thompson *et al.*, 2009). Additionally, regulations can target products that release microplastics during their use, such as microbeads in personal care products and synthetic fibers from textiles (Napper and Thompson, 2016). EPR schemes can hold manufacturers accountable for the environmental impacts of their products throughout their lifecycle. This can incentivize companies to design products that are easier to recycle and have a lower environmental footprint (European Commission, 2018).
b. **Technological Innovations:** Innovative technologies can play a crucial role in preventing and mitigating microplastic pollution. For example, researchers are developing biodegradable and compostable materials as alternatives to conventional plastics (Napper and Thompson, 2016). These materials can reduce the persistence of plastic waste in the environment and minimize the generation of microplastics. Advanced filtration systems can be installed in wastewater treatment plants and industrial facilities to capture microplastics before they are released into waterways (Conley et al., 2019). Similarly, washing machine filters can prevent synthetic fibers from entering the water system during laundry processes (Hernandez *et al.*, 2017).
c. **Behavioral Changes:** At the individual and community levels, behavioral changes are essential for reducing plastic waste and preventing microplastic pollution. Public awareness campaigns can educate people about the sources

and impacts of microplastics and encourage behaviors such as reducing plastic use, choosing eco-friendly products, and participating in recycling programs (Andrady, 2011). Initiatives like plastic-free movements and zero-waste lifestyles can promote sustainable consumption patterns and reduce the demand for plastic products. By making conscious choices, individuals can contribute to reducing the environmental burden of plastic pollution (Pandey *et al.*, 2023).

6.9.2 Improved Waste Management Systems

Effective waste management is crucial for preventing plastic waste from entering the environment and becoming microplastics. Improving waste management systems involves enhancing collection, sorting, recycling, and disposal processes.

a. **Collection and Sorting:** Efficient waste collection and sorting systems are essential for ensuring that plastic waste is properly managed and recycled. Investments in infrastructure and technology can improve the efficiency and effectiveness of waste collection and sorting processes (Pandey *et al.*, 2023). For example, automated sorting technologies can separate different types of plastics, making recycling more efficient (Lubongo et al., 2024; Olawade et al., 2024). Community-based waste management initiatives, such as waste segregation at source and local recycling programs, can also enhance waste collection and sorting efforts. Public participation in these initiatives can increase recycling rates and reduce the amount of plastic waste that ends up in landfills or the environment (UN Environment Programme, 2019).
b. **Recycling:** This is a key component of effective waste management. Enhancing recycling processes and increasing the capacity of recycling facilities can reduce the amount of plastic waste that needs to be disposed of (European Commission, 2018). Advanced recycling technologies, such as chemical recycling, can convert plastic waste into valuable resources and reduce the need for virgin plastic production (Hopewell *et al.*, 2009). Promoting the use of recycled materials in manufacturing can also create a market demand for recycled plastics, incentivizing recycling efforts and reducing the environmental impact of plastic production (Pandey *et al.*, 2023).
c. **Disposal:** Proper disposal of plastic waste is essential for preventing it from becoming microplastics. This includes ensuring that plastic waste is not littered or illegally dumped, as well as implementing safe disposal methods, such as incineration with energy recovery (Lubongo et al., 2024; Olawade et al., 2024). While incineration can reduce the volume of plastic waste, it must be carefully managed to minimize air pollution and other environmental impacts (Hopewell *et al.*, 2009). Landfills should be designed and managed to prevent plastic waste from escaping into the environment. Measures such as covering landfills and capturing landfill gas can reduce the risk of plastic pollution and mitigate environmental impacts (Pandey *et al.*, 2018).

6.9.3 Policy Frameworks and International Cooperation

Comprehensive policy frameworks and international cooperation are necessary to address the global nature of microplastic pollution. Policymakers need to implement regulations that cover the entire lifecycle of plastics, from production to disposal, and promote international collaboration to tackle transboundary pollution.

a. **Policy Frameworks:** National and regional policies should address the production, use, and disposal of plastics. This includes regulations to reduce plastic production, promote the use of sustainable materials, and enhance waste management systems (European Commission, 2018). Policies should also incentivize research and development of alternatives to conventional plastics and technologies for mitigating plastic pollution (UN Environment Programme, 2019). For instance, the European Union's Circular Economy Action Plan includes measures to reduce plastic waste, improve recycling, and promote the use of recycled materials (European Commission, 2018). Similar policies can be adopted by other countries to create a coordinated and comprehensive approach to plastic pollution.
b. **International Cooperation:** Plastic pollution is a global problem that requires international cooperation to address effectively. International agreements and frameworks, such as the Basel Convention on the Control of Transboundary Movements of Hazardous Wastes and Their Disposal, can facilitate cooperation and coordination among countries (UN Environment Programme, 2019). Collaborative efforts can include sharing best practices, technologies, and resources, as well as harmonizing policies and standards to ensure a unified response to plastic pollution (Lebreton *et al.*, 2017). For example, the UNEA has adopted several resolutions to address marine plastic litter and microplastics, encouraging member states to develop national action plans and engage in regional and international initiatives (UN Environment Programme, 2019).
c. **Global Governance:** Strengthening global governance mechanisms can enhance the effectiveness of international efforts to combat plastic pollution. This includes creating new international agreements or strengthening existing ones to address the full lifecycle of plastics, from production to disposal (UN Environment Programme, 2019). Global governance can also involve establishing monitoring and reporting mechanisms to track progress in reducing plastic pollution and holding countries accountable for their commitments. Transparent and robust governance frameworks can ensure that international efforts are effective and equitable (UN Environment Programme, 2019).

6.9.4 Public Awareness and Education

Public awareness and education are critical components of efforts to address microplastic pollution. Educating the public about the sources, impacts, and solutions to plastic pollution can drive behavioral changes and support policy initiatives.

a. **Public Education Campaigns:** These efforts can raise awareness about the impacts of microplastics on human health, wildlife, and the environment. These campaigns can provide information on how individuals can reduce their plastic use, properly dispose of plastic waste, and support recycling efforts (Andrady, 2011). For example, campaigns like the "break free from plastic" movement have successfully raised awareness about plastic pollution and mobilized public support for policy changes (Pandey *et al.*, 2023). Similar initiatives can be implemented at the local, national, and international levels to engage diverse audiences and promote sustainable behaviors.
b. **School and Community Programs:** Integrating environmental education into school curricula can help raise awareness among young people about the importance of reducing plastic waste and protecting the environment (UNESCO, 2024). Schools can also implement programs to reduce plastic use and promote recycling, such as using reusable materials and organizing waste reduction competitions. Community-based programs can engage local residents in efforts to reduce plastic pollution. Activities such as beach cleanups, recycling drives, and plastic-free campaigns can involve community members in direct actions to protect the environment (UN Environment Programme, 2019). These programs can also foster a sense of environmental stewardship and responsibility among participants.
c. **Corporate Social Responsibility:** Businesses and industries play a crucial role in addressing microplastic pollution through sustainable practices and corporate social responsibility (CSR) initiatives. Companies can adopt measures to reduce plastic use, improve waste management, and support environmental conservation efforts (Pandey *et al.*, 2023). For example, companies can eliminate single-use plastics from their operations, use recycled materials in their products, and invest in technologies to reduce plastic waste. Public pressure and consumer demand for eco-friendly products can drive industry change and innovation, leading to more sustainable business practices (Napper and Thompson, 2016).
d. **Media and Advocacy:** The media can play a powerful role in raising awareness about microplastic pollution and advocating for policy changes. Investigative journalism, documentaries, and news coverage can highlight the impacts of plastic pollution and hold policymakers and industries accountable for their actions (Andrady, 2011). Advocacy groups and non-governmental organizations (NGOs) can also mobilize public support for initiatives to reduce plastic pollution. By engaging in lobbying, campaigning, and community organizing, these groups can influence policy decisions and promote sustainable practices (Pandey *et al.*, 2023).

6.10 FUTURE DIRECTIONS

The ongoing battle against the global plastics pollution crisis requires proactive measures and innovative strategies to mitigate the pervasive impacts of microplastics. In the post COVID-19 era, it is essential to outline future directions that align with the

One Health perspective, focusing on research priorities, technological advancements, and holistic approaches to plastic pollution control.

6.10.1 Research Gaps and Priorities

Despite significant progress in understanding the prevalence and impacts of microplastics, several research gaps persist, necessitating focused efforts to address them. Priority areas for future research include:

a. Health Impacts: Further investigation into the long-term health effects of microplastic exposure on humans, including potential carcinogenicity, immunotoxicity, and neurotoxicity (Campanale *et al.*, 2020).
b. Ecological Consequences: Deeper understanding of the ecological impacts of microplastics on terrestrial and aquatic ecosystems, including effects on biodiversity, ecosystem services, and food web dynamics (Galloway *et al.*, 2017).
c. Fate and Transport: Enhanced knowledge of the fate and transport of microplastics in different environmental compartments, including air, soil, water, and biota, to assess their global distribution and environmental persistence (Lebreton *et al.*, 2017).
d. Standardization and Methodology: Standardization of sampling and analysis methodologies for microplastics to facilitate comparability and reproducibility of research findings across studies and regions (Karbalaei *et al.*, 2018).
e. Emerging Contaminants: Investigation into the interactions between microplastics and other emerging contaminants, such as pharmaceuticals, heavy metals, and persistent organic pollutants, to elucidate synergistic effects and cumulative risks (de Souza Machado *et al.*, 2018).

6.10.2 Technological Advancements in Detection and Remediation

Advancements in technology play a pivotal role in enhancing our capacity to detect and remediate microplastic pollution effectively. Key areas for technological innovation include:

a. Detection Techniques: Development of rapid, sensitive, and cost-effective detection techniques for microplastics across various environmental matrices, including water, sediment, soil, biota, and atmospheric aerosols (Lubongo et al., 2024; Olawade et al., 2024).
b. Remote Sensing: Integration of remote sensing technologies, such as satellite imagery and unmanned aerial vehicles, for large-scale monitoring and mapping of microplastic hotspots in marine and terrestrial environments (Lebreton *et al.*, 2017).
c. Nano-Scale Remediation: Exploration of nano-scale remediation technologies, such as nanomaterial-based adsorbents and filtration membranes, for efficient removal of microplastics from contaminated water sources (Sajid *et al.*, 2022).
d. Biodegradable Alternatives: Research and development of biodegradable polymers and eco-friendly materials as alternatives to conventional plastics to

minimize the accumulation of microplastics in the environment (Napper and Thompson, 2016).

e. Blockchain Solutions: Implementation of blockchain technology for tracking and tracing the life cycle of plastic products, from production to disposal, to promote transparency, accountability, and circular economy principles (Lubongo et al., 2024; Olawade et al., 2024).

6.10.3 Holistic Approaches to Plastic Pollution Control

Addressing the global plastics pollution crisis requires holistic approaches that transcend disciplinary boundaries and foster collaboration among diverse stakeholders. Key strategies for plastic pollution control include:

a. Circular Economy Transition: Promotion of circular economy principles, such as product redesign, reuse, repair, and recycling, to minimize plastic waste generation and foster a closed-loop system (European Commission, 2018).
b. Policy Integration: Integration of plastic pollution control measures into broader environmental policies and frameworks, encompassing waste management, pollution prevention, biodiversity conservation, and climate action (European Commission, 2018).
c. Community Engagement: Empowerment of local communities through education, capacity-building, and participatory decision-making processes to drive grassroots initiatives and behavioral change at the grassroots level (UN Environment Programme, 2019).
d. Public-Private Partnerships: Collaboration between governments, industry, academia, and civil society to catalyze innovation, investment, and collective action toward plastic pollution mitigation and sustainable development (Pandey *et al.*, 2018).
e. International Cooperation: Strengthening of international cooperation and coordination mechanisms to address the transboundary nature of plastic pollution and promote harmonized policies, regulations, and enforcement measures (UN Environment Programme, 2019).

6.11 CONCLUSION

The COVID-19 pandemic has underscored the intricate interdependence of human, environmental, and animal health, reinforcing the importance of a comprehensive One Health approach. The surge in plastic use, especially for PPE and packaging, has intensified the global plastic pollution crisis, with microplastics now recognized as an emerging contaminant of concern. These persistent pollutants can cause endocrine disruption, inflammation, and bioaccumulation, posing serious risks across species and ecosystems. As we navigate the post-COVID-19 era, there is a critical opportunity to realign strategies through One Health-aligned governance. This includes investing in sustainable technologies such as biodegradable PPE, advancing plastic degradation and recycling innovations, and strengthening waste management infrastructure. Furthermore, coordinated international policies are essential to tackle transboundary plastic pollution and mitigate its health and ecological consequences. A unified, science-informed, and collaborative global response is vital to addressing the long-term threats of microplastics and safeguarding health at all levels.

REFERENCES

An, R., Wang, X., Yang, L., and Zhang, S. (2021). Endocrine-disrupting effects of microplastics and their additives: From environmental occurrence to human exposure. *Environmental Pollution*, 275, 116603.

Andrady, A. L. (2011). Microplastics in the marine environment. *Marine Pollution Bulletin*, 62(8), 1596–1605.

Ardusso, M., Forero-López, A. D., Buzzi, N. S., Spetter, C. V., and Fernández-Severini, M. D. (2021). COVID-19 pandemic repercussions on plastic and antiviral polymeric textile causing pollution on beaches and coasts of South America. *Science of The Total Environment*, 763, 144365.

Bakir, A., Rowland, S. J., and Thompson, R. C. (2014). Enhanced desorption of persistent organic pollutants from microplastics under simulated physiological conditions. *Environmental Pollution*, 185, 16–23.

Batista de Castro, I., de Lima, L. R., Miranda de Sousa, A., de Souza e Mello, A., and de Castro, D. P. (2018). Bioaccumulation of microplastics in the marine environment. In P. Farrell (Ed.), *Marine Pollution and Microplastic Research: Solutions for Reducing Plastic Waste in the Ocean* (pp. 65–85). Elsevier.

Beaumont, N. J., Aanesen, M., Austen, M. C., Börger, T., Clark, J. R., Cole, M., and Wyles, K. J. (2019). Global ecological, social and economic impacts of marine plastic. *Marine Pollution Bulletin*, 142, 189–195.

Benson, N. U., Bassey, D. E., and Palanisami, T. (2021). COVID pollution: Impact of COVID-19 pandemic on global plastic waste footprint. *Heliyon*, 7(2), e06343.

Bergmann, M., Mutzel, S., Primpke, S., Tekman, M. B., Trachsel, J., and Gerdts, G. (2019). White and wonderful? Microplastics prevail in snow from the Alps to the Arctic. *Science Advances*, 5(8), eaax1157.

Bessa, F., Barría, P., Neto, J. M., Frias, J. P. G. L., Otero, V., Sobral, P., and Marques, J. C. (2018). Occurrence of microplastics in commercial fish from a natural estuarine environment. *Marine Pollution Bulletin*, 128, 575–584.

Bouwmeester, H., Hollman, P. C., and Peters, R. J. (2015). Potential health impact of environmentally released micro- and nanoplastics in the human food production chain: Experiences from nanotoxicology. *Environmental Science and Technology*, 49(15), 8932–8947.

Campanale, C., Massarelli, C., Savino, I., Locaputo, V., and Uricchio, V. F. (2020). A detailed review study on potential effects of microplastics and additives of concern on human health. *International Journal of Environmental Research and Public Health*, 17(4), 1212.

Carbery, M., O'Connor, W., and Palanisami, T. (2018). Trophic transfer of microplastics and mixed contaminants in the marine food web and implications for human health. *Environment International*, 115, 400–409.

Carr, S. A., Liu, J., and Tesoro, A. G. (2016). Transport and fate of microplastic particles in wastewater treatment plants. *Water Research*, 91, 174–182.

Chae, Y., and An, Y.-J. (2017). Effects of micro- and nanoplastics on aquatic ecosystems: Current research trends and perspectives. *Marine Pollution Bulletin*, 124(2), 624–632.

Chua, E. M., Shimeta, J., Nugegoda, D., Morrison, P. D., and Clarke, B. O. (2014). Assimilation of polybrominated diphenyl ethers from microplastics by the marine amphipod, Allorchestescompressa. *Environmental Science and Technology*, 48(14), 8127–8134.

Cole, M., Lindeque, P., Halsband, C., and Galloway, T. S. (2013). Microplastics as contaminants in the marine environment: A review. *Marine Pollution Bulletin*, 62(12), 2588–2597.

Conley, K., Clum, A., Deepe, J., Lane, H., and Beckingham, B. (2019). Wastewater treatment plants as a source of microplastics to an urban estuary: Removal efficiencies and loading per capita over one year. *Water Research X*, 3, 100030.

Cox, K. D., Covernton, G. A., Davies, H. L., Dower, J. F., Juanes, F., and Dudas, S. E. (2019). Human consumption of microplastics. *Environmental Science and Technology*, 53(12), 7068–7074.

Critchell, K., and Hoogenboom, M. O. (2018). Effects of microplastic exposure on the body condition and behavior of planktivorous reef fish (Acanthochromispolyacanthus). *PLoSOne*, 13(3), e0193308.

de Souza Machado, A. A., Lau, C. W., Till, J., Kloas, W., Lehmann, A., Becker, R., and Rillig, M. C. (2018). Impacts of microplastics on the soil biophysical environment. *Environmental Science and Technology,* 52(17), 9656–9665.

Dey, T. K., Uddin, M. E., and Jamal, M. (2021). Detection and removal of microplastics in wastewater: evolution and impact. *Environmental Science and Pollution Research International*, 28(14), 16925–16947.

Ellen MacArthur Foundation. (2017). *The New Plastics Economy: Rethinking the Future of Plastics and Catalysing Action*. Ellen MacArthur Foundation.

Emadian, S. M., Onay, T. T., and Demirel, B. (2017). Biodegradation of bioplastics in natural environments. *Waste Management*, 59, 526–536.

Eriksen, M., Lebreton, L. C., Carson, H. S., Thiel, M., Moore, C. J., Borerro, J. C., and Reisser, J. (2014). Plastic pollution in the world's oceans: More than 5 trillion plastic pieces weighing over 250,000 tons afloat at sea. *PLoSOne*, 9(12), e111913.

European Commission. (2018). *A European strategy for plastics in a circular economy*. European Commission Report. https://eur-lex.europa.eu/legal-content/EN/TXT/?uri=celex:52018DC0028

European Commission. (2019). Directive (EU) 2019/904 on the reduction of the impact of certain plastic products on the environment. *Official Journal of the European Union*. http://data.europa.eu/eli/dir/2019/904/oj

Fadare, O. O., and Okoffo, E. D. (2020). Covid-19 face masks: A potential source of microplastic fibers in the environment. *Science of The Total Environment*, 737, 140279.

Gall, S. C., and Thompson, R. C. (2015). The impact of debris on marine life. *Marine Pollution Bulletin*, 92(1–2), 170–179.

Galloway, T. S. (2015). Micro- and nano-plastics and human health. In M. Bergmann, L. Gutow, and M. Klages (Eds.), *Marine Anthropogenic Litter*(pp. 343–366). Springer.

Galloway, T. S., Cole, M., and Lewis, C. (2017). Interactions of microplastic debris throughout the marine ecosystem. *Nature Ecology and Evolution*, 1(5), 0116.

Guo, X., Pang, J., Chen, S., and Jia, H. (2020). Impact of plastic mulch film debris on soil physicochemical and hydrological properties. *Environmental Pollution*, 266, 115097.

Hartmann, N. B., Hüffer, T., Thompson, R. C., Hassellöv, M., Verschoor, A., Daugaard, A. E., and Wagner, M. (2017). Are we speaking the same language? Recommendations for a definition and categorization framework for plastic debris. *Environmental Science and Technology*, 53(3), 1039–1047.

Hernandez, E., Nowack, B., and Mitrano, D. M. (2017). Polyester textiles as a source of microplastics from households: A mechanistic study to understand microfiber release during washing. *Environmental Science and Technology*, 51(12), 7036–7046.

Hirt, N., and Body-Malapel, M. (2020). Immunotoxicity and intestinal effects of nano- and microplastics: A review of the literature. *Particle and Fibre Toxicology*, 17(1), 57.

Hopewell, J., Dvorak, R., and Kosior, E. (2009). Plastics recycling: challenges and opportunities. *Philosophical transactions of the Royal Society of London. Series B, Biological Sciences*, 364(1526), 2115–2126.

Igiebor, F. A., and Uwuigiaren, N. J. (2024). Influence of soil microplastic contamination on maize (*Zea mays*) development and microbial dynamics. *Discover Environment*, 2, 122.

Karbalaei, S., Hanachi, P., Walker, T. R., and Cole, M. (2018). Occurrence, sources, human health impacts and mitigation of microplastic pollution. *Environmental Science and Pollution Research*, 25(36), 36046–36063.

Kole, P. J., Löhr, A. J., Van Belleghem, F. G. A. J., and Ragas, A. M. J. (2017). Wear and tear of tyres: A stealthy source of microplastics in the environment. *International Journal of Environmental Research and Public Health,* 14(10), 1265.

Lebreton, L. C. M., Van Der Zwet, J., Damsteeg, J. W., Slat, B., Andrady, A., and Reisser, J. (2017). River plastic emissions to the world's oceans. *Nature Communications,* 8(1), 15611.

Lebreton, L., and Andrady, A. (2019). Future scenarios of global plastic waste generation and disposal. *Palgrave Communications*, 5(1), 1–11.

Lubongo, C., Daej, M. A. A. B., and Alexandridis, P. (2024). Recent developments in technology for sorting plastic for recycling: the emergence of artificial intelligence and the rise of the robots. *Recycling*, 9(4), 59.

Lusher, A. L., Hollman, P. C. H., and Mendoza-Hill, J. J. (2015). *Microplastics in fisheries and aquaculture: Status of knowledge on their occurrence and implications for aquatic organisms and food safety*. FAO Fisheries and Aquaculture Technical Paper, 615.

Napper, I. E., and Thompson, R. C. (2016). Release of synthetic microplastic plastic fibres from domestic washing machines: Effects of fabric type and washing conditions. *Marine Pollution Bulletin*, 112(1–2), 39–45.

Nizzetto, L., Futter, M., and Langaas, S. (2016). Are agricultural soils dumps for microplastics of urban origin? *Environmental Science and Technology*, 50(20), 10777–10779.

Obbard, R. W., Sadri, S., Wong, Y. Q., Khitun, A. A., Baker, I., and Thompson, R. C. (2014). Global warming releases microplastic legacy frozen in Arctic Sea ice. *Earth's Future*, 2(6), 315–320.

Olawade, D. B., Fapohunda, O., Wada, O. Z., Usman, S. O., Ige, A. O., Ajisafe, O., and Oladapo, B. I. (2024). Smart waste management: A paradigm shift enabled by artificial intelligence. *Waste Management Bulletin*, 2(2), 244–263.

Pandey, P., Dhiman, M., Kansal, A., and Subudhi, S. P. (2023). Plastic waste management for sustainable environment: techniques and approaches. *Waste Disposal and Sustainable Energy*, pp. 1–18.

Parashar, N., and Hait, S. (2021). Plastics in the time of COVID-19 pandemic: Protector or polluter? *Science of The Total Environment*, 759, 144274.

PlasticsEurope. (2020). *Plastics – the Facts 2020: An Analysis of European Plastics Production.* Demand and Waste Data. PlasticsEurope.

Pourebrahimi, S., and Pirooz, M. (2023). Microplastic pollution in the marine environment: A review. *Journal of Hazardous Materials Advances*, 10, 100327.

Prata, J. C. (2018). Airborne microplastics: Consequences to human health?*Environmental Pollution*, 234, 115–126.

Prata, J. C., da Costa, J. P., Lopes, I., Duarte, A. C., and Rocha-Santos, T. (2020). Environmental exposure to microplastics: An overview on possible human health effects. *Science of The Total Environment,* 702, 134455.

Prata, J. C., Silva, A. L. P., Walker, T. R., Duarte, A. C., and Rocha-Santos, T. (2020). COVID-19 pandemic repercussions on the use and management of plastics. *Environmental Science and Technology*, 54(13), 7760–7765.

Ragusa, A., Svelato, A., Santacroce, C., Catalano, P., Notarstefano, V., Carnevali, O., Papa, F., Rongioletti, M. C. A., Baiocco, F., Draghi, S., D'Amore, E., Rinaldo, D., Matta, M., and Giorgini, E. (2020). Plasticenta: First evidence of microplastics in human placenta. *Environment International*, 146, 106274.

Reisser, J., Shaw, J., Hallegraeff, G., Proietti, M., Barnes, D. K., Thums, M., and Pattiaratchi, C. (2015). Microplastics in the ocean: A review of the methods used for identification and quantification. *Environmental Science and Technology*, 49(10), 5939–5956.

Rillig, M. C., Ziersch, L., and Hempel, S. (2019). Microplastic transport in soil by earthworms. *Scientific Reports*, 7, 1362.

Rochman, C. M., Hoh, E., Hentschel, B. T., and Kaye, S. (2013). Long-term field measurement of sorption of organic contaminants to five types of plastic pellets: Implications for plastic marine debris. *Environmental Science and Technology*, 47(3), 1646–1654.

Rochman, C. M., Kross, S. M., Armstrong, J. B., Bogan, M. T., Darling, E. S., Green, S. J., and Veríssimo, D. (2016). Scientific evidence supports a ban on microbeads. *Environmental Science and Technology*, 50(3), 1031–1032.

Sajid, M., Ihsanullah, I., Khan, M. T., & Baig, N. (2022). Nanomaterials-based adsorbents for remediation of microplastics and nanoplastics in aqueous media: A review. *Separation and Purification Technology*, 305, 122453.

Schnurr, R. E., Alboiu, V., Chaudhary, M., Corbett, R. A., Quanz, M. E., Sankar, K., and Rehse, S. (2018). Reducing marine pollution from single-use plastics (SUPs): A review. *Marine Pollution Bulletin*, 137, 157–171.

Silva, A. L. P., Prata, J. C., Walker, T. R., Campos, D., Duarte, A. C., Soares, A. M. V. M., and Rocha-Santos, T. (2021). Rethinking and optimising plastic waste management under COVID-19 pandemic: Policy solutions based on redesign and reduction of single-use plastics and personal protective equipment. *Science of The Total Environment*, 742, 140565.

Silva, A. L. P., Prata, J. C., Walker, T. R., Duarte, A. C., Ouyang, W., Barcelò, D., and Rocha-Santos, T. (2021). Increased plastic pollution due to COVID-19 pandemic: Challenges and recommendations. *Chemical Engineering Journal*, 405, 126683.

Smith, M., Love, D. C., Rochman, C. M., and Neff, R. A. (2018). Microplastics in seafood and the implications for human health. *Current Environmental Health Reports*, 5(3), 375–386.

Sussarellu, R., Suquet, M., Thomas, Y., Lambert, C., Fabioux, C., Pernet, M. E. J., and Soudant, P. (2016). Oyster reproduction is affected by exposure to polystyrene microplastics. *Proceedings of the National Academy of Sciences*, 113(9), 2430–2435.

Thompson, R. C., Moore, C. J., vom Saal, F. S., and Swan, S. H. (2009). Plastics, the environment and human health: current consensus and future trends. *Philosophical Transactions of the Royal Society of London. Series B, Biological Sciences*, 364(1526), 2153–2166.

Thompson, R. C., Olsen, Y., Mitchell, R. P., Davis, A., Rowland, S. J., John, A. W., and Russell, A. E. (2004). Lost at sea: Where is all the plastic? *Science*, 304(5672), 838–838.

Torres, F. G., and De-la-Torre, G. E. (2021). Face mask waste generation and management during the COVID-19 pandemic: An overview and the Peruvian case. *Science of The Total Environment*, 786, 147523.

UN Environment Programme. (2019). *Addressing the plastic pollution crisis: The role of global governance*. UNEP Report.

UNESCO. (2024). *Greening Curriculum Guidance: Teaching and learning for climate action*. United Nations Educational, Scientific and Cultural Organization. Available online at www.unesco.org/sites/default/files/medias/fichiers/2024/09/Greening%20curriculum%20guidance%20Teaching%20and%20learning%20for%20climate%20action.pdf

Vanapalli, K. R., Sharma, H. B., Ranjan, V. P., Samal, B., Bhattacharya, J., Dubey, B., and Goel, S. (2021). Challenges and strategies for effective plastic waste management during and post COVID-19 pandemic. *Science of The Total Environment*, 750, 141514.

Wang, F., Wong, C. S., Chen, D., Lu, X., Wang, F., and Zeng, E. Y. (2021). Interaction of toxic chemicals with microplastics: A critical review. *Water Research*, 162, 263–279.

Wright, S. L., and Kelly, F. J. (2017). Plastic and human health: A micro issue? *Environmental Science and Technology*, 51(12), 6634–6647.

Wright, S. L., Thompson, R. C., and Galloway, T. S. (2013). The physical impacts of microplastics on marine organisms: A review. *Environmental Pollution,* 178, 483–492.

Wright, S. L., Ulke, J., Font, A., Chan, K. L., and Kelly, F. J. (2020). Atmospheric microplastic deposition in an urban environment and an evaluation of transport. *Environment International*, 136, 105411.

7 Antibiotics and Personal Care Products as Contributors to Antibiotic Resistance from a One Health Perspective

Damini and Rahul Vikram Singh

7.1 INTRODUCTION

Antibiotic resistance (ABR) has contributed to a critical global health concern in recent decades, posing a significant threat to human, animal, and environmental well-being. The widespread utilization of antibiotics in both medical and non-medical contexts has resulted in the exposure and spread of ABR to these drugs. While the overuse and misuse of antibiotics in medical practice have been well-documented, emerging evidence suggests that personal care products (PCPs) may also contribute to the development and spread of ABR. [1–3] Antibiotics are powerful drugs used to combat bacterial infections. However, their widespread and often inappropriate use has led to the development of antibiotic-resistant bacteria. These resistant strains can survive exposure to antibiotics, rendering them ineffective and posing a significant challenge to healthcare providers. [4, 5] ABR is primarily driven by genetic mutations or the acquisition of resistance genes through horizontal gene transfer. Over time, resistance-causing bacteria can spread within healthcare facilities, communities and even natural ecosystems, leading to severe infections that are difficult to treat. [6–9] PCPs encompass a wide range of consumer goods, including soaps, shampoos, cosmetics, and cleaning agents. [10] Many of these products contain antimicrobial agents such as triclosan and triclocarban, which are intended to kill or inhibit the growth of bacteria. However, their continuous and widespread use has raised concerns about the development of ABR. [11, 12] Studies have shown that prolonged exposure to these antimicrobial agents in PCPs can promote the emergence of antibiotic-resistant bacteria. Moreover, the presence of such resistant bacteria in the environment, including water bodies, can further contribute to the spread of resistance genes to other organisms. [13, 14] Adopting a One Health approach is crucial to addressing the complex issue of ABR. This perspective recognises the

DOI: 10.1201/9781032623801-7

interconnection of humans, animals and environmental health and emphasises the need for collaborative efforts across disciplines. [15, 16] Antibiotics used in human healthcare can find their way into the environment via various routes, such as inappropriate disposal of pharmaceutical waste or excretion by humans and animals. This environmental contamination can lead to the selection and dissemination of antibiotic-resistant bacteria. Consequently, resistant bacteria can be transmitted back to humans through direct contact or via the food chain, creating a vicious cycle of resistance. [17–19] ABR is a multifaceted problem that requires a comprehensive and integrated approach. Understanding the role of both antibiotics and PCPs as contributors to ABR is essential in mitigating its impact. By adopting a One Health perspective, stakeholders from human and veterinary medicine, environmental science and policymaking can collaborate to develop strategies that minimize the emergence and spread of ABR. This may include implementing prudent antibiotic use policies, promoting alternatives to antimicrobial agents in PCPs, and improving waste management practices.

7.2 ANTIBIOTICS

Antibiotics are one of the prominent medical innovations which have efficient performance in immunosuppressive therapies and infectious diseases that depend on antibiotic prophylaxis and are preferably used in surgeries. [8] Antibiotics are administered in sufficient doses to inhibit the microbial population, specifically bacteria, for various therapeutic applications in humans, and moreover have crucial use in agriculture to enhance crop yield and treat animals in livestock. [20, 21]

7.2.1 Source of Antibiotics

Antibiotics can be derived from various sources, which are as follows:

Natural Sources

The majority of antibiotics are produced by microorganisms like bacteria and fungi as a part of their defence mechanisms. These microorganisms produce antibiotics to inhibit the growth of competing bacteria in their environment. Examples include penicillin, which is derived from the fungus *Penicillium*, and *Streptomycin*, which is produced by the bacteria *Streptomyces*. [5, 9]

Semi-synthetic Antibiotics

Semi-synthetic antibiotics are derived from natural sources but are chemically modified to enhance their effectiveness or expand their spectrum of therapeutic index. For example, amoxicillin and ampicillin are semi-synthetic derivatives of the natural antibiotic penicillin. [22, 23]

Synthetic Antibiotics

Some antibiotics are entirely synthesized in the laboratory without being derived from natural sources. Synthetic antibiotics are designed to mimic the structure and mechanism of action of natural antibiotics. Examples include fluoroquinolones and sulphonamides. [24]

Combination Antibiotics

Combination antibiotics are formulations that contain two or more different antibiotics. These combinations can be designed to provide a broader spectrum of activity or synergistic effects, where the combined effect of the antibiotics is greater than the sum of their individual effects. Common examples include trimethoprim-sulfamethoxazole and amoxicillin-clavulanate. [25, 26]

Biosynthetic Engineering

Advances in genetic engineering and biotechnology have enabled the production of antibiotics through the engineering of microorganisms genetic material. Scientists can manipulate the genes responsible for antibiotic production in bacteria and fungi to increase yield or create novel antibiotics. This approach allows for the production of antibiotics that may not be easily obtained from natural sources. [27, 28]

7.2.2 Types of Antibiotics and Antibiotic Resistance

There are several types of antibiotics, each with its mode of action and spectrum of activity against different types of bacteria. Some common classes of antibiotics based on their action mechanism are shown in Figure 7.1. The efficient use of antibiotics is limited because of the development of resistance or tolerance since the first use of the compound due to certain biochemical as well as physicochemical mechanisms. [5] One of the major problems in healthcare systems globally is ABR, which has various contributing factors, but the major reason is the spread and evolution of resistant microbes. [29] A few factors are listed in Figure 7.2. [30]

7.3 MECHANISM OF ANTIBIOTIC RESISTANCE

ABR is the potential of bacteria or any microorganism to withstand the effects of antibiotics that were once effective in killing or inhibiting their growth. Numerous pathways exist through which bacteria can acquire ABR.

7.3.1 Enzymatic Inactivation

Enzymatic inactivation of antibiotics is a common resistance mechanism employed by bacteria to neutralise the effects of certain antibiotics. Antibiotics susceptible to enzymatic degradation, such as β-lactamases and aminoglycosides, are examples that are degraded due to the enzymes produced by bacteria. [30] There are several methods in the enzymatic inactivation of antibiotics, which are illustrated in Figure 7.3.

7.3.2 Biofilm Formation

Biofilm consists of a large bacterial population, which is protected by a robust exopolymer matrix that binds to the surface of the bacteria. The capability of biofilm synthesis depends on physical, genetic, and physiological mechanisms. Biofilm formation affects the therapeutic effect of antibiotics, and as a result, 65% 80% of

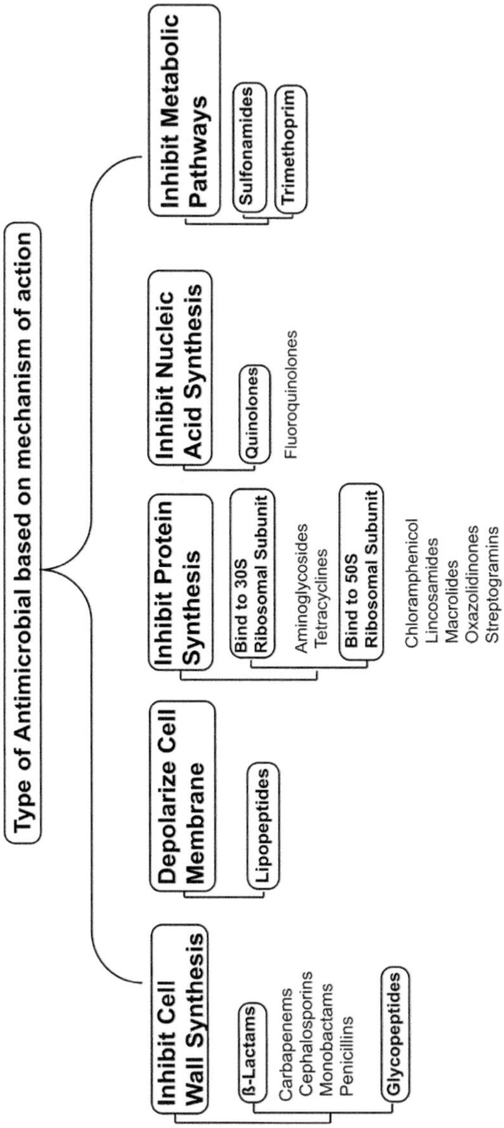

FIGURE 7.1 Different classes of antibiotics.

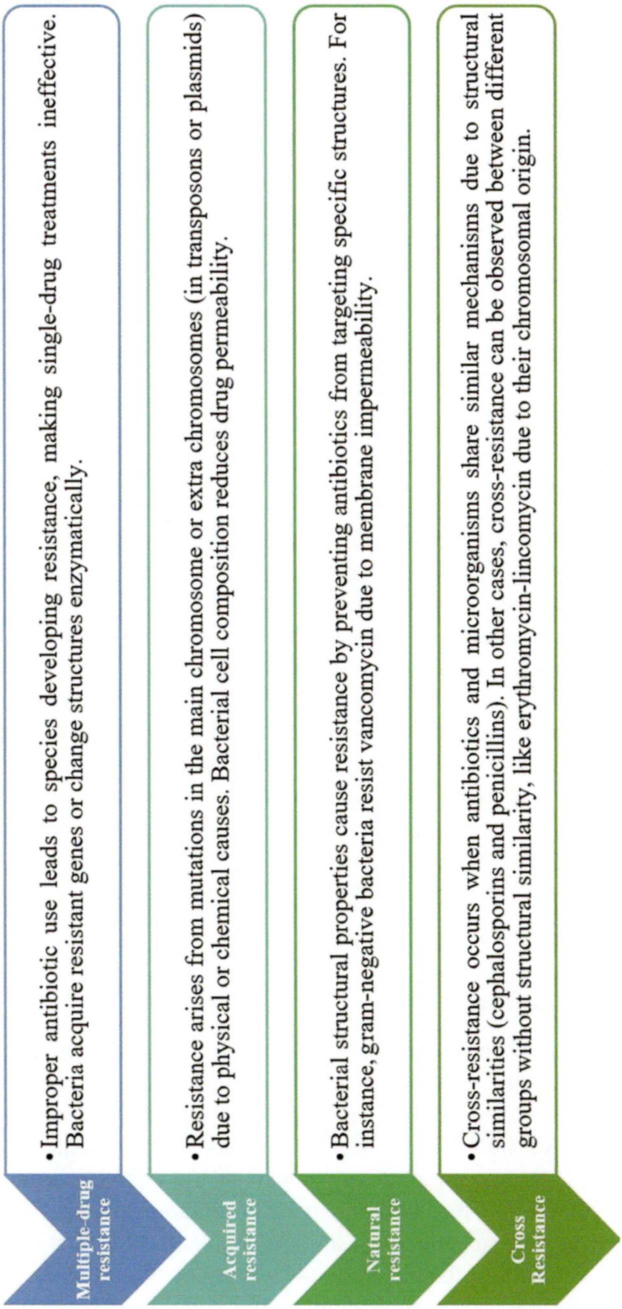

FIGURE 7.2 Major causes for antibiotic resistance.

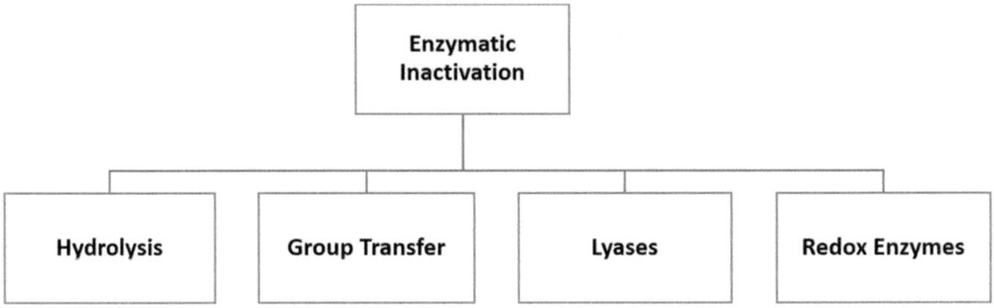

FIGURE 7.3 Different methods of enzymatic inactivation of antibiotics.

infections are a consequence of this. [31] Quorum sensing (QS), a communication biofilm mechanisms in bacterial cells, facilitates the formation of biofilms. It requires enough cell density to initiate QS signalling, which ultimately contributes to gene expression. The formation of biofilm is illustrated in Figure 7.4. [32]

7.3.3 Efflux Pump

The transporters that actively remove the diverse substrates from the cytoplasm are called bacterial efflux pumps (EPs), which are present in the plasma membrane of bacteria. In the bacterial repertoire of stress responses, efflux may be the fastest-acting and most highly potent resistance mechanism, depending on the antibiotic or toxin challenge. To prevent the build-up of poisonous compounds in the cells and to pump the dangerous molecules out of bacteria, Eps have evolved in bacteria. The resistant nodulation (RND) and cell division, which are particularly important in bacteria, the major facilitator superfamily (MFS), and the multidrug toxic compound extrusion (MATE) are some of the significant families of efflux transporters. Multiple drug resistance arises due to the broad specificity of various active efflux mechanisms. One illustrative example is *E. coli*'s AcrAB-TolC efflux system, which can confer resistance to a range of drugs such as tetracycline, florfenicol, erythromycin, and enrofloxacin. [33]

7.3.4 Alteration of Target Site

For antibiotics to have an antibacterial effect, they must combine with the target site; however, if the target site is altered or mutated, this will interfere with the usual combination and reduce the antibiotic's effectiveness. The majority of the mutations, which are harmful to the host bacterium and do not pass on to subsequent generations of cells or populations, happen randomly as part of the DNA replication process. On the bacterial cytoplasmic membrane are penicillin-binding proteins (PBPs) which serve as the target of β-lactam antibiotics and play a role in the formation of cell wall peptidoglycan. As a result of the mutation, the affinity between β-lactam antibiotics and their target PBPs will vanish, preventing the antibiotics from binding to the target and leading to bacterial resistance. [33]

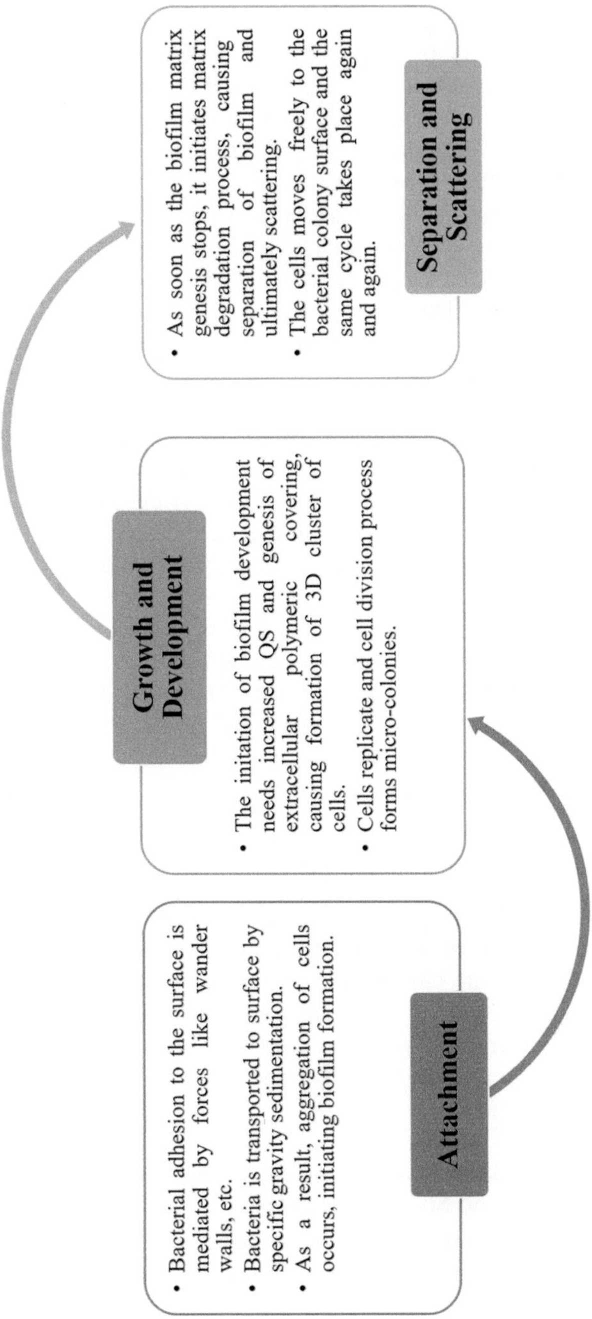

FIGURE 7.4 Bacterial biofilm formation.

7.4 ANTIBIOTICS IN PERSONAL CARE PRODUCTS

There are many different chemical classes included in pharmaceuticals and personal care products (PPCPs). Personal care goods, which include items like moisturizers, lipsticks, shampoos, hair colours, deodorants, and toothpaste, are intended to enhance daily living, while pharmaceuticals are largely used to prevent or treat human and animal sickness. [34] Additives play a big part in controlling the bacterial load of products that are made for cleaning and personal hygiene. Without them, the product would swiftly lose its effectiveness and, more crucially, expose the consumer to unforeseeable risks. Medical lotions, soaps, creams, and sprays that are used for daily hygiene or the treatment of localized bacterial infections also contain antimicrobial compounds.[12] PPCPs, which are pervasive and could be hazardous to the environment and human health, are also known as emerging pollutants. Antibiotics, hormones, and antimicrobial agents are examples of the various substances that are used to treat illnesses and improve quality of life. Additionally, PPCPs are utilized to improve the health or growth of farm animals. Currently, about 4,000 medicines are in use. Pharmaceutical prescription rates have nearly tripled over the past 14 years, and in 2013, over 9.1 million kg of antibiotics were utilized, of which 73.6% were used to increase the production of animals used for food. [35] PPCPs from the home and industrial sectors are frequently excreted in their unmetabolized form or as active metabolites into sewage and wastewater treatment plants (WWTPs). As a breeding ground for ABR, these bioengineered systems serve as a hub. Resistance spreads primarily through horizontal gene transfer, which occurs between microorganisms. The efficacy of these innovative compounds in treating bacterial infections is currently declining as a result of the unchecked use of antibiotics, and antimicrobial resistance is becoming more and more of a global problem. The lack of adequate access to fundamental public healthcare and sanitary facilities in underdeveloped nations accelerates the spread of resistance. [36]

7.4.1 Types of Personal Care Products

Modern society has an almost overwhelming selection of cosmetic and personal care products, both in terms of quantity and variety, as well as individual ingredients. In this section, personal care products are categorized in the subsections, and these different personal care products are formulated by various combinations of antibiotics and other active pharmaceuticals which are used in daily life; therefore, overexposure of these products creates a threat of ABR. Some of the daily-use products are given below with their formulation.

7.4.1.1 Haircare Products (Hair Colouring, Cleansing, Conditioning, Depilatories, and Styling Products)

One of the most reactive compounds in the cosmetic sector is the chemically active compounds of hair colouring or dye products. Direct (temporary or semi-permanent) dyes and oxidative (permanent) dyes are the two categories of hair colouring products. The hydrogen peroxide, which is an oxidiser, and dye intermediaries like paraphenylenediamine (PPD), resorcinol, or aminophenol are both found in oxidative hair dyes. Since PPD oxidizes with couplers to produce coloured reaction products, it is a common ingredient in practically all hair colouring formulations. Semi-permanent and transient dyes are examples of direct dyes. While semi-permanent hair colours comprise nitro-phenylenediamines, nitro-aminophenols, and some azo dyes, temporary hair dyes typically contain azo, triphenylmethane, anthraquinone, or indamine-like compounds. Permanent hair dyes are different from semi-permanent or temporary dyes. Permanent hair dyes are made of two ingredients that must be put together before usage for a chemical reaction to produce the dye on the hair. Although PPD has benefits for hair dyes and colours, studies suggest that this component is common, which causes an allergic reaction in people. PPD has been known to cause contact sensitization in certain cases. Some of these cases are related to the para group of PPDs' hypersensitivity or to the creation of oxidized products as a result of air exposure to PPD. PPD's sensitization potential is influenced by N-substitution. Additionally, the alkyl substituent's chain length frequently affects the sensitization potential. [37, 38] Shampoos normally contain an aqueous emulsion comprising several surfactants, which have significant detergent characteristics, as well as several compounds to boost the cleansing ability or sensory qualities, and that sometimes contain more than one preservative and aroma. Anti-dandruff shampoos contain different bioactive components that work in quite diverse ways. Coal tar, a component of certain conventional shampoos in this category, has been determined to be acceptable for use in anti-dandruff shampoos even though it is carcinogenic under some exposure circumstances. Although cationic surfactants are frequently the most efficient cleaners, they can irritate the scalp when used in large quantities. Sodium lauryl sulphate, sodium laureth sulphate, and ammonium laureth sulphate are examples of the most popularly used anionic surfactants with high washing and foaming qualities; however, because of their potential for skin irritation, they are typically combined with non-ionic surfactants, which are typically milder. [39, 40] Amphoteric or non-ionic surfactants, which are less severe, are frequently used in the formulation of baby shampoos. A large number of non-ionic surfactants fall into the category of polyethoxylated surfactants, which may also contain minute amounts of 1,4-dioxane or ethylene oxide, both of which have been classified as potentially carcinogenic substances. To reduce the quantity of these accidental contaminants, a procedure called vacuum stripping is performed during their production. In formulations containing other chemicals that break down to make nitrates, the surfactants diethanolamine and triethanolamine have also been recognized as potential contributors to the generation of nitrosamine. [41] Numerous sources advise limiting usage and avoiding surfactant-preservative combinations that can produce nitrosamine. Preservatives are present in shampoos to prevent bacteria growth. Diazolidinyl urea and Quaternium-1,5-imidiazolidinyl urea are examples of preservatives that work by releasing formaldehyde. Formaldehyde is a possible sensitizer and when inhaled it is proven to be carcinogenic. The use of formaldehyde-releasing preservatives should be limited, according to several expert organizations. In general, hair conditioners are designed to restore the beneficial features or elements of hair that have been lost or diminished during the washing process. The most popular product varieties often include cationic surfactants, fragrances, fatty acids, and fatty alcohols in a water-based emulsion that also includes a preservative system. The cationic surfactants alter the structure and appearance of the hair shaft by adhering to it or seeping into it. They are typically regarded as harmless and moderate. When compared to shampoos and other water-based emulsion solutions with similar water content, preservative systems are not significantly different. [42, 43] These hair removal products often include barium sulphide, thioglycolates, sodium, or calcium hydroxide (alkalis, which are highly hazardous), and are used to eliminate undesired hair. Due to their caustic nature, these substances should be used with caution to prevent skin irritation, especially of the mucous membranes. These products contain ingredients that cause chemical connections in hair to break or re-form to produce the desired results. They are therefore highly chemically aggressive; thus, caution must be exercised to prevent harm to the hair itself as well as to the nearby skin and body structures. To dissolve the disulfide bond present in the protein structure of the hair, when wrapped around style rods, permanent hair-waving solutions are used, which often include a reducing solution or lotion. Ammonium thioglycolate or other thioglycolate salts, glyceryl thioglycolate, and thioacetate salts are a few of the regularly utilised oxidising agents. Permanent hair treatments used in homes for a long time are made safely from ammonium thioglycolate, but they can occasionally cause sensitization in hairdressers. It has been demonstrated that glycerol monothioglycolate sensitizes hairdressers and, on rare occasions, their clientele. Skin contact with these compounds must be avoided, especially by hairdressers, as they are exposed more regularly in the workplace than private customers who use the products on an occasional basis. The neutralizing solutions containing hydrogen peroxide, sodium bromate, or perborate are also used in hair-waving products to re-form disulfide bonds and shape the hair to the same as style rods. Some fixatives for the permanent styling of hair contain 2%–8% (w/v) mercuric chloride. It is important to take precautions to keep neutralizing solutions and fixatives away from the eyes, it is advised to wear gloves, especially for hairdressers. [43] These products use synthetic polymers as the main functional component in water- or alcohol-based solvent vehicles, which act by forming films. Different additives are used to enhance the product's functionality and other application, function-related characteristics. Most of the time, the polymers, or resins used are acrylate-based copolymers composed of acrylamide, vinyl acetate, and methyl vinyl ether. Ethanol is frequently employed as a solvent in aerosol hairsprays, along with dimethyl ether, hydrocarbon, or carbon dioxide as a propellant. A compound such as Butane is used in typical hydrocarbon propellant systems. [43, 44]

7.4.1.2 Skincare Products (Moisturisers, Creams, Lotions, Sun Protection, Shaving Creams, Aftershave, and Toners)

These products are emulsions, such as oil in water (o/w), water in oil (w/o), aqueous gel, silicone in water, and combination emulsions of various sorts. Cold cream is a famous example of a w/o emulsion. Emulsifying agents include a variety of surfactants. The kind of surfactant used depends on the type of emulsion and the product's intended usage. The basic emulsion combination is then supplemented with several other compounds such as thickeners, preservatives, and moisturizing agents. Petrolatum and mineral oil are often used substances to help skin maintain moisture. Polycyclic aromatic hydrocarbons that have the potential to cause cancer, may be present in trace concentrations. When improperly treated to eliminate allergenic impurities, propylene glycol, used as a humectant, is a recognized contact irritant and as an outcome causes contact dermatitis. Lanolin can also do the same thing. Many face creams and lotions that market themselves as anti-ageing or wrinkle-reducing treatments include retinyl palmitate. [43, 45] Chemical sunscreens, which block UV rays, make up the majority of sunscreens, whereas mineral sunscreens physically block the sun's rays. Cinnamates, para-aminobenzoic acid and its derivatives, salicylates, camphor derivatives, avobenzone, triazines, and benzophenones are examples of common UV absorbers, often known as UV filters. Some nations regulate sunscreen products as medications, while others do so as cosmetics. Avobenzone and octyl methoxycinnamate, two UV filters, have been proven to produce negative skin effects, namely inflammation. Zinc oxide and titanium dioxide are the two most common chemically active components found in mineral sunscreens. These naturally occurring minerals serve as sunscreens and are typically regarded as neutral. However, the advent and widespread usage of titanium dioxide and zinc oxide nanoparticles in sunscreens present additional safety issues. It has been speculated that the use of nanoparticles may make it easier for these substances to permeate the skin, resulting in systemic exposure and potential toxicity. [43, 46] Conventional shaving creams are commonly formulated with a water base, encompassing scents, binders, foaming or lathering agents, thickeners, and additional components. Inhalation or exposure and toxic components are a critical consideration for aerosol types. A popular propellant component, isobutane may include trace levels of butadiene, a carcinogen. It is critical to reduce inadvertent inhalation. [43, 47] Aftershaves are generally water or alcohol based, and are intended to calm inflamed skin. They are, nevertheless, intensely perfumed and function as men's fragrances.

7.4.1.3 Skin Cleansing Products (Exfoliants or Scrubs, Soaps, Make-Up Removers, Liquid Skin Cleansers)

Exfoliating chemicals in scrubs remove dead skin cells. The exfoliating chemicals are carried in a liquid or paste that has a surfactant-based washing effect. Many scrubs use microbeads of polyethylene plastic as an exfoliating agent; others contain natural ingredients such as crushed nuts, seeds, or salt. Certain of these items contain chemical exfoliants such as alpha-hydroxy acids or beta-hydroxy acids. Because the goal effect is exfoliation, these chemicals must be irritants. Furthermore, they might make the skin more sensitive to sunlight. Therefore, the concentration of such exfoliating agents is minimized and designed for consumer use to avoid severe irritation and burning. [43] In its most basic form, soap is an alkali salt of a long-chain fatty acid. In the past, they were made by combining animal fats with lye derived from wood ashes. Bar soaps and cleaning bars are now created via a vast range of fats and oils using more sophisticated techniques, and they contain a variety of chemicals like surfactants, stabilizers, antioxidants, colourants, and aromas. Above, a brief overview of toxicological considerations concerning surfactants was presented; a similar discussion was offered for cosmetic colourants and aroma components. [43, 47] Make-up removers are used to eliminate the oily residues left behind by cosmetics such as eye make-up, foundation, and so on. They are water-in-oil emulsions and are frequently used in conjunction with the sort of liquid cleaner or bar soap as outlined above. They consist of preservatives (although at lesser concentrations than formulations having high water concentration), colourants, and aromas. [43] To eliminate filth and clean the skin and remove natural oils, commercial cleansers typically use alcohol and petroleum chemicals, and promote skin dryness. The popularly utilized non-ionic surfactants in these products in terms of their cleaning action are comparable to the ones found in shampoos but at lesser concentrations. Various anionic and other modifying surfactants are utilized to achieve greater 'mildness', particularly for usage on facial skin, as well as other sensory qualities relevant to customers when using face and body cleansers. It is worth noting that the growing demand of cleansers which make claims of being 'antibacterial' contain preservatives or other antibacterial chemicals at a higher amount than normal for preservation function. [43]

7.4.1.4 Fragrances

Fragrances are formulated using essential oils or other volatile aromatic chemicals, along with solvents and fixatives – substances that promote stability and slow down evaporation. Ethanol, or a combination of ethanol and water, is commonly used as a solvent. Essential oils are complex mixtures of ethers, hydrocarbons, esters, alcohols, and ketones. Examples of essential oils include eucalyptus, lemon, geranium, peppermint, and rosewood. Several ancient natural scent components, such as ambergris and natural musk oil, are no longer used – primarily due to environmental concerns, and in some cases, for safety reasons. The most common cause of allergic reactions in the cosmetic industry is the presence of certain chemicals in fragrance components. In particular, contact allergy to fragrance compounds can occur after significant skin exposure to these chemicals. This typically presents as localized contact dermatitis when perfume-type products, intensely scented products, antiperspirants, or deodorants are applied, as well as on the hands. Eczema (allergic contact dermatitis) may develop upon re-exposure to significant levels of certain fragrance components. [43, 48]

7.4.1.5 Antiperspirants and Deodorants

Deodorants include aluminium and zinc salts, as well as fragrance to mask the odour of sweat. A large number of classic antiperspirants rely on aluminium compounds, such as aluminium tri-chlorohydrin gly, to block sweat ducts and prevent sweat from reaching the skin's surface. Numerous aluminium compounds cause skin irritation, are neurotoxic, and interact with oestrogen, which increases the risk of breast cancer. Crystal deodorants are a popular alternative to antiperspirants and deodorants. They also contain potassium alum or aluminium alum as active ingredients, which are regarded as less allergenic than other compounds with equivalent activities. Many of the items in this category have a strong fragrance. [49]

7.4.1.6 Oral Care Products

Active compounds such as calcium phosphates, alumina, abrasives, and anionic surfactants are found in dental pastes, powders, and liquids. Mouthwashes are typically flavoured with essential oils and sweetened. Bicarbonates, borates, phosphates, and carbonates are all found in denture cleansers. While these products are not meant to be consumed, some ingestion is to be expected and therefore the product must be designed accordingly. For information about toxicity by consumption, the reader is directed to other topics in the present volume. [50]

7.4.1.7 Decorative Cosmetics

These are sometimes known as 'colour cosmetics'. Their primary function is to colour or hide, and they include some of the greatest quantities of colouring chemicals found in any cosmetic or personal care product. Although there are various other colouring compounds, these colouring substances are commonly categorized into two different classes, that is, synthetic organic pigments and mineral pigments.

7.4.1.8 Face, Eye, Lip, and Nail Make-Up

These products comprise foundation (also known as formation in some countries) and blusher, which are formulated in liquid, cream, or powder form, as well as additional finishing face powders. Suspensions or emulsions (mainly o/w) can be found in liquid and cream goods. These products come into contact with the face, and the emulsifying agents used must be carefully selected to avoid irritation. The oils used must have a minimal propensity to clog pores (comedogenicity), and may sometimes cause blackheads, whiteheads, and acne. [43, 51] This category includes items such as eye shadow, mascara, and eyeliner. Modern eyeliners and mascara are both formulated in liquid form. The most commonly used eye shadow formulations are pressed powders, while liquid (cream) forms are also available. Water, which is a gelling agent (gums, such as magnesium aluminium silica); polyols (propylene glycol or butylene glycol); colourants; wetting agents (water-soluble esters, such as polysorbate 80); alcohol; polymeric film former; and preservatives are the most common components of eyeliners. Mascara is available in the form of both types of emulsions (o/w and w/o) and also in anhydrous solvent-based suspensions. Polymers, other film formers, pigments, emulsifiers, waxes, and preservatives are among the active compound classes employed in all varieties. [43, 52] Classical lipstick has been used since ancient times to provide texture, colour, and protection to the lips. It is a cosmetic product which is composed of oils, waxes, pigments, and other emollients. Classical lipstick is still quite popular and comes in a variety of colours and surface appearance qualities. Lip pencils are a modern product that utilizes various combinations and particular components from the same fundamental constituent groups as conventional lipsticks, as well as fillers and other additions required to generate physical effects. [43, 53] The most popular nail beauty items are nail polishes and lacquers. These are formulated using modifying resins or polymers, polymeric film former, plasticizing agents, and solvents. The most popular film former is nitrocellulose, and the most popular modifying resin is para-toluene sulphonamide formaldehyde resin. Formaldehyde is a recognized sensitizer, and formaldehyde resin has been a major factor in allergic responses in certain users. Many products using different modifiers have been developed. Except for the dibutyl phthalate, which has been demonstrated to attack the immune system in animal tests, a range of plasticizers are utilized, most of which are thought to be rather innocuous. [43]

7.5 TYPES OF PRODUCTS CONTRIBUTING TO ANTIBIOTIC RESISTANCE

Well-known compounds including parabens, triclocarban, and triclosan are included in many items. Preservatives called parabens are added to several products (Table 7.1), including lotions, detergents, and cosmetics, to increase their stability over time. They are derivatives of benzoic acid, and the lengths of their side chains distinguish them from one another. Because of benzoic acid excessive use, it was released into the environment, contaminating several matrices. [12] A popular antibacterial agent that penetrates wastewater treatment plants (WWTPs) and the environment is triclosan (TCS). In the United States, WWTPs discharge between 1.1 and 4.2 105 kg of TCS annually. Various concerns have been raised about the effects of TCS's availability and antibacterial qualities on the spread of ABR in the environment. [54] Some of the major compounds are mentioned in Table 7.1 which contribute to ABR.

To maintain and enhance oral health and appearance, toothpaste is a paste or gel applied to the teeth with a toothbrush. The formulations of toothpaste have changed significantly since their invention many thousands of years ago, going from simple suspensions of crushed eggshells or ashes to complicated formulations with frequently more than 20 constituents. Compounds to fight dental cavities, gum disease, bad breath, calculus, erosion, and dentin hypersensitivity might be included in this list. Additionally, toothpaste includes abrasives to clean and whiten teeth, tastes to refresh breath, and colours for improved aesthetics. The three antiplaque/antigingivitis ingredients most frequently found in dental pastes and supported by science are triclosan, stannous chloride/fluoride, and zinc citrate/chloride. [12] Microorganisms are responsible for many artificial and biological processes, and triclosan (TCS) is made to affect microorganisms in hospitals and homes. The antimicrobial characteristics of TCS can affect intricate microbial communities present in constructed and natural systems after being released into the environment. TCS has

TABLE 7.1
Different types of skin-care products with antibiotics formulation

Agents	Compounds	Source	Reference
UV fillers, Sunscreen	2-Ethyl-hexyl-4-tri methoxycinnamate, 4-methyl-benzylidene-camphor	Skincare Products (Sun Protection)	36, 43
Preservatives	Parabens (alkyl-p-hydroxybenzoates), Triclocarban	Body washes, Antibacterial soaps, Detergents, and Toothpastes	12, 36
Chemical Stabilizer	Parabens (p-Hydroxybenzoates derivatives)	Shaving products, Hair care, Skin care (Moisturizers), Make-Up	12
Antimicrobial Agent	Triclosan	Dental care products, Cosmetics, and Soaps	12

been connected to changing the composition or performance of microbial communities in wastewater treatment processes including anaerobic digestion and activated sludge. The same is true for freshwater biofilms in receiving streams, where TCS can affect diversity and biofilm growth. TCS affects denitrification and respiration rates in soils, as well as enriches organisms that may dehalogenate. TCS causes reactions in microbial communities; however, these intricate microbial communities seldom contain TCS amounts that limit function. TCS is more likely to exert stress that promotes resistance than stress that functionally inhibits complex microbial communities at ambient concentrations. [55] Given the COVID-19 pandemic's infectious nature, it has become a huge public health problem and resulted in widespread usage of hand sanitizers. Hand sanitizers and soaps also possess TCS phenolic compound and show action by penetrating the cytoplasmic bilayer. Due to its negative effects on bacterial enzymes involved in the synthesis of fatty acids from cell walls and membranes, triclosan is bacteriostatic. TCS breaks down the bacteria's membrane in high doses, killing it. [56] Similarly, TCS is also present in disinfectants along with quaternary ammonium compounds, chlorine, iodine, alcohols, hydrogen peroxide, silver, and chlorhexidine; hence, antimicrobial resistance caused by the usage of disinfectants may make ABR worse. [57, 58]

7.6 ANTIBIOTIC RESISTANCE FROM A ONE HEALTH PERSPECTIVE

ABR is an important global health concern, and the One Health perspective is a crucial approach to addressing this issue. One Health is a collaborative and holistic approach that recognizes the interconnectedness of human health, animal health, and the environment. It emphasizes the importance of addressing health challenges through a unified effort that involves multiple sectors, including human medicine, veterinary medicine, agriculture, and environmental conservation. Antibiotic-resistant bacteria can be transmitted between animals and humans, and the excessive use of antibiotics in the agriculture sector can create a significant impact on resistance rates in other sectors like human health. Therefore, addressing ABR requires a joint effort and coordination across different sectors. A few common antibiotics are used in human medicine as well as in veterinary medicine and agriculture. The widespread use of antibiotics in these sectors can lead to the development and spread of antibiotic-resistant bacteria, making it essential to manage and monitor antibiotic use in all areas to minimize resistance. In the case of zoonotic diseases like rabies, blastomycosis (*Blastomyces dermatitidis*), psittacosis (*Chlamydophila psittaci, Chlamydia psittaci*), trichinosis (*Trichinella spiralis*), cat-scratch disease (*Bartonella henselae*), histoplasmosis (*Histoplasma capsulatum*), *coccidiomycosis* (Valley Fever), and so forth, these are infections which can be transmitted from animals to humans. ABR can complicate the treatment of zoonotic diseases, making it more challenging to control outbreaks and protect both human and animal populations. On the other hand, the use of antibiotics in agriculture can lead to the release of antibiotic residues and resistant bacteria into the environment through water run-off and manure. These resistant bacteria can then spread to other organisms, contributing to the development of resistance. The One Health approach calls for enhanced surveillance of ABR in both human and animal populations. This data can help researchers and policymakers to understand the dynamics of resistance transmission and devise appropriate strategies to combat it. By working together and adopting a multidisciplinary approach, we can develop more effective strategies to address ABR and ensure the continued efficacy of antibiotics for future generations.

7.7 CONCLUSION

In the current scenario, the demand for personal care products is increasing rapidly year by year, and the pharmaceutical industries are striving to fulfil consumers demands by developing a variety of such products. However, this has led to the worldwide use of antibiotics in personal care products, which is a contributor to the growing problem of ABR. To effectively address this issue, it is crucial to recognize the complex interplay between human health, animal health, and the environment. One approach that holds promise is the adoption of a One Health approach, which combines efforts from various sectors. By doing so, improved surveillance, regulation, and mitigation strategies can be developed to combat ABR. Responsible antibiotic use, the development of alternative products, and the implementation of sustainable wastewater management practices are essential steps to preserve the effectiveness of antibiotics and safeguard public and animal health for future generations. However, the situation remains concerning, especially in developing countries, where some pharmaceutical companies prioritize profits over people's health. For example, in India, bath soaps are required to have a total fatty matter (TFM) content of more than 76%, according to Bureau of Indian Standards guidelines. Yet, many bath soaps in the market have TFM content lower than the required percentage, but they are still advertised and sold as regular bath soaps. This lack of stringent rules and regulations raises concerns about the safety and efficacy of skincare products containing antibiotics. Without proper oversight, consumers may unknowingly use products that do not meet the necessary standards, potentially leading to adverse effects on their skin and overall health. Governments need to step in and implement strict regulations, and enforce quality control measures to ensure that pharmaceutical industries adhere to the appropriate standards when developing and marketing skincare products, especially those containing antibiotics. By doing so, consumers can be protected, and the integrity of the skincare industry can be maintained, ensuring safe and effective products for all.

ABBREVIATIONS

PCPs	Personal Care Products
ABR	Antibiotic Resistance
QS	Quorum Sensing
Eps	Efflux Pumps
RND	Resistant Nodulation
MFS	Major Facilitator Superfamily
MATE	Multidrug Toxic Compound Extrusion
PBPs	Penicillin-Binding Proteins

PPCPs	Pharmaceuticals and Personal Care Products
WWTPs	Wastewater Treatment Plants
HGT	Horizontal Gene Transfers
AMR	Antimicrobial Resistance
PPD	Paraphenylenediamine
TCS	Triclosan
TFM	Total Fatty Matter
BIS	Bureau of Indian Standards

REFERENCES

1. Prestinaci, Francesca, Patrizio Pezzotti, and Annalisa Pantosti. Antimicrobial resistance: A global multifaceted phenomenon. *Pathogens and Global Health* 109, no. 7 (2015): 309–318. https://doi.org/10.1179/2047773215Y.0000000030
2. Collignon, Peter J., and Scott A. McEwen. One health its importance in helping to better control antimicrobial resistance." *Tropical Medicine and Infectious Disease* 4, no. 1 (2019): 22. https://doi.org/10.3390/tropicalmed4010022
3. Velazquez-Meza, Maria Elena, Miguel Galarde-López, Berta Carrillo-Quiróz, and Celia Mercedes Alpuche-Aranda. "Antimicrobial resistance: One health approach." *Veterinary World* 15, no. 3 (2022): 743. https://doi.org/10.14202%2Fvetworld.2022.743-749
4. World Health Organization. (2018). *Antibiotic resistance*. Retrieved from www.who.int/news-room/fact-sheets/detail/antibiotic-resistance. (access on July 2023).
5. Aslam, Bilal, Wei Wang, Muhammad Imran Arshad, Mohsin Khurshid, Saima Muzammil, Muhammad Hidayat Rasool, Muhammad Atif Nisar et al. Antibiotic resistance: A rundown of a global crisis. *Infection and Drug Resistance* (2018): 1645–1658. https://doi.org/10.2147/IDR.S173867
6. Ventola, C. Lee.. "The antibiotic resistance crisis: Part 1: Causes and threats." *Pharmacy and Therapeutics* 40, no. 4 (2015): 277. https://pubmed.ncbi.nlm.nih.gov/25859123
7. Munita, Jose M., and Cesar A. Arias. Mechanisms of antibiotic resistance. Wiley Online Library , Editors name Indira T. Kudva, Nancy A. Cornick, Paul J. Plummer, Qijing Zhang, Tracy L. Nicholson, John P. Bannantine, Bryan H. Bellaire *Virulence mechanisms of bacterial pathogens* (2016): 481 511. https://doi.org/10.1128/9781555819286.ch17
8. Uddin, Tanvir Mahtab, Arka Jyoti Chakraborty, Ameer Khusro, BM Redwan Matin Zidan, Saikat Mitra, Talha Bin Emran, Kuldeep Dhama et al. Antibiotic resistance in microbes: History, mechanisms, therapeutic strategies and future prospects. *Journal of Infection and Public Health* 14, no. 12 (2021): 1750–1766. https://doi.org/10.1016/j.jiph.2021.10.020
9. Larsson, DG Joakim, and Carl-Fredrik Flach. Antibiotic resistance in the environment. *Nature Reviews Microbiology* 20, no. 5 (2022): 257–269. https://doi.org/10.1038/s41579-021-00649-x
10. Martín-Pozo, Laura, María del Carmen Gómez-Regalado, Inmaculada Moscoso-Ruiz, and Alberto Zafra-Gómez. Analytical methods for the determination of endocrine disrupting chemicals in cosmetics and personal care products: A review. *Talanta* 234 (2021): 122642. https://doi.org/10.1016/j.talanta.2021.122642

11. Weatherly, Lisa M., and Julie A. Gosse. Triclosan exposure, transformation, and human health effects. *Journal of Toxicology and Environmental Health, Part B* 20, no. 8 (2017): 447–469. https://doi.org/10.1080/10937404.2017.1399306
12. Caioni, Giulia, Elisabetta Benedetti, Monia Perugini, Michele Amorena, and Carmine Merola. Personal care products as a contributing factor to antimicrobial resistance: Current state and novel approach to investigation. *Antibiotics* 12, no. 4 (2023): 724. https://doi.org/10.3390/antibiotics12040724
13. Chaturvedi, Preeti, Parul Shukla, Balendu Shekher Giri, Pankaj Chowdhary, Ram Chandra, Pratima Gupta, and Ashok Pandey. Prevalence and hazardous impact of pharmaceutical and personal care products and antibiotics in environment: A review on emerging contaminants. *Environmental Research* 194 (2021): 110664. https://doi.org/10.1016/j.envres.2020.110664
14. Okoye, Charles Obinwanne, Raphael Nyaruaba, Richard Ekeng Ita, Samuel Ukpong Okon, Charles Izuma Addey, Chike C. Ebido, Adebayo Oluwole Opabunmi, Emmanuel Sunday Okeke, and Kingsley Ikechukwu Chukwudozie. Antibiotic resistance in the aquatic environment: Analytical techniques and interactive impact of emerging contaminants. *Environmental Toxicology and Pharmacology* (2022): 103995. https://doi.org/10.1016/j.etap.2022.103995
15. Conrad, Patricia A., Laura A. Meek, and Joe Dumit. Operationalizing a one health approach to global health challenges. *Comparative Immunology, Microbiology and Infectious Diseases* 36, no. 3 (2013): 211–216. https://doi.org/10.1016/j.cimid.2013.03.006
16. Shrestha, Pami, Shiying He, and Helena Legido-Quigley. Antimicrobial resistance research collaborations in Asia: Challenges and opportunities to equitable partnerships. *Antibiotics* 11, no. 6 (2022): 755. https://doi.org/10.3390/antibiotics11060755
17. Boxall, Alistair BA. The environmental side effects of medication: How are human and veterinary medicines in soils and water bodies affecting human and environmental health? *EMBO Reports* 5, no. 12 (2004): 1110–1116. https://doi.org/10.1038/sj.embor.7400307
18. Daughton, Christian G. Pharmaceuticals in the environment: Sources and their management. *Comprehensive Analytical Chemistry* 50 (2007): 1–58. https://doi.org/10.1016/S0166-526X(07)50001-2
19. Kumar, Vinay, Sivarama Krishna Lakkaboyana, Neha Sharma, Pritha Chakraborty, Mridul Umesh, Ritu Pasrija, Jithin Thomas et al. A critical assessment of technical advances in pharmaceutical removal from wastewater A critical review. *Case Studies in Chemical and Environmental Engineering* (2023): 100363. https://doi.org/10.1016/j.cscee.2023.100363
20. Tiedje, James M., Yuhao Fu, Zhi Mei, Andreas Schäffer, Qingyuan Dou, Wulf Amelung, Martin Elsner et al. Antibiotic resistance genes in food production systems support One Health opinions. *Current Opinion in Environmental Science & Health* (2023): 100492. https://doi.org/10.1016/j.coesh.2023.100492
21. Mann, Avantika, Kiran Nehra, J. S. Rana, and Twinkle Dahiya. Antibiotic resistance in agriculture: Perspectives on upcoming strategies to overcome upsurge in resistance. *Current Research in Microbial Sciences* 2 (2021): 100030. https://doi.org/10.1016/j.crmicr.2021.100030
22. Sambyal, Krishika, and Rahul Vikram Singh. Exploitation of E. coli for the production of penicillin G amidase: A tool for the synthesis of semisynthetic β-lactam antibiotics. *Journal of Genetic Engineering and Biotechnology* 19 (2021): 1–11. https://doi.org/10.1186/s43141-021-00263-7

23. Fernandes, Prabhavathi, Evan Martens, and David Pereira. Nature nurtures the design of new semi-synthetic macrolide antibiotics. *Journal of Antibiotics* 70, no. 5 (2017): 527 533. doi:10.1038/ja.2016.137
24. Mitcheltree, Matthew J., Amarnath Pisipati, Egor A. Syroegin, Katherine J. Silvestre, Dorota Klepacki, Jeremy D. Mason, Daniel W. Terwilliger et al. A synthetic antibiotic class overcoming bacterial multidrug resistance. *Nature* 599, no. 7885 (2021): 507–512. https://doi.org/10.1038/s41586-021-04045-6
25. Feldman, William, Theresa Sutcliffe, and Corinne Dulberg. Twice-daily antibiotics in the treatment of acute otitis media: Trimethoprim-sulfamethoxazole versus amoxicillin-clavulanate. *CMAJ: Canadian Medical Association Journal* 142, no. 2 (1990): 115. https://pubmed.ncbi.nlm.nih.gov/2403833
26. Dumitrescu, O., C. Badiou, M. Bes, M-E. Reverdy, F. Vandenesch, J. Etienne, and G. Lina. Effect of antibiotics, alone and in combination, on PantonValentine leukocidin production by a Staphylococcus aureus reference strain. *Clinical Microbiology and Infection* 14, no. 4 (2008): 384–388. https://doi.org/10.1111/j.1469-0691.2007.01947.x
27. Baltz, Richard H. Daptomycin: Mechanisms of action and resistance, and biosynthetic engineering. *Current Opinion in Chemical Biology* 13, no. 2 (2009): 144–151. https://doi.org/10.1016/j.cbpa.2009.02.031
28. Hutchings, Matthew I., Andrew W. Truman, and Barrie Wilkinson. Antibiotics: Past, present and future. *Current Opinion in Microbiology* 51 (2019): 72–80. https://doi.org/10.1016/j.mib.2019.10.008
29. Podolsky, Scott H. The evolving response to antibiotic resistance (1945 2018). *Palgrave Communications* 4, no. 1 (2018). https://doi.org/10.1057/s41599-018-0181-x
30. Hasan, Thualfakar Hayder, and Raad A. Al-Harmoosh. Mechanisms of antibiotics resistance in bacteria. *Systematic Reviews in Pharmacy* 11, no. 6 (2020): 817–823. https://doi.org/10.31838/srp.2020.6.118
31. Coenye, Tom, and Hans J. Nelis. In vitro and in vivo model systems to study microbial biofilm formation. *Journal of Microbiological Methods* 83, no. 2 (2010): 89–105. https://doi.org/10.1016/j.mimet.2010.08.018
32. El-Tarabily, Khaled A., Mohamed T. El-Saadony, Mahmoud Alagawany, Muhammad Arif, Gaber E. Batiha, Asmaa F. Khafaga, Hamada AM Elwan, Shaaban S. Elnesr, and Mohamed E. Abd El-Hack. Using essential oils to overcome bacterial biofilm formation and their antimicrobial resistance. *Saudi Journal of Biological Sciences* 28, no. 9 (2021): 5145–5156. https://doi.org/10.1016/j.sjbs.2021.05.033
33. Zhang, Fusheng, and Wei Cheng. The mechanism of bacterial resistance and potential bacteriostatic strategies. *Antibiotics* 11, no. 9 (2022): 1215. https://doi.org/10.3390/antibiotics11091215
34. Boxall, Alistair B. A., Murray A. Rudd, Bryan W. Brooks, Daniel J. Caldwell, Kyungho Choi, Silke Hickmann, Elizabeth Innes et al. Pharmaceuticals and personal care products in the environment: What are the big questions? *Environmental Health Perspectives* 120, no. 9 (2012): 1221–1229. https://doi.org/10.1289/ehp.1104477
35. Pennington, Marcus J., Jason A. Rothman, Michael B. Jones, Quinn S. McFrederick, Jay Gan, and John T. Trumble. Effects of contaminants of emerging concern on Megaselia scalaris (Lowe, Diptera: Phoridae) and its microbial community. *Scientific Reports* 7, no. 1 (2017): 8165. https://doi.org/10.1038/s41598-017-08683-7
36. Yadav, Shailendra, Niti B. Jadeja, Nishant A. Dafale, Hemant J. Purohit, and Atya Kapley. Pharmaceuticals and personal care products mediated antimicrobial resistance: Future challenges.

In *Pharmaceuticals and personal care products: Waste management and treatment technology* (2019): pp. 409 428. Butterworth-Heinemann. https://doi.org/10.1016/B978-0-12-816189-0.00017-2

37. Da França, Simone Aparecida, Michelli Ferrera Dario, Victoria Brigatto Esteves, André Rolim Baby, and Maria Valéria Robles Velasco. Types of hair dye and their mechanisms of action. *Cosmetics* 2, no. 2 (2015): 110–126. https://doi.org/10.3390/cosmetics2020110
38. He, Lin, Freideriki Michailidou, Hailey L. Gahlon, and Weibin Zeng. Hair dye ingredients and potential health risks from exposure to hair dyeing. *Chemical Research in Toxicology* 35, no. 6 (2022): 901–915. https://doi.org/10.1021/acs.chemrestox.1c00427
39. Trüeb, Ralph M. Shampoos: Ingredients, efficacy and adverse effects. *JDDG: Journal der Deutschen Dermatologischen Gesellschaft* 5, no. 5 (2007): 356–365. https://doi.org/10.1111/j.1610-0387.2007.06304.x
40. AlQuadeib, Bushra T., Eram K. D. Eltahir, Rana A. Banafa, and Lama A. Al-Hadhairi. Pharmaceutical evaluation of different shampoo brands in local Saudi market. *Saudi Pharmaceutical Journal* 26, no. 1 (2018): 98–106. https://doi.org/10.1016/j.jsps.2017.10.006
41. Cornwell, P. A. A review of shampoo surfactant technology: Consumer benefits, raw materials and recent developments. *International Journal of Cosmetic Science* 40, no. 1 (2018): 16 30. https://doi.org/10.1111/ics.12439
42. Dias, Maria Fernanda Reis Gavazzoni. Hair cosmetics: An overview. *International Journal of Trichology* 7, no. 1 (2015): 2. https://doi.org/10.4103%2F0974-7753.153450
43. Manayi, A., and Saeidnia, S. "Cosmetics and personal care products." In *Encyclopedia of toxicology: Third Edition* (Vol. 1), (Editor: Wexler, P). (2014) Elsevier. https://doi.org/10.1016/B978-0-12-386454-3.00979-9
44. Hahladakis, John N., Costas A. Velis, Roland Weber, Eleni Iacovidou, and Phil Purnell. An overview of chemical additives present in plastics: Migration, release, fate and environmental impact during their use, disposal and recycling. *Journal of Hazardous Materials* 344 (2018): 179–199. https://doi.org/10.1016/j.jhazmat.2017.10.014
45. Kang, Seok-Young, Ji-Young Um, Bo-Young Chung, So-Yeon Lee, Jin-Seo Park, Jin-Cheol Kim, Chun-Wook Park, and Hye-One Kim. Moisturizer in patients with inflammatory skin diseases. *Medicina* 58, no. 7 (2022): 888. https://doi.org/10.3390/medicina58070888
46. Gabros, Sarah, Trevor A. Nessel, and Patrick M. Zito. Sunscreens and photoprotection. (2021): 202. StatPearls Publishing. PMID: 30725849. Bookshelf ID: NBK537164.
47. Savary, G., M. Grisel, and C. Picard. Cosmetics and personal care products. natural polymers: Industry techniques and applications. *Natural Polymers Industry Techniques and Applications* (2015): 219–261. https://dx.doi.org/10.1007/978-3-319-26414-1_8
48. Sharmeen, Jugreet B., Fawzi M. Mahomoodally, Gokhan Zengin, and Filippo Maggi. Essential oils as natural sources of fragrance compounds for cosmetics and cosmeceuticals. *Molecules* 26, no. 3 (2021): 666. https://doi.org/10.3390/molecules26030666
49. Abrutyn, Eric S. "Editor: Zoe Diana Draelos. Antiperspirants and Deodorants. In *Cosmetic dermatology: Products and procedures* (2022): 213–222. Publisher: Blackwell Publishing Ltd. http://dx.doi.org/10.1002/9781444317657.ch20
50. Colijn, Ivanna, Anthony Ash, Marie Dufauret, Céline Loussert-Fonta, Martin E. Leser, Peter J. Wilde, and Tim J. Wooster. Colloidal dynamics of emulsion droplets in mouth. *Journal of Colloid and Interface Science* 620 (2022): 153–167. https://doi.org/10.1016/j.jcis.2022.03.117

51. Yurdasiper, Aysu, Aslı Şahiner, and Evren Homan Gökçe. Preparation of thermoresponsive triclosan poly (N-isopropylacrylamide) nanogels and evaluation of antibacterial efficacy on Cutibacterium acnes. *Journal of Drug Delivery Science and Technology* 76 (2022): 103734. https://doi.org/10.1016/j.jddst.2022.103734
52. Yazdani, Mazyar, Katja Benedikte Prestø Elgstøen, and Tor Paaske Utheim. Eye make-up products and dry eye disease: A mini review. *Current Eye Research* 47, no. 1 (2022): 1–11. https://doi.org/10.1080/02713683.2021.1966476
53. Mawazi, Saeid Mezail, Nurul Aqilah Binti Azreen Redzal, Noordin Othman, and Sultan Othman Alolayan. Lipsticks history, formulations, and production: A narrative review. *Cosmetics* 9, no. 1 (2022): 25. https://doi.org/10.3390/cosmetics9010025
54. Heidler, Jochen, and Rolf U. Halden. Mass balance assessment of triclosan removal during conventional sewage treatment. *Chemosphere* 66, no. 2 (2007): 362–369. https://doi.org/10.1016/j.chemosphere.2006.04.066
55. Carey, Daniel E., and Patrick J. McNamara. The impact of triclosan on the spread of antibiotic resistance in the environment. *Frontiers in Microbiology* 5 (2015): 780. https://doi.org/10.3389/fmicb.2014.00780
56. Jing, Jane Lee Jia, Thong Pei Yi, Rajendran J. C. Bose, Jason R. McCarthy, Nagendran Tharmalingam, and Thiagarajan Madheswaran. Hand sanitizers: A review on formulation aspects, adverse effects, and regulations. *International Journal of Environmental Research and Public Health* 17, no. 9 (2020): 3326. https://doi.org/10.3390/ijerph17093326
57. van Dijk, Harrie F. G., and Henri A. Verbrugh. Resisting disinfectants. *Communications Medicine* 2, no. 1 (2022): 6. https://doi.org/10.1038/s43856-021-00070-8
58. www.healthline.com/health/beautyskincare/chemicalexfoliation#product-recomendations (Access on July 2023)

8 Potential Ecological Impacts of Pesticides and Herbicides on the One Health System

Palashpriya Das, Sakshi Ranjan, and Ulrich Vasconcelos

8.1 INTRODUCTION

Since the dawn of civilization, agricultural activities have been the driver of human life, and various technologies have been implemented to enhance productivity with the aim of attaining United Nations Sustainable Development Goal 2, i.e. SDG-2, zero hunger. But with the introduction of one such practice in this sector, namely the application of pesticides, environmental degradation was initiated, which became an environmental concern. In order to make this process sustainable, the concept of One Health entered the limelight. This concept addresses the intimate link between the health of humans, animals and the environment and aligns them with modern development in a collaborative interdisciplinary manner (Zhang et al., 2024). Implementation of this approach not only empowers the scientific community to have an understanding of global health issues and strategize to combat the same but also to sustainably balance and optimize health.

8.2 CONCEPT OF ONE HEALTH

The approach of One Health has been practiced for centuries, well before its official naming. The term has been well known for several decades but under various names (Pettan-Brewer et al., 2024; Atlas, 2013). The approach of One Health (OH) suggests that the health of humans, wild animals, domestic animals and ecosystems are in an alignment. In order to be in good shape, researchers and scholars across disciplines and sectors need to join hand in hand to address threats to the ecosystem and to promote sustainable development. For implementing the OH approach, organizations like the World Health Organization, the World Organisation for Animal Health, the Food and Agriculture Organization, the World Bank and UNICEF have developed a pentapartite strategy. Capacity building is needed for future generations for which decision-making bodies must invest in the direction of One Health research and development.

DOI: 10.1201/9781032623801-8

8.3 BRIEF DETAILS OF PESTICIDES AND HERBICIDES

Any living organism, whether a bacterium, fungus, insect or nematode, which disturbs the well-being of humans or animals is known as a pest, and the substances used to destroy pests are referred to as pesticides. The term "pesticide" encompasses a multitude of chemicals, namely, herbicides that destroy undesired weeds; insecticides that kill insects; fungicides that annihilate fungi; rodenticides that kill rodents like rats; avicides that control the population of birds; and so on. Pesticides increase agricultural productivity. They are a saviour in a scenario where a substantial reduction in crop production can occur due to infestation of pests, insects and diseases. They play a vital role in plant protection, in augmentation of yield and in the security of food crops, ensuring their quality as well as quantity. Organophosphorus pesticides make up 40% of the world's pesticides. The most commonly used pesticides are herbicides with a targeted application on plants that compete with desired crops. They are very effective and economical, and hence are a major tool in places where no-till practices are implemented. Despite their positive roles, however, the concomitant effect on non-target organisms is a matter of concern. In fact, herbicides are so hazardous that they have been identified as one of the topmost priority chemicals of environmental concern. These compounds can exert acute or chronic toxicity. If transient exposure to a pesticide causes a marked poisonous result, the toxicity of the pesticide is said to be highly acute. In contrast, if a pesticide exerts its toxic effects over a considerable period of time, the toxicity of the pesticide is said to be chronic. It is quite alarming that pesticide residues have been detected in human breast milk samples for the first time in India, indicating a sharp rise in risk for individuals in urban areas as compared to those in semi-urban agricultural areas (Anand et al., 2021). Various pesticide residues were also detected in blood and urine samples from people in the US (Karwal et al., 2022). In fact, the application of synthetic pesticides does not always ensure a marked reduction in pest densities; rather, the pesticides kill them and in so doing, these chemicals reinvigorate the pests (Janssen and van Rijn, 2021).

The strategy used for the classification of pesticides involves looking at their chemical constitution, the target organism they kill, and their path of entry into the host. Systemic pesticides are those which are ingested and can spread to different systems of the body, while non-systemic pesticides are those which require an intimate contact to exert their action. Farmers and gardeners are most vulnerable to the action of pesticides because they directly handle these compounds. Their exposure route is not only through dermal contact, but they may also inhale these toxic compounds, which can lead to respiratory problems. They may mistakenly consume food without cleaning their hands, thereby poisoning themselves, which might be the genesis of intestinal disturbances. Moreover, people who are not directly involved in farming activities but dwell at nearby farms, as well as those living far away from these areas, are also exposed through agricultural merchandise and pesticide-contaminated water.

Toxicokinetic or toxicodynamic interactions of pesticides might take place in a mixture of pesticides, and this might result in higher or lower levels of toxicity as compared to the additive effect of the individual components. Due to long-term application of pesticides, pests often become resistant to the action of these substances. Hence, higher doses of the same pesticide or pesticides with higher toxic effects need to be applied in order to obtain the same effect, thereby directly affecting the environment through incorporation of increased quantities of these chemicals (Demirak and Canpolat, 2022).

8.4 IMPACT OF PESTICIDES ON HUMAN BEINGS

8.4.1 Exposure Routes and Risk of Exposure

One of the major pesticide exposure routes of human beings is through inhalation, whether direct occupational or even direct non-occupational. The occupational exposure of humans to pesticides is most dangerous because it is linked to heightened risk of pulmonary disease. This exposure intensifies if pesticides are used in places with no ventilation and if the temperature is high because at elevated temperatures, a pesticide can volatilize and will soon saturate the closed area, resulting in coughing, sneezing and respiratory irritation. When growing crops, the farmers are actually exposed to a mixture of pesticides, and hence effects can range from skin inflammation and respiratory problems to the development of male sterility and even cancer (Plaat et al., 2018). Farmers as well as gardeners who employ pesticides without personal protective equipment can experience several undesired complications. They may suffer from skin problems like itching and skin reddening, eye problems, breathing issues like gasping for breath, nose burning, dry throat, headache and digestive issues.

Children are more sensitive than adults as they are delicate and their body is still in a developing state. Pesticides applied indoors may enter their systems through oral, dermal or nasal routes. Due to pesticide exposure, the tissues of their nose, throat and lungs may be affected and they may develop asthmatic symptoms, cough, a respiratory tract infection and even lung cancer (Tarmure et al., 2020). Passive exposure of children can occur through visiting places located close to pesticide-application sites. It is also quite obvious that children will inevitably be exposed to pesticides at least passively when they come in contact with their parents who are somehow engaged in handling pesticides. Pregnant women tend to be at high risk as well, because even mild maternal alterations may affect foetal neurological development. Foetal death or abortion might also take place. Prenatal exposure can also lead to lower IQ, lower perceptual reasoning skills and poor memory in children (Engel et al., 2011).

8.4.2 Effect of Pesticides on Different Organs of the Human Body

Pesticides can have adverse effects on the whole human body because exposure to pesticides results in excessive reactive oxygen production, damage to DNA, disruption in balance between pro- and anti-oxidant molecules, reduced immunity, disturbance in hormone balance and reduction in intelligence. A major organ of the human body that is affected by pesticides is the eyes, because ocular tissues absorb pesticides and get damaged much faster compared to other tissues of the body. This can also be direct occupational or direct non-occupational. Symptoms begin with itching continuing up to corneal damage, increased incidence of myopia, constriction of pupils and finally to loss of vision which might occur (Kempuraj et al., 2023).

Pesticides can be absorbed through the skin as well and may exert toxicity on nails and hair. Changes in colour or growth pattern of nails have been observed due to exposure to pesticides. Pesticides can also get introduced into the hair shaft. There are blood vessels at the root of the hair follicles, and thus there is an exchange of substances between the hair and other body parts. It is interesting to note that either through occupational exposure or through the handling of pesticide-treated objects like agricultural produce, pesticides accumulated in the hair of pregnant women, or effects were observed on newborn birth parameters like birth weight and length (Béranger et al., 2020).

Pesticide exposure might also lead to decreased fertility due to sperm DNA damage, resulting in problems in reproduction. Several malignancies like cancer of the head, neck, breast, thyroid, brain, colon, rectum, pancreas, ovaries and so on might even occur as reported in various studies. Pesticide exposure and leukaemia development have been studied and found to be linked. Exposure to pesticides has also been reported to make people vulnerable to cardiovascular disease (Ahmad et al., 2024). Neurological problems might result from exposure to pesticides because these chemicals can impair brain development. Serious diseases like Alzheimer's as well as Parkinson's might also be a result of pesticide application and subsequent exposure. The liver is adversely affected through eating pesticide-contaminated food, especially fat-rich foods, which might increase the exposure level to pesticides. Consumption of such contaminated foods over prolonged periods might cause perturbation of healthy gut microbiota, leading to conditions like non-alcoholic fatty liver disease, diabetes, neurotoxicity, reproductive disorders and so forth, and also behavioural changes in the host (Yuan et al., 2019; Rives et al., 2020; Matsuzaki et al., 2023).

8.4.3 Urban versus Rural Scenario with Reference to Pesticide Exposure

In urban areas, the density of population is high in a particular area, and application of pesticides for urban agriculture presents a serious threat to the environment. Homeowners often use pesticides when growing fruits and vegetables in their household gardens, and when they pick or consume these products, residues of pesticides may be ingested. These pesticides often volatilize or get washed off, and might then contaminate air and water resources. It so happens that the fishes become contaminated, and when consumed by humans, health effects are observed (Bhandari et al., 2020). There are even instances when pesticides are applied to lawns, gardens and so forth for controlling weeds, fungi and various types of insects. Urban pesticides being used in increased quantities in comparison to rural agricultural pesticides, their toxic effects get enhanced and this is becoming a matter of environmental relevance. During outdoor recreational activities of children, pesticide exposure may easily occur through dermal contact, contact with eyes, inhalation or ingestion (Islam et al., 2020).

All the human health issues mentioned here that occur due to pesticide exposure fall under the One Health approach. Thus, all these findings emphasize how intricately humans are affected by pesticide application in agriculture, which in turn cascades across public and environmental health systems, thus exemplifying the core of One Health.

8.5 IMPACT OF PESTICIDES ON PLANTS AND ANIMALS

8.5.1 Impact on Plants

Pesticides are applied to enhance production, but if this is done in an indiscriminate manner for a prolonged period, they can accumulate considerably in soil. They get absorbed by the plants through their roots and hinder the absorption of essential nutrients from the soil. This results in a decreased rate of seed germination, stunted plant growth and decreased plant yield. Absorbed pesticides get translocated to various parts of the plant body, and if they are consumed by other organisms, negative effects are imparted to consumers (Sharma et al., 2019).

8.5.2 Impact on Insects

Insects, for example the honeybee, butterfly, beetle and so forth are involved in many ecosystem services like pollination, pest control, regulation of soil quality and decomposition. Honeybees not only pollinate flowers but also produce honey, which is a valuable product of the food market. The breeding season of arthropods (phylum of animal kingdom, including insects) often coincides with the crop-growing season, during which pesticides are applied to plants. Therefore, insects get directly exposed. They often ingest pesticides like malathion, which reaches their stomach, followed by absorption into the body, resulting in death. If honeybees are impacted by the application of pesticides, the apiculture industry suffers a big loss. It so happens that insects are exposed to low doses of pesticides which they bioaccumulate. In this condition of reduced vigour, they are preyed upon by predators, and with the passage of time this exerts effects on the entire ecosystem, bringing about an imbalance.

8.5.3 Impact on Animals

Beneficial organisms like earthworms, often referred to as natural tillers, aerate the soil and recycle soil nutrients, thus enhancing plant growth. Therefore, they are also called a farmer's friend. Their ability to transform organic pollutants is largely dependent on their gut microbial flora. Substantial use of pesticides for a long period of time has been reported to cause a decline in this microbial community, thus hampering the concept of sustainable agriculture. Pesticides like imidacloprid, fenitrothion, malathion, monocrotophos and phorate have been studied and found to influence physiological reactions and hence the life cycle of earthworms. Another insect that forms the basis of the silk industry, the silkworm, is also reported to be damaged by pesticides. This is because the immune function of these insects, post exposure to pesticides, is impaired due to the change in gut microbial flora. It was noted that *Staphylococcus* sp. became the major member of the microbial population in the insect gut, replacing genera like *Methylobacterium* and *Aurantimonadaceae*. Interestingly, antimicrobial peptide formation was also blocked, thus making the insects more susceptible to pesticides (Houda et al., 2021; Li et al., 2020).

It is quite alarming that pesticide application locations are found to overlap with wild animals' activity areas because of the expansion of the zones of agrofarming activities. It was noted in a study that gradual accumulation of pesticide residues presented a perilous situation for the survival of leopard cat, an endangered animal (Jing-Yu et al., 2020). Pesticides may also affect non-target animals like pets through the inhalation of pesticide mists.

Thus, all these findings emphasize how intricately pesticide application in agriculture has cascading effects not only on plants but also on insects and animals, hence exemplifying the core of One Health.

8.6 IMPACT OF PESTICIDES ON THE ENVIRONMENT

8.6.1 Effect of Pesticides on the Aquatic Environment

Pesticides, after they are exposed to the environment, often enter adjoining water bodies either through atmospheric precipitation or release from chemical and pesticide manufacturing industries, which deteriorates the quality of the adjoining aquatic habitat, making it unhygienic for the inhabitants. Pesticides can increase nutrient availability in the aquatic environment, support algal blooms, decrease the biological oxygen demand and deteriorate water quality. Co-exposure to pesticides along with increased nutrient levels or elevated temperature can synergistically affect algal natural defence and community organization of algae (Xiang-Wu et al., 2024; Elersek et al., 2021).

Due to the reduction in the number of phytoplankton and zooplankton, small fish in the aquatic environment get deprived of their food sources. Apart from this, the oxygen content of the water body also decreases due to pesticide contamination, posing a great threat to water dwellers. Fish ingests pesticides, and though a small portion can be regulated by metabolic activities, a major part bioaccumulates. These chemicals get hoarded in their gills, skin and alimentary canal. Evidence suggests that pesticide exposure in fish has an impact on the neurological system; can promote the generation of free radicals, which have an impact on their body defence strategies, thereby increasing their susceptibility to diseases; promotes eggshell weakness; and exhibits teratogenic and intergenerational effects (Tahir et al., 2021; Raibeemol and Chitra, 2020; Salbego et al., 2020).

Filter feeders like oysters filter large volumes of water, thereby retaining pesticide residues in their tissues. Hence, they can act as biological indicators to give an estimation of pesticide contamination in the water body from where they were harvested (Vera-Candioti et al., 2021). Pesticide-laden water is often used for agricultural purposes, and therefore these residues may get incorporated into food products, which can be threatening to food safety (Islam et al., 2020). Pesticides accumulate in stormwater ponds in urban areas, and these residues get bioaccumulated in non-target fauna like frogs and bats (Julie et al., 2022; Ozgur et al., 2021). Pesticides can be lethal to the larvae of various arthropods because this stage of the arthropod life cycle is spent in water. However, if they survive and emerge as adults, they become greatly toxic to their predators with the accumulated high concentrations of pesticides within them (Alexis et al., 2023). Coral reefs, the most vital marine ecosystem, are highly receptive to pesticide accumulation in the marine environment. Pesticide-contaminated water affects the growth rates of algae, poses problems for photosynthesis, and even causes the death of the algal population (Creusot et al., 2022).

8.6.2 Effect of Pesticides on the Terrestrial Environment

Soil microbiome plays an indispensable role in biogeochemical cycles and the breakdown of organic compounds, thereby assisting crop growth. Any perturbation in the number and structure of the corresponding microbes affects the ecosystem. Large-scale application of pesticides badly impacts the inhabitant microbial communities in the soil. Certain pesticides have been reported to affect the symbiotic and free-living nitrogen-fixing bacteria in the soil. These bacteria promote the addition of nitrogenous compounds into soil, thus enhancing fertility. There has been a precipitous decrease in the population of microorganisms due to non-judicious pesticide application for increasing crop production. Hence, if this bacterial population is endangered, there will be a decline in soil fertility, requiring the addition of fertilizers, which definitely augments plant growth but in the long run affects soil health and also imperils human health (Sharma et al., 2019; Yengkokpam and Mazumder, 2020). Farmers currently use a mixture of pesticides, which poses a greater risk to soil microbiota than the single pesticide used previously. Excess use of pesticides leads to contamination of air, surface and groundwater (Figure 8.1). It also leads to the loss of rotational crops

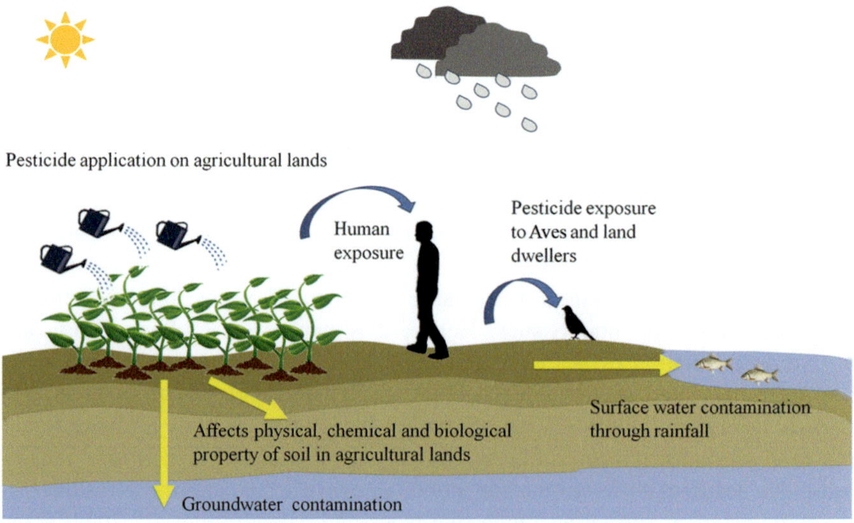

FIGURE 8.1 Impact of pesticides on environment. (Source: Rajak et al., 2023, CC-BY 4.0 license.)

and promotes persistence in the environment (Egbe et al., 2021; Mu et al., 2023; Nitin et al., 2023).

Hence, all the issues mentioned here, as related to the effect of pesticides on the environment, focus on the intensification of sustainable agriculture. This means that agricultural productivity has to be invariably attained but not at the cost of the environment, thereby exemplifying the core of One Health.

8.7 IMPACT OF HERBICIDES ON HUMAN BEINGS, PLANTS, ANIMALS AND THE ENVIRONMENT

8.7.1 Impact of Herbicides on Human Beings

Herbicides exert acute to chronic toxicities on human health, and the effects range from skin irritation and blurred vision, to body dehydration, difficulty in breathing, nausea, waist pains and eventual death (Ghazi et al., 2023). The herbicide glyphosate is administered most commonly and was initially considered safe for humans. Later, it was reported to affect human health significantly and has been grouped as a "probable human carcinogen" (Peillex and Pelletier, 2020; Galli et al., 2024). Glyphosate was found to enter the body mostly through consumption of water and agricultural products. It was found to interfere with the production and action of the feminizing hormone oestrogen (Gasnier et al., 2009). This herbicide chelates metals like manganese, which influences sperm motility, and also influences the process of sperm production, therefore leading to reproductive issues. It may cause neurological issues and has been reported to enhance the risk of breast cancer (Rajwinder et al., 2024; Islam et al., 2018). Reports have indicated the ability of herbicides to induce problems related to reproduction, birth defects, oxidative stress and damage to the liver, especially if there is an incidence of exposure during childhood (Eskenazi et al., 2023; Pathak and Dixit, 2011).

Another herbicide in frequent use is atrazine, which finds its way into the human body through ingestion or inhalation. Following ingestion, various organs of the alimentary system like the salivary glands, the liver and various parts of the small intestine might be severely damaged. Other sites of damage as noted through mammalian studies include the endocrine system and various portions of the central nervous system. Impairment of the kidneys along with the increased incidence of heart attacks might also occur (Das et al., 2023).

8.7.2 Influence of Herbicides on Plants

Herbicide applications alter plant communities because such plants are often affected which are not the exactly targeted ones, and thus the biodiversity of a particular region is greatly disturbed. Herbicides inhibit production of plant pigments, hormones and major metabolites like amino acids, fatty acids and so on. They influence photosynthetic machinery, wither leaves, cause chromosomal aberrations and weaken plant health, thereby making them inefficient in fighting against pathogenic attacks (Feng et al., 2023; Hashimi et al., 2020). One study revealed that increased herbicide application affected many characteristics like plant biomass, total arbuscular myccorhizal fungi colonization, nodule mass and phosphorous content (Carriquiry et al., 2024).

8.7.3 Influence of Herbicides on Terrestrial and Aquatic Animals

Apart from plants, various animals like earthworms, fish, bees, beetles and butterflies are affected by herbicides. Earthworms enrich soil fertility and promote nutrient recycling. Herbicides are reported to harm their cells and DNA, and also reduce their reproduction. The gut bacterial diversity of earthworms was also reported to become altered with an increase in bacterial species like *Comamonas testosteroni*, known to degrade organic pollutants (Gaupp-Berghausen et al., 2015; Iteb et al., 2024). Chickens often roam and peck in fields where they might ingest herbicide-contaminated particles. Hence, these birds often face gut-related problems. There can be reproductive issues in the exposed chickens and they can die as well. Herbicide residues might remain on plant parts like flowers when sprayed. Nectar-seeking insects like bees and even birds like the hummingbird get exposed to herbicides when they visit flowers. This view is substantiated by the detection of herbicide residues in beehives. However, bee foragers while collecting honey from beehives may also unintentionally introduce herbicide residues. It has been reported that exposing bees to glyphosate changes their behaviour, alters gut community composition and makes them susceptible to pathogen attack. Beetles have been reported to be affected by triazine herbicides when exposed during sexual maturation. A few insects feed upon weeds, which can be damaged by the overuse of herbicides, thereby leading to a shortage of food for the insects. Indiscriminate use of herbicides may have a significant ecological impact on some insects due to inhibition of the caterpillar stage of their life cycles (Erick et al., 2018; Naccarato et al., 2023; Mallick et al., 2023). A few insects secrete pheromones to attract their mates to begin their reproduction process, and herbicides have been reported to affect this process too. Often prey–predator relationships are also affected (Wrinn et al., 2012). It is very interesting to note how the microbiome in wasps gets modified upon exposure to the herbicide atrazine. Wasps, exposed to atrazine at concentrations not toxic to them, gradually developed an atrazine-resistant microbial population in their guts. Few bacteria in their microbiome metabolized this toxic herbicide, and they also passed on this trend to their offspring, which become resistant to the herbicide without even being exposed to the same compound.

Herbicides inhibit vital metabolic enzymes of fishes, affect the electron transport system, and generate free radicals which can oxidize lipids, proteins and DNA (Das et al., 2023). These compounds were also reported to reduce fertility in annual fish, disturb oxidative balance and promote histopathological damage (Reichert et al., 2022; Lopes et al., 2022). Fishes get maximum exposure to herbicides through the application of these compounds into water bodies to control aquatic weeds, especially in rice fields. One study indicated that this scenario poses a dreadful situation to the native fish population, making them anaemic and causing certain alterations in blood

(Uchenna et al., 2022). Due to exposure to herbicides, malformation in amphibians was reported. Upon 2.4-D exposure, the central nervous system of a few tadpoles was reported to be affected (Pavan et al., 2021; Gaaied et al., 2020).

8.7.4 Influence of Herbicides on Aquatic Environment

Herbicides applied to agricultural fields often get washed away from land into water. This is lethal to aquatic vegetation, as a result of which the oxygen level drops in water. Aquatic plants often act as food for fish, and their reduction may result in fish death. Also, the drop in the oxygen level might lead to fish death or impairment in their reproduction capacities (Helfrich et al., 2009). A decrease in the fish population can be problematic as fish are a good nutrient source for humans. It may be likely that fishes do not die but accumulate toxic compounds within their bodies and after being consumed by humans, the herbicide residues wreak havoc on the human physiological system. In an aquatic environment, apart from herbicides, heavy metals are present, and their synergistic effect may create a dangerous situation for the aquatic inhabitants. Often, a mixture of herbicides and other compounds present in the water body creates a toxic environment for a non-target species.

Herbicides target algal photosynthetic machinery and reduce photosynthesis by either blocking chlorophyll formation or blocking photosystems, thereby disrupting light harvesting and subsequent electron transport (Cao et al., 2023; Machado et al., 2021; Du et al., 2023; Monika et al., 2021). Herbicides like atrazine and triazine, when released into the aquatic environment, harm animals, leading to severe ecological damage, and also reduce the abundance of the toxic dinoflagellate *Alexandrium minutum*. Due to the decrease in the algal population, massive algal bloom formation is blocked, but saxitoxin gene expression is enhanced (Eyal and Barak, 2023; Chen et al., 2022; Zhang et al., 2022). A herbicide like glyphosate, due to land run-off, can result in an adverse situation for microalgae, aquatic bacteria and freshwater sediment microbial communities (Gaupp-Berghausen et al., 2015). Certain species metabolize this pollutant and accumulate it in their bodies, which has implications for the whole ecosystem. This herbicide can change the algal community composition and nutrient dynamics by favouring certain cyanobacteria like *Planktothrix* sp. while adversely affecting species like *Microcystis* sp., thereby suggesting that the phytoplankton can use glyphosate as a nutrient (Saxton et al., 2011). Another herbicide, metolachlor (MET), negatively affects non-target aquatic organisms like fish, crustaceans and phytoplanktons (Maazouzi et al., 2016). It impacts chytrid fungi, which parasitize bloom-forming cyanobacteria, inducing an anoxic state affecting all aquatic organisms. Thus, MET might have an impact on the trophic interactions in aquatic ecosystems (Martínez-Ruiz et al., 2024).

8.7.5 Influence of Herbicides on Terrestrial Environment

Herbicides, after application, infiltrate surface and groundwater, thereby contaminating drinking water sources (Guillon et al., 2019; Majewski et al., 2014). Continued usage leads to their persistence in the ecosystem, thereby affecting soil ecosystems, soil fertility, causing alterations in diversity, abundance and movement of the microbial population (Figure 8.2); hence affecting fundamental ecological relationships (Ruuskanen et al., 2023).

Application of the herbicide glyphosate can cause non-target toxicity to soil flora and fauna and promote antimicrobial resistance in soil bacteria through the overexpression of antibiotic-resistant genes. One of the several mechanisms bacteria use to evade antibiotics is efflux pump, and overactivity of these efflux pumps has been observed due to herbicide treatment. It is highly alarming that abundance in antibiotic-resistant genes is correlated with mobile genetic elements like transposons and plasmids, which may be taken up by different bacterial population in the soil through the process of conjugation, resulting in further worsening the menace of antimicrobial resistance. Herbicides may also affect ecosystem functions mediated by other microbes (Liao et al., 2021; Xiao et al. 2023).

Therefore, all the issues mentioned here regarding the various implications of herbicides for humans, plants, animals and the environment associated with all of them, are considered within the One Health domain. Adoption of a One Health

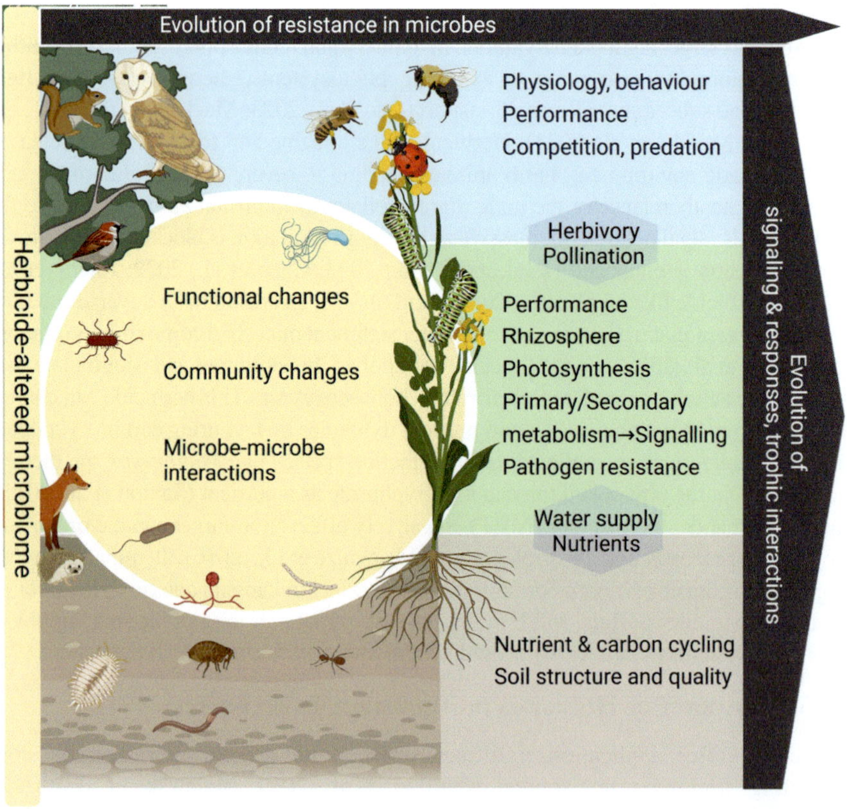

FIGURE 8.2 Impact of herbicides on plants and environment. (Source: Ruuskanen et al., 2023, CC-BY 4.0 license.)

strategy across various sectors can help in the mitigation of problems caused by the overuse of pesticides and herbicides, thereby leading to practicing agricultural activities in a sustainable manner.

8.8 CONCLUSION

The need of the hour is a sustainable increase in crop production for enhanced food security through the utilization of better technologies as permitted by the One Health concept, with the aim to attaining SDG 3, that is, Sustainable Development Goal 3, good health and well-being. Researchers across diverse fields related to One Health need to collaborate to promote a synergistic relationship across various sectors and develop strategies in line with One Health, for example, plant health clinics established in Bangladesh, Bolivia, Nicaragua and Uganda. Farmers and gardeners should be made aware of the pros and cons of pesticide and herbicide usage. They must learn how to apply these plant growth-promoting chemicals in a judicious manner so as not to contaminate the surrounding environment through run-off from the application sites. International organizations need to work in close association with different countries and different sectors like healthcare professionals, farmers and so on, and develop a joint plan of action for the proper implementation of technologies for sustainable development.

REFERENCES

Ahmad, M.F., Ahmad, F.A., Alsayegh, A.A., Zeyaullah, M., AlShahrani, A.M., Muzammil, K., Saati, A.A., Wahab, S., Elbendary, E.Y., Kambal, N., Abdelrahman, M.H., Hussain, S. 2024. Pesticides impacts on human health and the environment with their mechanisms of action and possible countermeasures. *Heliyon* 10: e29128.

Alexis, P.R., Maike, H., Martin, H.E., Ralf, S. 2023. Aquatic-terrestrial transfer of neonicotinoid insecticides in riparian food webs. *J. Haz. Mat.* 455: 131635.

Anand, N., Chakraborty, P., Ray, S. 2021. Human exposure to organochlorine, pyrethroid and neonicotinoid pesticides: Comparison between urban and semi-urban regions of India. *Env. Poll.* 270: 116156.

Atlas, R.M. 2013. One Health: Its origins and future. *Curr. Top. Microbiol. Immunol.* 365: 1–13.

Béranger, R., Hardy, E.M., Binter, A.C., Charles, M.A., Zaros, C., Appenzeller, B.M.R., Chevrier, C. 2020. Multiple pesticides in mothers' hair samples and children's measurements at birth: Results from the French national birth cohort (ELFE). *Int. J. Hygiene Env. Health.* 223: 22–33.

Bhandari, G., Kishor, A., Paul, T.J., Scheepers, V.G. 2020. Concentration and distribution of pesticide residues in soil: Non-dietary human health risk assessment. *Chemosphere.* 253: 126594.

Cao, L., Halpern, B.S., Troell, M., Short, R., Zeng, C., Jiang, Z., Liu, Y., Zou, C., Liu, C., Liu, S., Liu, X., Cheung, W.W.L., Cottrell, R.S., DeClerck, F., Gelcich, S., Gephart, J.A., Godo-Solo, D., Kaull, J.I., Micheli, F., Naylor, R.L., Payne, H.J., Selig, E.R., Sumaila, U.R., Tigchelaar, M. 2023. Vulnerability of blue foods to human-induced environmental change. *Nat. Sustain.* 6: 1186–1198.

Carriquiry, I.G., Silva, V., Raevel, F., Harkes, P., Osman, R., Bentancur, O., Fernandez, G., Geissen, V. 2024. Effects of mixtures of herbicides on nutrient cycling and plant support considering current agriculture practices, *Chemosphere.* 349: 140925.

Chen, M., Jin, X., Liu, Y., Guo, L., Ma, Y., Guo, C., Wang, F., Xu, J. 2022. Human activities induce potential aquatic threats of micropollutants in Danjiangkou Reservoir, the largest artificial freshwater lake in Asia. *Sci. Total Env.* 850: 157843

Creusot, N., Chaumet, B., Eon, M. 2022. Metabolomics insight into the influence of environmental factors in responses of freshwater biofilms to the model herbicide diuron. *Env. Sc. Poll. Res.* 29: 29332–29347.

Das, S., Sakr, H., Al-Huseini, I., Jetti, R., Al-Qasmi, S., Sugavasi, R., Sirasanagandla, S.R. 2023. Atrazine toxicity: The possible role of natural products for effective treatment. *Plants (Basel)*. 12: 2278.

Demirak, M.S.S., Canpolat, E. 2022. Plant-based bioinsecticides for mosquito control: Impact on insecticide resistance and disease transmission. *Insects.* 13: 162.

Du, J., Izquierdo, D., Xu, H.F., Beisner, B., Lavaud, J., Ohlund, L., Sleno, L., Juneau, P. 2023. Responses to herbicides of Arctic and temperate microalgae grown under different light intensities. *Env. Poll.* 333: 121985.

Egbe, C.C., Oyetibo, G.O., Ilori, M.O. 2021. Ecological impact of organochlorine pesticides consortium on autochthonous microbial community in agricultural soil. *Ecotoxicol. Env. Saf.* 207: 111319.

Elersek, T., Notersberg, T., Kovačič, A. 2021. The effects of bisphenol A, F and their mixture on algal and cyanobacterial growth: From additivity to antagonism. *Env. Sc. Pollut. Res.* 28: 3445–3454.

Engel, S.M., Wetmur, J., Chen, J., Zhu, C., Barr, D.B., Canfield, R.L., Wolff, M.S. 2011. Prenatal exposure to organophosphates, paraoxonase 1 and cognitive development in childhood. *Env. Health Perspect.* 119: 1182–1188.

Erick, V.S., Motta, K.R., Nancy, A.M. 2018. Glyphosate perturbs the gut microbiota of honey bees. *Proc. Nat. Ac. Sci.* 115: 10305–10310.

Eskenazi, B., Gunier, R.B., Rauch, S., Kogut, K., Perito, E.R., Mendez, X., Limbach, C., Holland, N., Bradman, A., Harley, K.G., Mills, P.J., Mora, A.M. 2023. Association of lifetime exposure to glyphosate and aminomethylphosphonic acid with liver inflammation and metabolic syndrome at young adulthood: Findings from the CHAMACOS Study. *Env. Health Pers.* 131: 037001.

Eyal, R., Barak, H. 2023. Impact of combined seawater warming and triazine-type herbicide pollution on the physiology and potential toxicity of the dinoflagellate Alexandrium minutum. *Mar. Poll. Bull.* 196: 115612.

Feng, W., Wang, T., Zhu, Y., Sun, F., Giesy, J.P., Wu, F. 2023. Chemical composition, sources and ecological effect of organic phosphorus in water ecosystems: A review. *Carbon Res.* 2: 12.

Gaaied, S., Oliveira, M., Domingues, I., Banni, M. 2020. 2,4-Dichlorophenoxyacetic acid herbicide effects on zebrafish larvae: Development, neurotransmission and behavior as sensitive endpoints. *Env. Sc. Poll. Res.* 27: 3686–3696.

Galli, F.S., Mollari, M., Tassinari, V., Alimonti, C., Ubaldi, A., Cuva, C., Marcoccia, D. 2024. Overview of human health effects related to glyphosate exposure. *Front. Toxicol.* 6: 1474792.

Gasnier, C., Dumont, C., Benachour, N., Clair, E., Chagnon, M.C., Séralini, G.E. 2009. Glyphosate-based herbicides are toxic and endocrine disruptors in human cell lines. *Toxicol.* 262: 184–191.

Gaupp-Berghausen, M., Hofer, M., Rewald, B., Zaller, J.G. 2015. Glyphosate-based herbicides reduce the activity and reproduction of earthworms and lead to increased soil nutrient concentrations. *Sci. Rep.* 5: 12886.

Ghazi, R.M., Yusoff, N.R.N., Halim, N.S.A., Wahab, I.R.A., Latif, N.A., Hasmoni, S.H., Zaini, M.A.A., Zakaria, Z.A. 2023. Health effects of herbicides and its current removal strategies. *Bioengineered.* 14: 2259526.

Guillon, A., Videloup, C., Leroux, C., Bertin, H., Philibert, M., Baudin, I., Bruchet, A., Esperanza, M. 2019. Occurrence and fate of 27 triazines and metabolites within French drinking water treatment plants. *Water Supply*. 19: 463–471.

Hashimi, M.H., Hashimi, R., Ryan, Q. 2020. Toxic effects of pesticides on humans, plants, animals, pollinators and beneficial organisms. *Asian Plant Res. J.* 5: 37–47.

Helfrich, L.A., Weigmann, D.L., Hipkins, P.A., Stinson, E.R. 2009. Pesticides and aquatic animals: A guide to reducing impacts on aquatic systems. *Virginia Tech Pub*. 420-013: 1–24.

Houda, B., Basma, K., Rabeh, R., Kaouter, H. 2021. Toxicological and histopathological impact of insecticide acetamiprid and agricultural fertilizer activeg on earthworm (*Aporrectodea giardi*). *Appl. Biol. Res*. 23: 341–348.

Islam, F., Wang, J., Farooq, M.A., Khan, M.S.S., Xu, L., Zhu, J., Zhao, M., Muños, S., Li, Q.X., Zhou, W. 2018. Potential impact of the herbicide 2, 4-dichlorophenoxyacetic acid on human and ecosystems. *Env. Int*. 111: 332–351.

Islam, M.M., Kadiyala, V., Rajarathnam, D., Prasath, A., Mallavarapu, M. 2020. Pesticides in the urban environment: A potential threat that knocks at the door. *Sci. Total Env*. 711: 134612.

Iteb, B., Filippo, V., Leilei, Z., Francesca, B., Begona, MM, Omayma, M., Sabrine, H., Marouane, M., Luigi, L., Edoardo, P., Mohamed, B. 2024. Co-exposure to environmental microplastic and the pesticide 2,4-dichlorophenoxyacetic acid induce distinctive alterations in the metabolome and microbial community structure in the gut of the earthworm *Eisenia andrei*. *Env. Poll*. 344: 123213.

Janssen, A., van Rijn, P.C.J. 2021. Pesticides do not significantly reduce arthropod pest densities in the presence of natural enemies. *Ecol. Lett*. 24: 2010–2024.

Jing-Yu, L., Chihhao, F., Ya-Zhen, H., Kurtis, J.C.P. 2020. Distribution of residual agricultural pesticides and their impact assessment on the survival of an endangered species. *J. Haz. Mat*. 389: 121871.

Julie, C.B., María, J.D., Dante, E.R., Diego, C., Claudia, V., María, B.P., Adrián, E.A. 2022. Concentration of current-use pesticides in frogs from the Pampa region and correlation of a mixture toxicity index with biological effects. *Env. Res*. 204: 112354.

Karwal, P., Mittal, P., Nagar, G., Singh, A., Singh, I.K. 2022. Effects of pesticides on human physiology, genetics, and evolution. In Sarma, H., Dominguez, D.C., Lee, W.Y. (eds), *Emerging Contaminants in the Environment* (pp. 287–310). Elsevier.

Kempuraj, D., Zhang, E., Gupta, S., Gupta, R.C., Sinha, N.R., Mohan, R.R. 2023. Carbofuran pesticide toxicity to the eye. *Exp. Eye Res*. 227: 109355.

Li, F., Li, M., Mao, T., Wang, H., Chen, J., Lu, Z., Qu, J., Fang, Y., Gu, Z., Li, B. 2020. Effects of phoxim exposure on gut microbial composition in the silkworm, *Bombyx mori*. *Ecotoxicol. Env. Saf*. 189: 110011.

Liao, H., Li, X., Yang, Q., Bai, Y., Cui, P., Wen, C., Liu, C., Chen, Z., Tang, J., Che, J., Yu, Z., Geisen, S., Zhou, S., Friman, V.P., Zhu, Y.G. 2021. Herbicide selection promotes antibiotic resistance in soil microbiomes. *Mol. Biol. Evol*. 38: 2337–2350.

Lopes, A.R., Moraes, J.S., Martins, C.M.G. 2022. Effects of the herbicide glyphosate on fish from embryos to adults: A review addressing behavior patterns and mechanisms behind them. *Aquatic Toxicol*. 25: 106281.

Maazouzi, C., Coureau, C., Piscart, C., Saplairoles, M., Baran, N., Marmonier, P. 2016. Individual and joint toxicity of the herbicide S-metolachlor and a metabolite, deethylatrazine on aquatic crustaceans: Difference between ecological groups. *Chemosphere*. 165: 118–125.

Machado, M.D., Soares, E.V. 2021. Exposure of the alga *Pseudokirchneriella subcapitata* to environmentally relevant concentrations of the herbicide metalochlor: Impact on the redox homeostasis. *Ecotoxicol. Env. Safety*. 207: 111264.

Majewski, M.S., Coupe, R.H., Foreman, W.T., Capel, P.D. 2014. Pesticides in Mississippi air and rain: A comparison between 1995 and 2007. *Env. Toxicol. Chem*. 33: 1283–1293.

Mallick, B., Rana, S., Ghosh, T.S. 2023. Role of herbicides in the decline of butterfly population and diversity. *J. Exp. Zool. A Ecol. Integr. Physiol.* 339: 346–356.

Martínez-Ruiz, E.B., Agha, R., Spahr, S., Wolinska, J. 2024. Widely used herbicide metolachlor can promote harmful bloom formation by stimulating cyanobacterial growth and driving detrimental effects on their chytrid parasites. *Env. Poll.* 344: 123437.

Matsuzaki, R., Gunnigle, E., Geissen, V., Clarke, G., Nagpal, J., Cryan, J.F. 2023. Pesticide exposure and the microbiota-gut-brain axis. *Int. Soc. Microbial Ecol. J.* 17: 1153–1166.

Monika, M., Darya, H., Wojciech, P., Agnieszka, B.R., Szymon, T., Anna, A. 2021. Does diclofenac act like a photosynthetic herbicide on green algae? *Chlamydomonas reinhardtii* synchronous culture-based study with atrazine as reference. *Ecotoxicol. Env. Safety* 208: 111630.

Mu, H., Yang, X., Wang, K., Tang, D., Xu, W., Liu, X., Ritsema, C.J., Geissen, V. 2023. Ecological risk assessment of pesticides on soil biota: An integrated field-modelling approach. *Chemosphere.* 326: 138428.

Naccarato, A., Vommaro, M.L., Amico, D., Sprovieri, F., Pirrone, N., Tagarelli, A., Giglio, A. 2023. Triazine herbicide and NPK fertilizer exposure: Accumulation of heavy metals and rare earth elements, effects on cuticle melanization, and immunocompetence in the model species *Tenebrio molitor*. *Toxics.* 11: 499.

Nitin, K.S., Gaurav, S., Manish, Y., Hirendrasinh, P., Johnson, C., Vijai, S. 2023. Fate of pesticides in agricultural runoff treatment systems: Occurrence, impacts and technological progress. *Env. Res.* 237: 117100.

Ozgur, K., Ilker, S., Tarkan, Y., Begum, Y.D., Ozcan, O., Ayhan, F. 2021. Multiresidues of environmental contaminants in bats from Turkey. *Chemosphere.* 282: 131022.

Pathak, R.K., Dikshit, A.K. 2011. Atrazine and human health. *Int. J. Ecosys.* 1: 14–23.

Pavan, F.A., Samojeden, C.G., Rutkoski, C.F., Folador, A., Da Fr´e, S.P., Müller, C., Hartmann, P.A., Hartmann, M.T. 2021. Morphological, behavioral and genotoxic effects of glyphosate and 2,4-D mixture in tadpoles of two native species of South American amphibians. *Env. Toxicol. Pharmacol.* 85: 103637.

Peillex, C., Pelletier, M. 2020. The impact and toxicity of glyphosate and glyphosate-based herbicides on health and immunity. *J. Immunotoxicol.* 17: 163–174.

Pettan-Brewer, C., Penn, G., Biondo, A.W., Jaenisch, T., Grützmacher, K., Kahn, L.H. 2024. Who coined the term "One Health"? Cooperation amid the siloization. *One Health.* 18: 100678.

Plaat, V.D.A., deJong, K., deVries, M., vanDiemen, C.C., Nedeljković, I., Amin, N. 2018. Occupational exposure to pesticides associated with differential DNA methylation. *Occup. Env. Med.* 75: 427–435.

Raibeemol, K.P., Chitra, K.C. 2020. Induction of immunological, hormonal and histological alterations after sublethal exposure of chlorpyrifos in the freshwater fish, Pseudetroplus maculatus. *Fish Shellfish Immunol.* 102: 1–12.

Rajak, P., Roy, S., Ganguly, A., Mandi, M., Dutta, A., Das, K., Nanda, S., Ghanty, S., Biswas, G. 2023. Agricultural pesticides – friends or foes to biosphere? *J. Haz. Mat. Adv.* 10: 100264.

Rajwinder, K., Diksha, C., Bali, S., Bandral, S.S., Singh, V., Ahmad, M.A., Rani, N., Singh, T.G., Chandrasekaran, B. 2024. Pesticides: An alarming detrimental to health and environment. *Sc. Total Env.* 915: 170113.

Reichert, L.M.M., de Oliveira, D.R., Papaleo, J.L., Valgas, A.A.N., Oliveira, G.T. 2022. Biochemical and body condition markers in *Rhinella icterica* tadpoles exposed to atrazine, glyphosate, and quinclorac based herbicides in ecologically relevant concentrations. *Env. Toxicol. Pharmacol.* 93: 103884.

Rives, C., Fougerat, A., Ellero-Simatos, S., Loiseau, N., Guillou, H., Gamet-Payrastre, L., Wahli, W. 2020. Oxidative stress in NAFLD: Role of nutrients and food contaminants. *Biomolecules*. 10: 1702.

Ruuskanen, S., Fuchs, B., Nissinen, R., Puigbò, P., Rainio, M., Saikkonen, K., Helander, M. 2023. Ecosystem consequences of herbicides: The role of microbiome. *Trends Ecol. Evol.* 38: 35–43.

Salbego, J., Seben, D., Sippert, L.R., Gressler, L.T., da Cunha, J.A., Zanella, R., Vaucher, R.A., Marchesan, E., Baldisserotto, B., Loro, V.L., Golombieski, J.I. 2020. Toxicological response of silver catfish (*Rhamdia quelen*) after acute exposure to a commercial insecticide containing thiamethoxam. *J. Env. Sci. Heal. B.* 55: 749–755.

Saxton, M.A., Morrow, E.A., Bourbonniere, E.A., Wilhelm, S.W. 2011. Glyphosate influence on phytoplankton community structure in Lake Erie. *J. Gt. Lakes Res*. 37: 683–690.

Sharma, A., Kumar, V., Shahzad, B., Tanveer, M., Sidhu, G.P.S., Handa, N. 2019. Worldwide pesticide usage and its impacts on ecosystem. *SN Appl. Sci*. 1: 1–16.

Tahir, R., Ghaffar, A., Abbas, G., Turabi, T.H., Kausar, S., Xiaoxia, D. 2021. Pesticide induced hematological, biochemical and genotoxic changes in fish: A review. *Agrobiol. Rec.* 3: 41–57.

Tarmure, T.G., Alexescu, O.O., Negrean, V., Sitar-taut, A.V., Coste, S.C., Todea, D.A. 2020. Influence of pesticides on respiratory pathology – A literature review. *Ann. Agric. Env. Med*. 27: 194–200.

Uchenna, U.B., Uka, A., Obiahu, O.H. 2022. The impact of sub-lethal concentrations of glyphosate on growth and haematology of African catfish under aquatic ecological microclimate. *Env. Chem. Ecotoxicol*. 4: 164–170.

Vera-Candioti, J., Araujo, P.I., Huerga, I.R., Rojas, D.E., Cristos, D.S., Malmantile, A.D. 2021. Pesticides detected in surface and groundwater from agroecosystems in the Pampas region of Argentina: Occurrence and ecological risk assessment. *Env. Monit. Assess.* 193: 689.

Wrinn, K.M., Evans, S.C., Rypstra, A.L. 2012. Predator cues and an herbicide affect activity and emigration in an agrobiont wolf spider. *Chemosphere.* 87: 390–396.

Xiang-Wu, C., Heng, C., Hai-Ling, Z., Da-Wei, L., Lin-Jian, O. 2024. Triazine herbicide reduced the toxicity of the harmful dinoflagellate K. mikimotoi by impairing its photosynthetic systems. *Ecotoxicol. Env. Safety.* 269: 115740.

Xiao, R., Kuang, B., Hu, Y., Wang, Y., Bai, J., Zhang, L., Wang, C., Wei, Z., Zhang, K., Jorquera, M.A., Campos, M., Acuña, J.J., Pan, W. 2023. Ecological risk assessment of glyphosate and its possible effect on bacterial community in surface sediments of a typical shallow Lake, northern China. *Ecotoxicol. Env. Safety.* 249: 114445.

Yengkokpam, P., Mazumder, P.B. 2020. Phytotoxicity of malathion (PM) and tatafen (PTF) towards *Solanum melongena* L. cv. Longai: A case study. *Plant Phys. Rep.* 25: 149–156.

Yuan, X., Pan, Z., Jin, C., Ni, Y., Fu, Z., Jin, Y. 2019. Gut microbiota: An underestimated and unintended recipient for pesticide-induced toxicity. *Chemosphere*. 227: 425–434.

Zhang, G., Qiu, Y., Boireau, P., Zhang, Y., Ma, X., Jiang, H., Xin, T., Zhang, M., Tadesse, Z., Wani, N.A., Song, J., Ding, J. 2024. Modern agriculture and One Health. *Infect Dis Poverty.* 13: 74.

Zhang, S., Wang, J., Li, Y., Li, X., Zheng, Y., Arif, M., Ru, S. 2022. Environmental relevant herbicide prometryn induces developmental toxicity in the early life stages of marine medaka (*Oryzias melastigma*) and its potential mechanism. *Aquatic Toxicol*. 243: 106079.

9 Management of Endocrine-Disrupting Compounds with Reference to One Health Framework

Saurabh Shukla and Ramsha Khan

9.1 INTRODUCTION

Endocrine-disrupting chemicals (EDCs) are exogenous chemical mixtures that disrupt the normal functioning of the endocrine system in humans as well as animals. EDCs are typically classified into three groups: androgenic compounds (EDsA), thyroid compounds (EDsT), and estrogen compounds (EDsE). Some other examples include industrial chemicals such as polychlorinated biphenyls (PCBs), dioxins, and alkylphenols; agricultural pesticides such as dichlorodiphenyltrichloroethane (DDT) and atrazine; plasticizers such as bisphenol A (BPA) and phthalates; pharmaceuticals such as synthetic hormones; and personal care products with parabens and UV filters (You and Song 2021). EDCs have the capability to block, or interfere with the stimulation, secretion, transfer, metabolization, or discharge of endogenous hormones. This may lead to disturbances in homeostasis, reproduction, development, and behavioural effects as well (Vieira et al. 2021). These effects are reported even at low EDC concentrations, and may lead to delayed, transgenerational effects, making them difficult to detect, to apply risk analysis, as well as to manage (Bilal et al. 2022).

EDCs can enter the environment via different routes. Agricultural and municipal wastewater, municipal and industrial effluent, leachates from sanitary landfill, and agricultural field run-off are among the most common routes of entry. Because of their persistence, EDCs are found in surface water, groundwater, sediments, and soils. These EDCs have often been reported far away from their initial emission point of release, for example in remote areas like the Arctic (Marlatt et al. 2022). These chemicals can bioaccumulate in the food chain as well, leading to dietary ingestion as the primary pathway for human as well as animal exposure. Their persistence and ubiquitousness present critical concerns regarding their long-term and overall impacts. Moreover, most of these chemicals are not removed in conventional wastewater treatment systems, and hence they can re-enter the environment, leading to chronic exposures for both humans and animals (Table 9.1).

 DOI: 10.1201/9781032623801-9

TABLE 9.1
Some commonly reported EDCs, their uses and characteristics with the associated ecotoxicological effects towards exposed organisms

Common EDCs	Category	Ecotoxicological effects	Uses	Exposed organisms
Di-n-butyl phthalate	Phthalates	Androgen and oestrogen receptors are interfered with, infertility, reproductive issues	Personal care products, vinyl flooring, lubricating oil, medical devices	Mice, guinea pigs, wistar rats, and ferrets
Bisphenol A	Plasticizers	Histone acetylation, haematological, biochemical, and hormonal changes	Storage containers, computer appliances, eyeglass lenses	Invertebrates, molluscs, amphibians
Acephate	Pesticide	Abrupt hormonal functioning of hypothalamus	Control sucking and biting insects	–
Cadmium chloride	Inorganic elements	Testicular tumours, prostate cancer, respiratory issues, pancreatic cancer, affect adrenal and pituitary glands	Commercial industries, baby products, packaging	Wistar rats, aquatic plants
Tetrabromobisphenol A	Brominate flame retardants	Thyroid hormone deiodinase	Building materials, electronics	Fish, algae, and invertebrates
DDT & metabolites	Pesticide (organochlorine)	Stimulation of oestrogen production and androgen-sensitive cell proliferation	Popular pesticide, additionally used against malaria, typhus, and bubonic plague	–
Butylbenzyl phthalate	Phthalates	Reproductive disorders, carcinogenicity	Vinyl flooring, synthetic leather, and toys, food packaging	Aquatic organisms

(continued)

TABLE 9.1 (Continued)
Some commonly reported EDCs, their uses and characteristics with the associated ecotoxicological effects towards exposed organisms

Common EDCs	Category	Ecotoxicological effects	Uses	Exposed organisms
Chlorpyrifos	Pesticide (organophosphorus)	Genotoxic, neurological, reproductive, and haematological effects	Works against multiple pests	Aquatic flora and fauna, soil organisms
Lead (lead acetate)	Inorganic elements	Reproductive system disorders, human granulosa cells altered	Batteries, chemical additives, cables and pigments, mining and foundry industries	Rats, dogs
Fluoxetine	Pharmaceuticals	Autophagic cell death, cancer, cytogenotoxic effects	Acute stroke, depression, autism	In liver/muscle of wild fish (wastewater outfalls)
Triclosan	Antimicrobials (preservatives)	Sublethal changes, reproductive disorders	Antibacterial activity, Consumer goods, personal care products,	Invertebrates, algae, flora and dolphins
Perfluorononanoic acid	Perflourochemicals	Alter spermatogenesis and steroidogenesis, oxidative stress	Commercial industries, surfactants	Aquatic organisms
Tributyltin	Antifoulants	Changes in bone modelling, suppresses immune system	Used for painting	Oysters, guppies, pond snails
Prochloraz	Pesticide (imidazole)	Activation of Pregnane X cellular receptor. Antagonist to cellular androgen and oestrogen receptors, agonist to Ah receptor and inhibition of aromatase activity	Fungicide in gardening and agriculture.	–

Ethinyl estradiol	Contraceptives (synthetic steroid)	Impairs foetal growth, reproductive system disorders	Birth control	Fish (plasma and muscle)
Paraben	Preservatives	Genotoxic effects, childhood allergies, respiratory issues	Personal care products like cosmetics and cleaning, pharmaceuticals	Wildlife living near the outfalls of sewage treatment
Cyclotetrasiloxane	Emollient and solvent	Nervous issues (drowsiness, dizziness, hypotension headache)	Cosmetics, hair conditioning and skin	–
Benzophenone-3	UV absorbent and filter	Oestrogenic effects and hormonal imbalance	Clothing, laundry detergents, sunscreens, and other personal care products	–
Nonylphenols	Alkylphenols	Neuroimmune, reproductive, and estrogenic effects	Detergents and surfactants	Aquatic organisms (microalgae), rot fungi

Source: Modified and produced with permission from Surana et al. (2022).

9.1.1 The Need for a One Health Perspective

The multitude of challenges posed by EDCs highlights the inadequacy of a conventional health management framework. Environmental and health management within the conventional framework tends to be focused on the end point, addressing either human health or environmental pollution but not the interfaces between them. However, endocrine disruption is an interface problem. Discharges of chemicals into the environment can interfere with wildlife populations and environmental processes, and subsequently affect human health and the ecosystem (Prata et al. 2021).

The One Health approach is seen as a possible holistic solution for addressing the risks associated with EDCs. One Health promotes the interrelation between human, animal, and environmental health, facilitating interdisciplinary collaborations. It also supports the sharing of data, knowledge, and policy across medicine, environmental science, and public health. Regarding EDCs, this implies acknowledging that:

- Wildlife as well as livestock may act as an indicator of environmental pollution.
- Both aquatic life and human population are impacted by contaminated water bodies.
- Agricultural activities as well as municipal waste management impact the spread of EDCs to different habitats.

Employing EDC management within a One Health framework makes it possible for interventions to be holistic, preventive, and sustainable. It also promotes improved surveillance, enhanced risk communication, as well as intensive action at multiple disciplinary levels. This approach also allows for the addition of social and economic aspects, such as occupational hazards (e.g. agricultural workers), and hazards from chemical pollution. Therefore, One Health is not just a technical and scientific approach but is considered an ethical and inclusive platform for governance as well (Wee et al. 2024).

9.2 SOURCES AND PATHWAYS OF EDCS

EDCs originate from various anthropogenic activities. Identifying the sources is an important step in planning an effective monitoring and control system. Agricultural run-off, industrial wastewater, pharmaceutical waste, and domestic wastewater are the major sources of EDCs. Pesticides, herbicides, and fungicides have active ingredients or metabolites that impact endocrine function (Wilk et al. 2019). Pesticides like atrazine, used extensively as an herbicide, have been linked with effects on amphibian and fish reproduction. Livestock growth promoting agents like synthetic oestrogens or androgens can lead to excretion of hormonally active compounds that move into surface waters and groundwater with run-off, or leach out from fields where manure is used as soil amendment (Wilk et al. 2019).

Industrial discharge is another significant source of EDCs. Industrial processes include chemicals which are known to have endocrine-disrupting activity. For example, phthalates in the production of plastics, nonylphenol ethoxylates in detergents, and flame retardants such as polybrominated diphenyl ethers (PBDEs) are all found in industrial effluent. Bu et al. (2018) reported that median dermal uptake of phthalates from skin wipes were in the range of 0.18–0.34 μg/day/kg-bw, which were ~7 times higher than the inhaled quantity. Industrial wastewater in most parts of the world is inadequately treated for such emerging contaminants; hence, they are able to enter aquatic environments without restriction (Czernych et al. 2017).

Pharmaceuticals and personal care products (PPCPs) are an expanding group of EDCs that include prescription medicines like synthetic hormones (e.g. ethinylestradiol), over-the-counter drugs, and numerous chemicals used in cosmetics, shampoos, sunscreens, and antiperspirants. These chemicals usually reach wastewater treatment plants via domestic sewage (Chakraborty et al. 2023). Conventional treatment technologies are generally inadequate for their complete removal; consequently, they are discharged into surface waters. Besides these point sources, non-point sources like atmospheric deposition, landfill leachate, and the application of biosolids on land exist. It is difficult to monitor and manage EDCs due to the variability and complexity of the sources.

9.2.1 Pathways into Water Bodies, Soil, and Food Chains

Once released into the environment, EDCs take different routes, and most of them facilitate a wider dispersion and persistence. Water and soil, as well as biological entities, act as a carrier and reservoir for these harmful chemical compounds. Aquatic environments are among the most widely impacted environmental matrices affected by EDCs. Stormwater drainage, agricultural drainage, and wastewater treatment plants discharge EDCs into coastal sites, lakes, and rivers. These compounds are either lipophilic (they can persist in water), or hydrophilic (they have an affinity for accumulating in biota or sediments). For instance, hydrophobic chemicals like PCBs and DDT adsorb on sediments as organic compounds, acting as long-term exposure sources in fish and benthos (Ghose and Mitra 2022).

Soil is exposed to EDCs via numerous mechanisms. Sewage sludge or biosolids applied to agricultural fields are often found to leach to different pharmaceuticals and personal care product residues into the soil. Irrigating with contaminated water can also result in the sorption of EDCs in the root zone, where plants can uptake them (Choudhury et al. 2025). Soil particles can adsorb some EDCs, further affecting their mobility, degradation, and bioavailability. Atmospheric transport is another significant pathway which is often undermined. Semi-volatile or volatile EDCs may evaporate into the atmosphere, transport over long distances, and get deposited on land surfaces or water bodies. This also highlights the primary reason for the presence of EDCs in remote regions like the Arctic, where local emissions are negligible (Khan et al. 2021).

EDCs have a tendency towards bioaccumulation and biomagnification in the food chain as well. Lipophilic chemicals are especially likely to accumulate in the fatty tissues of animals. This transfer in the trophic level is one of the major concerns for human health, especially for people whose dietary intake includes fish, or

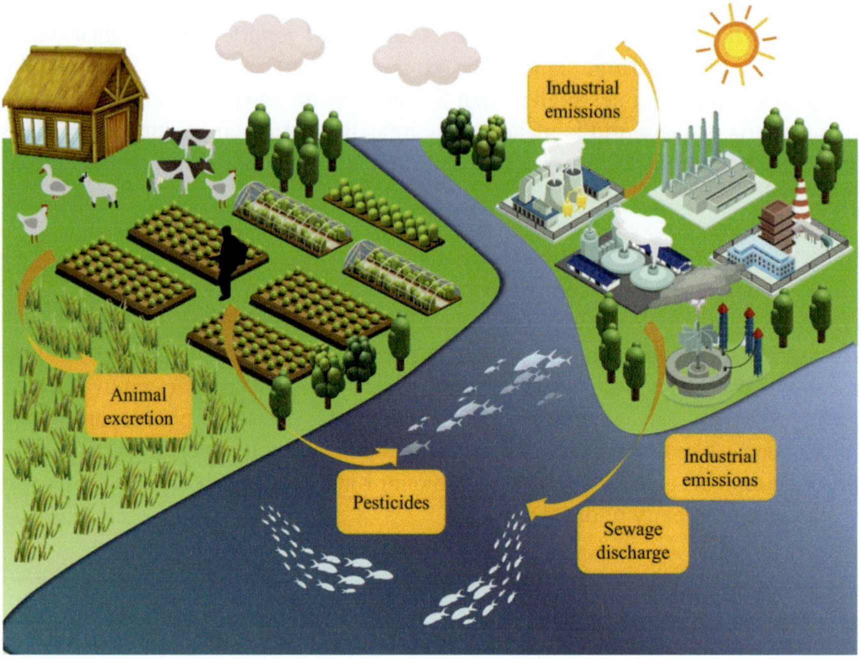

FIGURE 9.1 Primary sources and fate of EDCs in the environment. Reproduced with permission from Chen et al. (2022).

undomesticated animals as staples (rabbits, bison etc.). The primary sources and the fate of EDCs are illustrated in Figure 9.1.

Another critical concern with EDCs is their persistence and capability of forming transformation by-products. While certain compounds rapidly degrade under environmental settings, others are resistant to photolytic or microbial degradation. Even if the compounds are degraded, the transformation products can cause greater endocrine activity than the parent compound. Moreover, even with advanced treatment options like chlorination with ultraviolet radiation, which were reported to be successful for the degradation of studied EDCs, there were certain by-products which were found to be toxic (Saggioro et al. 2019). In the recent times, microplastics have emerged as possible and efficient carriers for many emerging contaminants including EDCs. Most plastic matrices include additives like phthalates as well as bisphenol A. Microplastics can adsorb other hydrophobic EDCs present in aquatic environments and can also serve as carriers, leading to their ingestion by aquatic animals. Yang et al. (2020) studied the combined toxicity of microplastics (MPs) with nonylphenol (NP). They found that the presence of MPs had a positive effect and increased the NP toxicity. They also found that adsorption capacity of MPs to NP was the dominant contributing factor for this increased toxicity. Table 9.2 summarizes the contamination pathways of some common EDCs and their impact on aquatic life.

TABLE 9.2
Prominent EDCs and their contamination pathways along with their harmful effects on aquatic life

EDC	Contamination path	Effects on aquatic life
Phthalates	Reach the aquatic bodies via improper disposal/sewage network of industrial solid waste	Reproductive toxicity, infertility, and adverse impacts on metabolism
17α-Ethinylestradiol	Used as contraceptive. Reach the aquatic bodies (surface water) via excreta and inappropriate drug disposal	Alterations in reproductive behaviour, changes in mating behaviour
Flame retardants	Waste generated out of textile and electrical industries/household waste	Disrupt oestrogen and androgen signalling, increased fish toxicity, altered thyroid gland
Perfluorinated chemicals	Polish wax and packaging waste/sewage networks	Suppresses hormonal action and impedes fertility
Bisphenol A	Xenoestrogen, reach the aquatic bodies via sewage networks/ inappropriate dumping of industrial solid waste	Offspring deformities, reproduction is hampered
Diethylstilbestrol	Oestrogen/discarded as excreta	Infertility, abortions, and premature birth
Nonylphenol	Xenoestrogen/reach the aquatic bodies via sewage networks/ inappropriate dumping of industrial solid waste	Decrease in male fertility, feminization of aquatic life, and hampering of the survival of juveniles
Triclosan	Agricultural waste/inappropriate dumping of cosmetic waste	Disturbed metamorphosis and sex ratio, irregular foetal development, and chances of infertility
Paraben	Wastewater industries, pharmaceuticals (preservatives)	Decreases fish fertility and increases risk of carcinogenicity
Primidone	Pharmaceutical waste drainage to aquatic streams	Altered reproductive functions and risk of carcinogenic effects
DDT (2,2 bis-*p*-chlorophenyl-1,1,1-trichloroethane)	Xenoestrogen pesticide/reach the aquatic bodies via inappropriate dumping of solid waste/surface run-off and generate leachate	Chronic effects on nervous system and immune system
Atrazine	Agriculture waste discharge into streams	Reproductive tumours, morphology abnormalities, feminization of male frogs
Dibenzofuran	Industrial waste	Polyhalogenated compounds impede hormonal functioning
17β-estradiol	Natural oestrogen used in contraception/excreted in urine and faeces and/or inadequately dumped through sewage networks	Hermaphroditism, VTG synthesis in blood, feminization of the reproductive tube, change in gonad weight

Source: Reproduced with permission from Surana et al. (2022).

9.3 MECHANISMS OF ENDOCRINE DISRUPTION

The endocrine system is a complex network of glands which produces hormones in both humans and animals. These hormones are responsible for regulating various vital physiological processes such as growth, reproduction, metabolism, and development. These glands communicate through the secreted hormones, which act as chemical messengers. These hormones travel through the bloodstream and reach the receptors of the target cells, and produce a biological response, maintaining the normal functioning of the body (Kasonga et al. 2021). There are different glands, and some produce multiple types of hormones while others produce simpler hormones, but they all are very important. However, these hormones must remain within the regulated concentrations, as either excesses or deficiencies can lead to serious health disorders, including reproductive dysfunction, metabolic syndromes, and endocrine-related cancers (Vieira et al. 2021).

EDCs are exogenous chemical compounds that interfere with the normal functioning of hormones. There are several possible mechanisms through which EDCs are able to do this. EDCs can mimic the natural hormones or block their actions; they can alter the hormone synthesis process. They can also affect the transport and metabolism of natural hormones, which in turn affects their binding capacity to the receptors and gene expression. One of the most critical concerns is that EDCs can have these effects even at very low concentrations and that too during the critical growth phases such as the foetal phase or the puberty phase.

9.3.1 Interaction with Hormone Receptors

One of the most direct mechanisms by which EDCs make an impact is through interaction with hormone receptors. Many EDCs either mimic the action of endogenous hormones or block the receptor function. For example, BPA, nonylphenol, and certain phytoestrogens have been shown to bind to oestrogen receptors (ERα and ERβ), and subsequently activate the activities which are normally triggered by the natural hormone oestradiol (Katchy et al. 2014). This can lead to inappropriate gene expression, abnormal cell production, and problems in reproductive development. On the other hand, compounds such as fungicides (e.g. vinclozolin) act as anti-androgens. That is, they inhibit the binding of testosterone to its receptor and result in shifts in gender and reproductive irregularities in exposed male animals (Draskau et al. 2021). EDCs may also cause disruption in signalling for other receptors as well, for example, thyroid hormone receptors and glucocorticoid receptors. In this regard, chemicals that can interfere with thyroid signalling are of particular medical concern. Thyroid plays a very important role in neurodevelopment, especially during gestation and early childhood. Disruption in this pathway can lead to cognitive deficits and growth retardation.

9.3.2 Disruption of Synthesis and Metabolism

EDCs can also affect the endocrine system by interfering with the synthesis of certain hormones, their transport, and metabolism. Some pesticides have been found to inhibit key enzymes in steroidogenesis, such as aromatase. Aromatase is responsible for converting androgens into oestrogens (a key enzyme for reproductive health). Inhibition of such enzymes alters the hormonal balance and has ill effects on reproductive health and secondary sexual characteristics as well (You and Song 2021). Obesity in children and adolescents has become one of the most common diseases in modern times, a public health concern at a global level. EDCs commonly are fat soluble, and hence they can easily accumulate in the adipose tissue, and in turn, can directly affect the metabolism of humans. Cano-Sancho Salmon and La Merrill (2017) reported a positive correlation between exposure to DDT, dichlorodiphenyldichloroethylene (DDE, a metabolite of DDT), and increasing obesity.

Some EDCs like polychlorinated biphenyls (PCBs) and dioxins have the capability to bind themselves with the hormone transport proteins (e.g. thyroxine-binding globulin), thus reducing the bioavailability of hormones in circulation. Furthermore, these chemicals may induce or inhibit hepatic enzymes responsible for hormone metabolism, affecting the rate at which hormones are broken down and eliminated. Such disruptions in the metabolic actions may either prolong hormone action or lead to the accumulation of inactive or harmful metabolites (Corsini, Ruffo, and Racchi 2018). Furthermore, a combination of even low-dose EDCs may have a compounding effect (also known as "cocktail effect"), which increases the complexity, as the toxic effects may not be predicted through the individual toxicological profiles. This presents a significant challenge for regulatory toxicology, which often assesses the risk based on exposure to a single compound.

While the basic mechanism of endocrine disruption is similar across vertebrates, its physiological impacts may vary significantly between different species due to varying hormone receptor structure, difference in development phases, and also the way environmental interactions happen (Teta et al. 2018). Exposure to EDCs in humans has been linked to effects such as reduced fertility, altered puberty onset, menstrual irregularities, and hormone-dependent cancers like breast and prostate cancer. Moreover, in wildlife, the effects of EDCs are equally concerning and even more apparent in certain cases. For example, the fish population, exposed to estrogenic compounds downstream from wastewater treatment plants, frequently displays intersex characteristics, that is, the presence of both male and female gonadal tissues. Similarly, certain amphibians (e.g. frogs) when exposed to atrazine display feminization and in some cases showed abnormal gonadal development as well. In birds, exposure to organochlorine pesticides such as DDT has led to eggshell thinning, decreased hatchability, and population declines in recent times (Marlatt et al. 2022).

What is more concerning is the fact that these effects may not be limited to the exposed population only. There have been studies where multi- and transgenerational impacts have been reported, which were facilitated through epigenetic mechanisms. In most of those cases, chemical exposure was found to have induced inherited changes in gene expression without altering the underlying DNA sequence. For example, studies in rodents which were exposed to vinclozolin or BPA have shown altered reproductive traits in multiple generations, even in the absence of direct exposure in later generations (Robaire et al. 2022).

9.4 IMPACTS ON HUMAN, ANIMAL, AND ENVIRONMENTAL HEALTH

9.4.1 Human Health

One of the biggest challenges posed by EDCs is their harmful effect on the human reproductive system. Multiple studies have reported that exposure to chemicals such as BPA, phthalates, and pesticides causes lower fertility rates, alterations in hormonal levels, and abnormal functioning of the reproductive organs. For instance, exposure to phthalates has been reported as linked to lower levels of testosterone as well as poor sperm quality in men. In women, EDCs can impair menstrual cycles as well as lead to endometriosis and polycystic ovary syndrome (PCOS).

Moreover, research also shows that EDCs have a high correlation with hormonally mediated cancers. For example, phthalates were detected in pre-diagnosis and pre-treatment urine samples from the breast cancer patients in India (Mukherjee Das et al. 2022). Similarly, EDC exposure has been found to cause higher risks of prostate as well as testicular cancer, particularly in cases of prolonged or foetal/neonatal exposure.

Prenatally as well as in the earliest postnatal periods there are developmental problems in particular of significant concern. Exposure in utero to EDCs like PCBs and dioxins is linked with birth weight reduction, changed neurodevelopment, as well as intellectual impairments. A few studies have also reported that the presence of BPA may cause an increase in several proteins which are associated with Alzheimer's disease (AD). It is correlated that BPA may cause the triggering of AD, especially in people who have type-2 diabetes. Moreover, exposure to lead and mercury may also cause an increase in certain other pathological proteins which are related to AD, hence aggravating its onset (Wang et al. 2017).

9.4.2 Animal Health

Existing research indicates that EDCs are strongly associated with a variety of hormonal and physiological disturbances in wildlife as well. These compounds often enter the aquatic ecosystems untreated, which leads to chronic exposure across several species. The impacts of EDCs on wildlife are usually assessed through biomarkers, tissue analysis, and physiological assessments of individual organisms. Among the most well-documented effects are alterations in the reproductive system, reduced fertility, eggshell thinning, disruptions in endocrine hormone levels, impaired hatchability, and changes in plasma vitellogenin (VTG) concentrations. These disruptions can also lead to decreased survival and viability in offspring, potentially influencing population dynamics. One of the earliest cases of the impacts of EDCs on wildlife was reported in alligator populations in Florida during the 1980s (Guillette et al. 1994). Young male alligators exhibited abnormally low testosterone levels and reduced genital size, while females showed morphological changes in ovarian tissue and elevated plasma hormone concentrations. These abnormalities were later found to be linked to the exposure to DDE.

EDCs also have a widespread impact of both natural and synthetic hormones (e.g. estrone, estradiol, and ethinylestradiol) on aquatic species (including fish, turtles, and molluscs). Changes in VTG levels have been consistently reported in these organisms, due to disruptions in estrogenic signalling pathways. In addition, contamination of the marine food web with organic pollutants, such as polychlorinated biphenyls (PCBs) and derivatives of DDT, has been associated with reproductive and immune dysfunction in aquatic animals. Studies have also reported that in Arctic ecosystems, exposure to these contaminants has caused hormonal imbalances in polar bears, including masculinization in females and suppressed testosterone levels in the male population (Ciesielski et al. 2023).

Similarly, feminization of the male fish population and masculinization in females have been widely reported in areas exposed to oestrogenic compounds and plasticizers like phthalates and BPA. The changes in endocrine levels have frequently caused reproductive damage, and consequently skewed sex ratios, and progressive anomalies in offspring (Gross-Sorokin, Roast, and Brighty 2006). Also, structural changes to eggshells in bird populations have been observed, suggesting disruptions in calcium metabolism potentially linked to endocrine interference (Gosler and Wilkin 2017). Based on the clarity of results and the impacts of EDCs on animals, these studies have become a critical resource for understanding the mechanisms through which EDCs operate. Insights gained from these studies offer valuable results to predict the impact and outcome of EDC exposure in humans as well. A summary of major health impacts of EDCs on human and animal populations are illustrated through Figure 9.2.

9.4.3 Ecosystem health

Other than the impacts on humans and wildlife, the presence of EDCs has harmful effects on the ecosystem as well, particularly significant effects on soil microbial communities. EDCs often persist in soil matrices and interact with microbial communities in more than one way. Disruption of microbial community structure and enzymatic activity has been reported, which suggests that EDCs are able to interfere with critical soil functions such as conversion and utilization of organic matter, and nutrient cycling. Moreover, many soil bacteria which have been exposed to compounds like BPA, or phthalates exhibit stress responses which reduce their metabolic flexibility (Moukhtari, Martín-Pozo, and Zafra-Gómez 2023). The subsequent imbalance can also affect plant–microbe interactions in a negative manner, and may even facilitate the propagation of resistance genes within the microbial community. Unfortunately, these ecological impacts are not easy to predict as EDC mixtures more often than not act in a synergetic manner. Overall, the impacts of EDCs extend beyond immediate toxicity, causing broader changes in the structure and functioning of soil ecosystems, and may have long-term consequences for agricultural productivity, ecological balance, and public health.

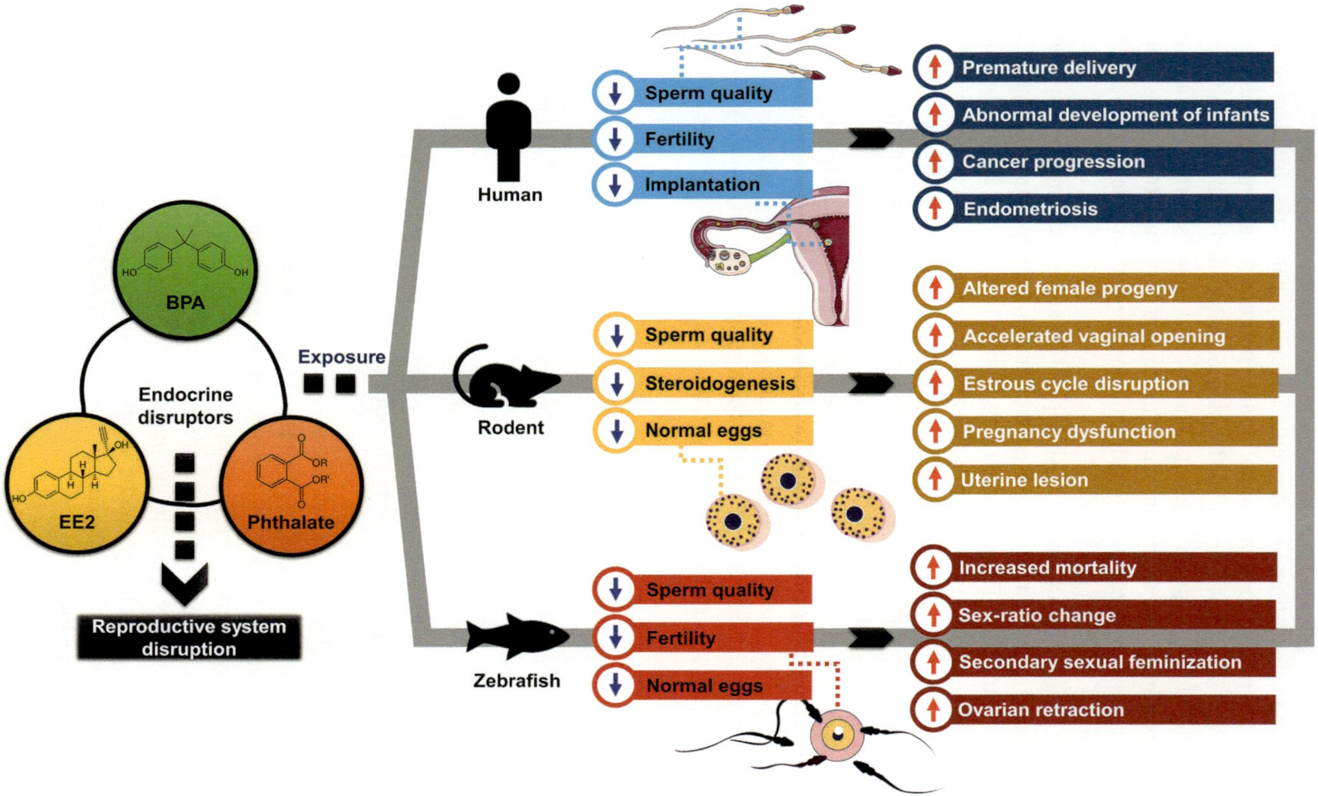

FIGURE 9.2 Schematic representation of the effects of EDCs reported in human and animal health. Reproduced with permission from You and Song (2021).

9.5 THE ONE HEALTH FRAMEWORK: AN INTEGRATIVE APPROACH

One Health is an adaptable, transdisciplinary, and multisectoral framework that acknowledges human, animal, and environmental health as interconnected. The concept of One Health has gained popularity in recent times, especially after the three-way engagement between the World Health Organization (WHO), the Food and Agriculture Organization (FAO), and the World Organization for Animal Health (WOAH) in 2010. The agreement defines One Health as

> An approach to address a health threat at the human-animal-environment interface based on collaboration, communication, and coordination across all relevant sectors and disciplines, with the ultimate goal of achieving optimal health outcomes for both people and animals; a One Health approach is applicable at the subnational, national, regional, and global level.
>
> **FAO-OIE-WHO, 2010**

EDCs fits well within this One Health framework as they are available at multiple interfaces, being applied in agriculture (through use of pesticides), being used in different industries, having domestic uses, and once released in the environment, polluting air, water, and food resources. EDCs, as discussed in previous sections, have a harmful impact on humans and animals alike, and their persistence in the environment underlines the need for intensive action.

Implementation of the One Health approach for EDCs involves consistent activities among public health, veterinary medicine, environmental science, and last but not the very least, regulatory authorities. For example, investigation campaigns must monitor both human biomarkers as well as wildlife populations and environmental contamination. Moreover, the risk assessments should be based on evidence collected from clinical, toxicological, and ecological studies as well. Some key components of a One Health approach include:

- **An integrated monitoring system:** Performing joint monitoring of EDCs across human, animal, and environmental domains.
- **Integrated data analysis:** Linking epidemiological, ecotoxicological, and chemical monitoring data for more holistic and comprehensive risk assessments.
- **Collaborative policies:** Involving diverse stakeholders before policies are formed, including health departments, environmental agencies, and agricultural sectors.
- **Capacity building:** Training professionals to understand interdisciplinary connections and communicate the associated risks to the general public and increase their awareness of EDCs.

One Health takes preventive steps rather than suggesting measures which can only be implemented after the damage has been done. By identifying hotspots of EDC exposure, their use and source, specific preventive steps can be implemented to reduce their ill impacts. Harmonization of legislation and provision of resources can be improved by a concerted global effort, particularly for middle- and low-income countries.

9.6 MONITORING AND RISK ASSESSMENT

9.6.1 Analytical Techniques

Regular monitoring and continuous risk assessment are very important in the One Health framework; hence, there is a need for both biological assays and chemical analyses for effective monitoring for EDCs. The yeast oestrogen screen (YES) assay has been commonly used as a cost-effective approach to detect and quantify the oestrogenic activity (EA) induced by the presence of EDCs in environmental samples and to screen for mixtures of high complexity as well (Azevedo et al. 2023). Moreover, chemical analysis using instruments like LC-MS and GC-MS can be employed for quantification of EDCs in different kinds of environmental samples (biological or environmental) at parts-per-trillion levels with high sensitivity. However, these are costly, both in terms of equipment (operational and maintenance, reagents etc.), and also skilled personnel are required for their operation. Pre-analysis requirements like preconcentration procedures are also needed in most cases since EDC concentration in samples is very low. New methods such as high-resolution mass spectrometry (HRMS) and non-target screening are also being applied for the detection of novel and unregulated EDCs, which can help in better understanding of exposure sources (Neto et al. 2024).

9.6.2 Risk Assessment Models

Conventional risk frameworks of hazard identification, dose response, exposure, and risk characterization have several challenges for them to be applied in case of contamination with EDCs. Most EDCs exhibit non-monotonic dose-response relations, in the sense that effects at low doses are observed which cannot be predicted from the outcomes of high-dose studies. Response is addressed through the development of new models such as adverse outcome pathways (AOPs) for connecting molecular-change with organism- and population-level effects. In silico models, for example, quantitative structure-activity relationships (QSARs) are used with increasing interest and frequency for the prediction of endocrine activity based on chemical structure. Cumulative risk assessments are increasingly being accepted, acknowledging that individuals and wildlife are exposed to numerous EDCs at once. Integrated exposure models and improved knowledge on the toxicity of mixtures are required in this case (Evangelista, Chirico, and Papa 2024).

9.6.3 Challenges in Risk Assessment

Several critical challenges exist during the comprehensive assessment and regulation of EDCs in environmental and public health contexts. One of the primary challenges is the lack of toxicological data for the majority of known and suspected EDCs. This limitation is faced particularly with respect to the endocrine-specific effects of these EDCs. Subsequently, this data gap restricts the development of evidence-based regulatory frameworks. Furthermore, EDCs often exhibit biological activity at low doses that challenge the traditional toxicological assumptions, and question the appropriateness of current exposure thresholds (Ho et al. 2022). This challenge is further complicated because of the matrix effect. That is, real-world samples have a combination of different chemicals, and their interaction and effect are hard to predict. Most of the lab-scale studies do not take this 'matrix effect' into account and fail in terms of scalability. Exposure patterns also vary widely across populations. That is, the most vulnerable groups such as children, pregnant women, and subsistence communities may experience disproportionately high exposures owing to their physiological and behavioural factors. Another critical concern is related to transgenerational impacts of the EDCs. Certain EDCs are able to cause epigenetic modifications which can potentially affect offspring across multiple generations. Addressing these challenges requires a more precautionary, integrative, and adaptive approach to risk assessment. The One Health framework accounts for these uncertainties, adapts well to system complexities, and gives priority to the protection of sensitive populations and ecological health.

9.7 CASE STUDIES AND SUCCESS STORIES

9.7.1 Case Study 1: Ban on DDT and Wildlife Recovery in the United States

DDT, which is one of the earliest and most renowned EDCs, was used extensively as a pesticide during the mid-20th century, and it was considered a miraculous and revolutionary chemical at that time. However, its persistence and bioaccumulation led to severe ecological consequences, particularly in bird populations. After the famous book by Rachel Carson, *Silent Spring*, and subsequent ecological studies, the United States enforced a ban on the use of DDT in 1972. Following this ban, the affected bird species (e.g. bald eagle, peregrine falcon, brown pelican etc.) showed significant recovery at the population level (Grier 1982). Improvement in reproduction rates was also observed owing to a reduction in eggshell thinning (caused by DDT exposure).

Relevance to One Health approach:
This case study highlights the interconnectedness between the ecosystem (wildlife health), human health (bioaccumulation in food chains), and the regulatory framework. It illustrates a One Health approach where policy decisions based on environmental evidence translated into benefits across ecosystems and human exposure.

9.7.2 Case Study 2: European Union (EU) Ban on BPA in Baby Bottles and Subsequent Policy Change

BPA, which is widely used in polycarbonate plastics and epoxy resins, has been widely studied for its oestrogenic activity. However, various concerns were raised because of its use in feeding bottles and other items for infants, which could cause neonatal and infant exposure (Vandenberg et al. 2007). Hence, in 2011, the EU implemented a precautionary ban on BPA in baby bottles, citing potential health risks despite the ongoing debate about the effects observed at low doses. Despite those debates, the ban prompted multiple reforms. One of the most significant was the implementation of strict limits for the use of BPA in food packaging and the inclusion of BPA in the EU REACH (Registration, Evaluation, Authorisation and Restriction of Chemicals) list of substances of very high concern (SVHC). Subsequently, many manufacturers voluntarily phased out the commercial use of BPA, and replaced it with less hazardous alternatives.

Relevance to One Health approach:
This case study highlights the integration of human health protection, especially in the case of highly vulnerable population such as neonates, with chemical safety and regulatory reforms. It aligns with One Health principles as the stakeholders took positive steps after recognizing the long-term consequences of chronic low-dose exposure of BPA and implemented a ban.

9.8 FUTURE DIRECTIONS

9.8.1 Research Needs

In recent years, there has been an increase in research with respect to the health impacts of EDCs on humans and the environment. However, despite these efforts critical gaps exist in the scientific understanding about the fate and effects of EDCs, especially in real-world settings. More research is required which can help highlight the action mechanism of EDCs after exposure at low doses. Also, the matrix effect, or the combined effects of chemical mixtures present in different environmental matrices, must be taken into consideration to develop more precise risk assessments. Expanding the research into transgenerational effects is also important, particularly in the context of epigenetic modifications that may persist across generations without being directly exposed. Novel research fields such as neurodevelopmental impacts, and gene-environment interactions are also being investigated in detail. Moreover, ecotoxicological studies, should try to look beyond vertebrate models and include invertebrate species. This approach can play a significant role in ecological frameworks, but remain less explored in research with respect to EDCs. Technological advancements, such as high-throughput screening, multi-omics technologies, and artificial intelligence-enabled data integration hold promise for identifying previously unrecognized EDCs and improving predictive models of their biological effects.

9.8.2 Policy Development

Many of the effects of EDC exposure are very hard to identify and predict, and some of them can even remain inactive for a long time. Hence, regulatory frameworks must adopt a precautionary approach to reflect these effects. Policymaking should also prioritize preventive actions over responsive remediation actions. Preventive measures, such as incentivizing green chemistry towards the synthesis of safer chemicals, enforcing clear labelling of product contents, and restricting non-essential uses of high-risk

EDCs, are essential. Equally, it is very important that international regulations are coordinated. Inconsistent policies across regions can result in the relocation of hazardous substances to jurisdictions with weaker controls and economic needs, which will hamper the collective progress.

9.9 CONCLUSION

EDCs represent a complex and universal challenge to human, animal, and environmental health systems. EDCs originate from a variety of sources, including industrial processes, agricultural run-off, pharmaceutical residues, and household products. EDCs have been reported to be capable of interfering with hormonal signalling mechanisms. Their harmful impacts range from developmental challenges, reproductive issues in humans, to population declines, alteration in behaviours, and reproductive dysfunction in wildlife. EDCs also have an additive effect when present along with other chemicals in different matrices in real-world settings. Owing to these characteristics, many of the conventional toxicological frameworks become inadequate for risk assessment. Moreover, these EDCs can affect different ecosystems, while being undetected, and this highlights the need for an integrated approach to their management.

Considering these challenges, the One Health framework offers a holistic and realistic path forward towards addressing the complicated risks posed by EDCs. The One Health approach adopts a combined effort between disciplines that have operated in isolation, by highlighting the significance of the interconnectedness of human, animal, and environmental health. This approach enables the application of a more responsive surveillance systems and ensures evenly coordinated policy interventions, while promoting shared accountability among stakeholders. The case studies discussed in Section 9.7 establish the effectiveness of coordinated actions which are an integral aspect of this framework. However, these success stories are only the tip of the iceberg. Continued research into EDC management will require stricter regulatory mechanisms, incentives, and investment in safer/greener chemical alternatives, public engagement, and promotion of interdisciplinary research. Importantly, adopting a One Health approach not only highlights the management of EDCs as a technical or regulatory necessity. It also advocates for a moral and societal responsibility to environmental planetary health across generations.

REFERENCES

Azevedo, Taiza dos Santos, Juliana Palermo Evangelista dos Santos, João Monteiro Neto, Fernanda Bento Rosa Gomes, Sue Ellen Costa Bottrel, Ann Honor Mounteer, and Renata de Oliveira Pereira. 2023. "Removal of Endocrine-Disrupting Chemical Mixtures in Water Using Chlorination and Photolysis." *Water Supply* 23 (3): 1284–96. https://doi.org/10.2166/ws.2023.022.

Bilal, Muhammad, Komal Rizwan, Muhammad Adeel, Damià Barceló, Youssef Ahmed Awad, and Hafiz M. N. Iqbal. 2022. "Robust Strategies to Eliminate Endocrine Disruptive Estrogens in Water Resources." *Environmental Pollution* 306 (August): 119373. https://doi.org/10.1016/j.envpol.2022.119373.

Bu, Zhongming, Jiahui Wang, Wei Yu, and Baizhan Li. 2018. "Dermal Exposure to Phthalates in Home Environment: Handwipes, Influencing Factors and Implications." *Building and Environment* 133 (April): 1–7. https://doi.org/10.1016/j.buildenv.2018.02.013.

Cano-Sancho, German, Andrew G. Salmon, and Michele A. La Merrill. 2017. "Association between Exposure to p,p'-DDT and Its Metabolite p,p'-DDE with Obesity: Integrated Systematic Review and Meta-Analysis." *Environmental Health Perspectives* 125 (9): 096002. https://doi.org/10.1289/EHP527.

Chakraborty, Aritra, Satadal Adhikary, Suchandra Bhattacharya, Sohini Dutta, Sovona Chatterjee, Diyasha Banerjee, Abhratanu Ganguly, and Prem Rajak. 2023. "Pharmaceuticals and Personal Care Products as Emerging Environmental Contaminants: Prevalence, Toxicity, and Remedial Approaches." *ACS Chemical Health & Safety* 30 (6): 362–88. https://doi.org/10.1021/acs.chas.3c00071.

Chen, Yuxin, Jian Yang, Bin Yao, Dan Zhi, Lin Luo, and Yaoyu Zhou. 2022. "Endocrine Disrupting Chemicals in the Environment: Environmental Sources, Biological Effects, Remediation Techniques, and Perspective." *Environmental Pollution* 310 (October): 119918. https://doi.org/10.1016/j.envpol.2022.119918.

Choudhury, Moharana, Manab Deb Adhikari, Sangita Agarwal, Palas Samanta, Anu Sharma, Debajyoti Kundu, and Sunil Kumar. 2025. "Pharmaceuticals and Personal Care Products in Soil: Sources, Impacts and Myco-Remediation Strategies." *Emerging Contaminants* 11 (2): 100488. https://doi.org/10.1016/j.emcon.2025.100488.

Ciesielski, Tomasz M., Christian Sonne, Eli I. Smette, Gro Dehli Villanger, Bjarne Styrishave, Robert J. Letcher, Daniel J. Hitchcock, Rune Dietz, and Bjørn M. Jenssen. 2023. "Testosterone and Persistent Organic Pollutants in East Greenland Male Polar Bears (Ursus Maritimus)." *Heliyon* 9 (3): e13263. https://doi.org/10.1016/j.heliyon.2023.e13263.

Corsini, Emanuela, Federica Ruffo, and Marco Racchi. 2018. "Steroid Hormones, Endocrine Disrupting Compounds and Immunotoxicology." *Current Opinion in Toxicology* 10 (August): 69–73. https://doi.org/10.1016/j.cotox.2018.01.006.

Czernych, Radosław, Milena Chraniuk, Paweł Zagożdżon, and Lidia Wolska. 2017. "Characterization of Estrogenic and Androgenic Activity of Phthalates by the XenoScreen YES/YAS in Vitro Assay." *Environmental Toxicology and Pharmacology* 53 (July): 95–104. https://doi.org/10.1016/j.etap.2017.05.010.

Draskau, Monica Kam, Aurélie Lardenois, Bertrand Evrard, Julie Boberg, Frédéric Chalmel, and Terje Svingen. 2021. "Transcriptome Analysis of Fetal Rat Testis Following Intrauterine Exposure to the Azole Fungicides Triticonazole and Flusilazole Reveals Subtle Changes despite Adverse Endocrine Effects." *Chemosphere* 264 (Pt 1): 128468. https://doi.org/10.1016/j.chemosphere.2020.128468.

Evangelista, Marco, Nicola Chirico, and Ester Papa. 2024. "In Silico Models for the Screening of Human Transthyretin Disruptors." *Journal of Hazardous Materials* 480 (December): 136188. https://doi.org/10.1016/j.jhazmat.2024.136188.

FAO-OIE-WHO. 2010. "The FAO-OIE-WHO Collaboration. A Tripartite Concept Note." September 28, 2010. www.who.int/publications/m/item/the-fao-oie-who-collaboration.

Ghose, Anamika, and Sudip Mitra. 2022. "Spent Waste from Edible Mushrooms Offers Innovative Strategies for the Remediation of Persistent Organic Micropollutants: A Review." *Environmental Pollution* 305 (July): 119285. https://doi.org/10.1016/j.envpol.2022.119285.

Gosler, Andrew G., and Teddy A. Wilkin. 2017. "Eggshell Speckling in a Passerine Bird Reveals Chronic Long-Term Decline in Soil Calcium." *Bird Study* 64 (2): 195–204. https://doi.org/10.1080/00063657.2017.1314448.

Grier, J. W. 1982. "Ban of DDT and Subsequent Recovery of Reproduction in Bald Eagles." *Science* 218 (4578): 1232–35. https://doi.org/10.1126/science.7146905.

Gross-Sorokin, Melanie Y., Stephen D. Roast, and Geoffrey C. Brighty. 2006. "Assessment of Feminization of Male Fish in English Rivers by the Environment Agency of England and Wales." *Environmental Health Perspectives* 114 (Suppl 1): 147–51. https://doi.org/10.1289/ehp.8068.

Guillette, L. J., T. S. Gross, G. R. Masson, J. M. Matter, H. F. Percival, and A. R. Woodward. 1994. "Developmental Abnormalities of the Gonad and Abnormal Sex Hormone Concentrations in Juvenile Alligators from Contaminated and Control Lakes in Florida." *Environmental Health Perspectives* 102 (8): 680–88. https://doi.org/10.1289/ehp.94102680.

Ho, V., L. Pelland-St-Pierre, S. Gravel, M. F. Bouchard, M. A. Verner, and F. Labrèche. 2022. "Endocrine Disruptors: Challenges and Future Directions in Epidemiologic Research." *Environmental Research* 204 (Pt A): 111969. https://doi.org/10.1016/j.envres.2021.111969.

Kasonga, Teddy Kabeya, Martie A. A. Coetzee, Ilunga Kamika, Veronica M. Ngole-Jeme, and Maggy Ndombo Benteke Momba. 2021. "Endocrine-Disruptive Chemicals as Contaminants of Emerging Concern in Wastewater and Surface Water: A Review." *Journal of Environmental Management* 277 (January): 111485. https://doi.org/10.1016/j.jenvman.2020.111485.

Katchy, Anne, Caroline Pinto, Philip Jonsson, Trang Nguyen-Vu, Marchela Pandelova, Anne Riu, Karl-Werner Schramm, et al. 2014. "Coexposure to Phytoestrogens and Bisphenol a Mimics Estrogenic Effects in an Additive Manner." *Toxicological Sciences* 138 (1): 21–35. https://doi.org/10.1093/toxsci/kft271.

Khan, Muhammad Tariq, Izaz Ali Shah, Ihsanullah Ihsanullah, Mu. Naushad, Sharafat Ali, Syed Hassan Ali Shah, and Abdul Wahab Mohammad. 2021. "Hospital Wastewater as a Source of Environmental Contamination: An Overview of Management Practices, Environmental Risks, and Treatment Processes." *Journal of Water Process Engineering* 41 (June): 101990. https://doi.org/10.1016/j.jwpe.2021.101990.

Marlatt, V. L., S. Bayen, D. Castaneda-Cortès, G. Delbès, P. Grigorova, V. S. Langlois, C. J. Martyniuk, et al. 2022. "Impacts of Endocrine Disrupting Chemicals on Reproduction in Wildlife and Humans." *Environmental Research* 208 (May): 112584. https://doi.org/10.1016/j.envres.2021.112584.

Moukhtari, Fatine El, Laura Martín-Pozo, and Alberto Zafra-Gómez. 2023. "Strategies Based on the Use of Microorganisms for the Elimination of Pollutants with Endocrine-Disrupting Activity in the Environment." *Journal of Environmental Chemical Engineering* 11 (1): 109268. https://doi.org/10.1016/j.jece.2023.109268.

Mukherjee Das, Ankan, Ajay Gogia, Manoj Garg, Arun Elaiyaraja, Priyadarshini Arambam, Sandeep Mathur, Ramaswamy Babu-Rajendran, et al. 2022. "Urinary Concentration of Endocrine-Disrupting Phthalates and Breast Cancer Risk in Indian Women: A Case-Control Study with a Focus on Mutations in Phthalate-Responsive Genes." *Cancer Epidemiology* 79 (August): 102188. https://doi.org/10.1016/j.canep.2022.102188.

Neto, João Monteiro, Lorena Gotelip Costalonga, Nathacha Oliveira Pires, Thais Girardi Carpanez, Fernanda Bento Rosa Gomes, Sue Ellen Costa Bottrel, Jonathas Batista Gonçalves Silva, Marcelo Henrique Otenio, and Renata de Oliveira Pereira. 2024. "Yeast Estrogen Screen Assay Applied in the Assessment of Estrogenic Activity Removal from Dairy Cattle Wastewater Treated by Anaerobic Digestion." *Science of The Total Environment* 908 (January): 168243. https://doi.org/10.1016/j.scitotenv.2023.168243.

Prata, Joana C., João P. da Costa, Isabel Lopes, Anthony L. Andrady, Armando C. Duarte, and Teresa Rocha-Santos. 2021. "A One Health Perspective of the Impacts of Microplastics on Animal, Human and Environmental Health." *The Science of The Total Environment* 777 (July): 146094. https://doi.org/10.1016/j.scitotenv.2021.146094.

Robaire, Bernard, Geraldine Delbes, Jessica A. Head, Vicki L. Marlatt, Christopher J. Martyniuk, Stéphane Reynaud, Vance L. Trudeau, and Jan A. Mennigen. 2022. "A Cross-Species Comparative Approach to Assessing Multi- and Transgenerational Effects of Endocrine Disrupting Chemicals." *Environmental Research* 204 (Pt B): 112063. https://doi.org/10.1016/j.envres.2021.112063.

Saggioro, Enrico M., Fernanda P. Chaves, Louise C. Felix, Giselle Gomes, and Daniele M. Bila. 2019. "Endocrine Disruptor Degradation by UV/Chlorine and the Impact of Their Removal on Estrogenic Activity and Toxicity." *International Journal of Photoenergy* 2019 (May): 1–9. https://doi.org/10.1155/2019/7408763.

Surana, Deepti, Juhi Gupta, Satyawati Sharma, Sunil Kumar, and Pooja Ghosh. 2022. "A Review on Advances in Removal of Endocrine Disrupting Compounds from Aquatic Matrices: Future Perspectives on Utilization of Agri-Waste Based Adsorbents." *Science of The Total Environment* 826 (June): 154129. https://doi.org/10.1016/j.scitotenv.2022.154129.

Teta, C., B. F. Holbech, L. Norrgren, and Y. S. Naik. 2018. "Occurrence of Oestrogenic Pollutants and Widespread Feminisation of Male Tilapia in Peri-Urban Dams in Bulawayo, Zimbabwe." *African Journal of Aquatic Science* 43 (1): 17–26. https://doi.org/10.2989/16085914.2017.1423269.

Vandenberg, Laura N., Russ Hauser, Michele Marcus, Nicolas Olea, and Wade V. Welshons. 2007. "Human Exposure to Bisphenol A (BPA)." *Reproductive Toxicology* 24 (2): 139–77. https://doi.org/10.1016/j.reprotox.2007.07.010.

Vieira, Wedja Timóteo, Marina Barbosa de Farias, Marcela Pires Spaolonzi, Meuris Gurgel Carlos da Silva, and Melissa Gurgel Adeodato Vieira. 2021. "Endocrine-Disrupting Compounds: Occurrence, Detection Methods, Effects and Promising Treatment Pathways—A Critical Review." *Journal of Environmental Chemical Engineering* 9 (1): 104558. https://doi.org/10.1016/j.jece.2020.104558.

Wang, Xueping, Jingchao Han, Chunjuan Bi, Xing Huang, Jinpu Jia, and Zhenlou Chen. 2017. "Distribution, Sources, and Risk Assessment of Polychlorinated Biphenyls in Surface Waters and Sediments of Rivers in Shanghai, China." *Frontiers in Earth Science* 11 (2): 283–96. https://doi.org/10.1007/s11707-016-0590-3.

Wee, Sze Yee, Ahmad Zaharin Aris, Fatimah Md. Yusoff, and Sarva Mangala Praveena. 2024. "Public Perception on Human Exposure Risk: A Case Study on Endocrine Disrupting Compounds in the Environment." *Ecotoxicology and Environmental Safety* 270 (January): 115830. https://doi.org/10.1016/j.ecoenv.2023.115830.

Wilk, Barbara K., Sylwia Fudala-Ksiazek, Małgorzata Szopińska, and Aneta Luczkiewicz. 2019. "Landfill Leachates and Wastewater of Maritime Origin as Possible Sources of Endocrine Disruptors in Municipal Wastewater." *Environmental Science and Pollution Research International* 26 (25): 25690–701. https://doi.org/10.1007/s11356-019-05566-4.

Yang, Wenfeng, Xinxin Gao, Yixiao Wu, Liang Wan, Lichen Tan, Shaoman Yuan, Huijun Ding, and Weihao Zhang. 2020. "The Combined Toxicity Influence of Microplastics and Nonylphenol on Microalgae Chlorella Pyrenoidosa." *Ecotoxicology and Environmental Safety* 195 (June): 110484. https://doi.org/10.1016/j.ecoenv.2020.110484.

You, Hyekyoung Hannah, and Gwonhwa Song. 2021. "Review of Endocrine Disruptors on Male and Female Reproductive Systems." *Comparative Biochemistry and Physiology. Toxicology & Pharmacology* 244 (June): 109002. https://doi.org/10.1016/j.cbpc.2021.109002.

10 Potential of One Health-Based Interventions towards Addressing Persistent Organic Pollutants

Luca Baccini and Luigi Marino

10.1 INTRODUCTION

10.1.1 Persistent Organic Pollutants (POPs)

Persistent organic pollutants (POPs) are chemical substances widely recognized for their extreme toxicity, persistence and ability to accumulate in living organisms over time due to their high lipophilicity. Compared to other pollutants, POPs are much more toxic even at very low concentrations (Kumar et al., 2022). Halogenated POPs are known for their chemical stability, in which the strong bond between carbon and halogens (e.g. chlorine/bromine/fluorine) makes them resistant to environmental degradation via chemical, biological and photolytic reactions (Guo et al., 2019). These pollutants can enter air, soil and water through a range of pathways. Some natural sources of POPs are volcanic eruptions and forest fires, but the main sources are of anthropogenic origin. In fact, inappropriate waste disposal, run-off from landfills, industrial and agricultural sites, industrial and domestic wastewater discharge, solid waste transport, waste incineration, fumes from vehicles, and improper combustion are considered the main POPs sources (Kumar et al., 2022; Wang et al., 2024) (Figure 10.1). Additionally, once emitted, POPs tend to bind to organic matter in soils and sediments, effectively favouring their transport in areas far from the source (El-Shahawi et al., 2010; Mojiri et al., 2019) (Figure 10.1).

Their persistence and low water solubility allow POPs to disperse over vast distances through a repeated process of evaporation and deposition, making it difficult to trace the original source of contamination. Therefore, mechanisms such as atmospheric transport, water currents and migratory species make POPs a global environmental concern. As a result, POPs can contaminate remote ecosystems such as the Arctic and the Antarctic, where they have been detected in the tissues of wildlife (Hung et al., 2022; Tenenbaum, 2004; Yuan et al., 2021). POPs, and products containing POPs, were introduced for various industrial, agricultural and commercial purposes. They were extensively used as pesticides to control agricultural pests

DOI: 10.1201/9781032623801-10

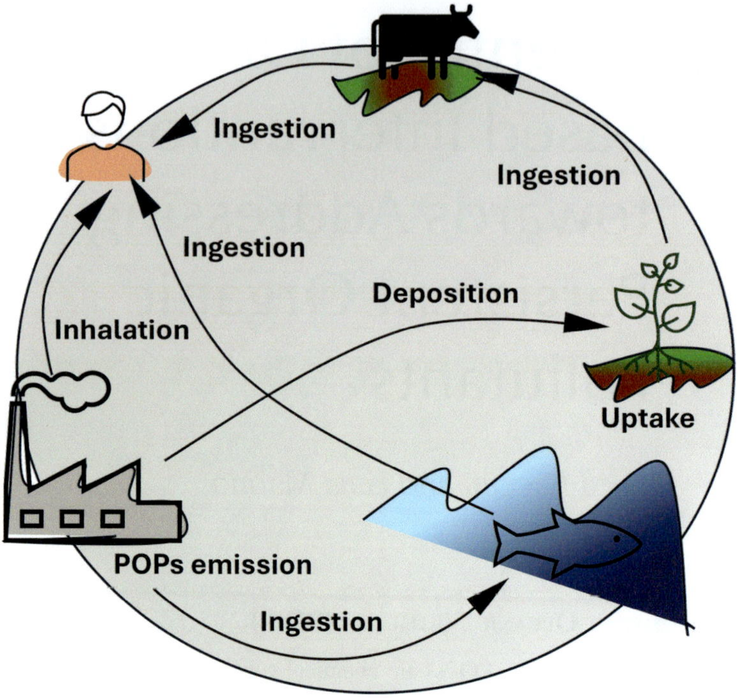

FIGURE 10.1 POP sources, pathways and targets.

and as flame retardants (e.g. polybrominated diphenyl ethers, PBDEs) in manufacturing. In addition, certain POPs, such as polychlorinated biphenyls (PCBs), were utilized as dielectric and coolant fluids in electrical equipment like transformers and capacitors. The chemical properties of these substances have provided significant benefits, leading to their widespread adoption in the market. On the other hand, the same chemical properties contributed to their accumulation in the environment and in the tissues of living organisms. However, their persistent nature and bioaccumulation/biomagnification potential, due to their lipophilic behaviour, have raised significant concerns regarding their adverse effects on both environmental and human health. For instance, POPs such as PCBs and DDT have been found in alarming concentrations in top predators like polar bears, killer whales and predatory birds such as ospreys and eagles. These high concentrations can disrupt endocrine systems, impair reproduction and increase vulnerability to disease, posing significant threats to biodiversity. For humans, more than 90% of exposure to POPs occurs primarily through the consumption of contaminated food, particularly animal products such as fish, dairy and meat (Guo et al., 2019). POPs have been linked to numerous adverse health effects, including endocrine disruption, developmental delays in children, immune system suppression and an increased risk of cancer (Gaur et al., 2022; Rickard et al., 2022).

Recognizing the global scale of the problem, the international community took a significant step in 2001 with the adoption of the Stockholm Convention on Persistent

Organic Pollutants (Stockholm Convention, 2023). The Convention's primary goal is to eliminate or restrict the production and use of POPs, while promoting the safe management of existing stockpiles and contaminated sites. The Convention, entered into force in 2004 and signed by over 90 countries, has focused on an initial list of 12 substances. The POPs were classified into three categories – by-products, industrial chemicals, pesticides – and listed into three annexes of the Convention text based on the measures that Parties have to adopt (i.e. elimination, restriction, reduction of unintentional release). Some of the initial POPs listed in the Convention are reported below:

- Polychlorinated dibenzo-p-dioxins and polychlorinated dibenzofurans (PCDD and PCDF respectively) are known by-products of many industrial processes involving chlorinated chemical compounds. The waste incineration is a major source of dioxins and furans;
- PCBs were widely used in the 20th century as coolants and lubricants in transformers and capacitors and as hydraulic and heat exchange fluids in electrical apparatuses;
- Pesticides, including dichloro-diphenyl-trichloroethane (DDT), hexachlorobenzene (HCB), and Aldrin, have been widely used in the agriculture industry since the Second World War to protect plants from fungi and insects.

Polycyclic aromatic hydrocarbons (PAHs) are generated mostly because of the incomplete combustion of either natural sources such as coal and wood or man-made sources such as automobile emissions and cigarette smoke. Although PAHs are unintentional toxic by-products, they are not listed in the Convention and are regulated by the Aarhus Protocol (Guo et al., 2019).

Ongoing revisions and scientific reviews have expanded the list, reflecting new research on emerging contaminants and extending protection to additional ecosystems and populations. Therefore, additional emerging POPs (ePOPs) were incorporated into the Convention via amendments implemented between 2009 and 2023 (Lorenzo et al., 2018; Stockholm Convention, 2023). Most of these emerging contaminants encompass substances widely used in industrial processes and everyday products:

- Polybrominated diphenyl ethers (PBDEs) are organobromine compounds that were extensively used from the beginning of the 1960s as flame retardants in electrical appliances, computers, televisions, insulation wires, cables and building materials. The waste of electrical and electronic equipment has been released to the environment and resulted in high concentrations of PCBs and PBDEs in fatty foods such as meat, fish and poultry;
- α-Hexachlorocyclohexane (α-HCH) and β-Hexachlorocyclohexane (β-HCH) are industrial by-products of the production of the insecticide lindane (γ-HCH);
- PFAS were largely used in everyday products, including food packaging, non-stick pans, waterproof textiles, firefighting foams and so on. Perfluorooctane

sulfonic acid (PFOS), its salts, and perfluorooctane sulfonyl fluoride (PFOSF), perfluorooctanoic acid (PFOA), its salts, and PFOA-related compounds were added as ePOPs.

Non-PBDEs, novel brominated flame retardants (NDFRs), organophosphate flame retardants (PFRs), short-chain chlorinated paraffins (SCCPs), and short-chain PFAS have been extensively adopted as substitutes for restricted or banned compounds (Brunn et al., 2023; Lorenzo et al., 2018). Nonetheless, the short-chain perfluorohexane sulfonic acid (PFHxS), its salts, and PFHxS-related compounds, as well as SCCPs were recently introduced into the list of the Convention. However, the ePOPs present additional challenges since they are less studied. In general, addressing POPs remains a complex and evolving challenge. Due to their high resistance, advanced treatments need to be employed for their remediation. Several well-established technologies have been developed for POPs remediation from all the environmental matrices, based on separation/concentration (e.g. phytoremediation, adsorption) and destruction (advanced/reduction processes, high-temperature incineration). However, drawbacks including high-energy demand, costs for disposal/regeneration of the adsorbent, and production of toxic by-products represent the main limitation for the scale-up of these technologies. The introduction of emerging technologies, involving sustainable materials, energy-efficient and cost-effective solutions, as well as using a multi-barrier approach can help to address the POPs remediation. However, reducing/restricting the production of these substances, preventing POPs pollution before they enter the environment, may be more efficient.

POPs are an increasing global threat. Addressing this issue requires coordinated efforts across multiple fronts, including stricter regulatory controls, advances in remediation technologies, and increased public awareness. Continued research is essential to develop sustainable and effective strategies to prevent POPs contamination and reduce existing environmental burdens, with particular emphasis on understanding the long-term ecological and health impacts of these persistent pollutants (Marican and Durán-Lara, 2018; Sun et al., 2020). Additionally, a holistic vision recognizing the complex interdependence of human, animal and environmental health is highly desirable. The One Health (OH) approach acknowledges that the health of human beings is strongly connected to the health of animals and a shared environment. This view supports the necessity of cooperation in addressing POPs challenges. Integrating research from multiple disciplines, including public health, veterinary medicine, environmental science, and medicine may mitigate POPs-related hazards and promote a new key perspective for a sustainable future (Schaffner et al., 2024; Wang et al., 2024).

10.1.2 Pills of One Health

Actions in the framework of the OH approach are proposed and advocated by the Food and Agriculture Organization of the United Nations (FAO), the United Nations Environment Programme (UNEP), the World Organisation for Animal Health (WOAH, founded as OIE), and the World Health Organization (WHO), making up the Quadripartite Organizations. The OH approach is applicable to many issues, such as: antimicrobial resistance (AMR), zoonotic diseases, vector-borne diseases and food safety and foodborne diseases, as well as the environment. The One Health Joint Plan of Actions (OHJPA) draws interventions to provide adequate guidance and tools for the effective implementation of multisectoral approaches (Action 1); to reduce the risk and minimize the local and global impacts of zoonotic epidemics and pandemics adopting upstream prevention measures and strengthening One Health surveillance, early warning and response systems (Action 2); to reduce the burden of endemic zoonotic, neglected tropical and vector-borne diseases by supporting countries in implementing community-centric, risk-based solutions, strengthening policy and legal frameworks, and increasing political commitment and investment (Action 3); to promote awareness, policy changes and action coordination among stakeholders (Action 4); to take joint action to preserve antimicrobial efficacy and ensure sustainable and equitable access to antimicrobials for responsible and prudent use in human, animal and plant health (Action 5); and to protect and restore biodiversity and to prevent the degradation of ecosystems and the wider environment (Action 6). Likewise, the Policy Brief (PB) advocates the need for strengthening surveillance through monitoring, risk assessment and control to prevent chain reactions through animals, humans and environment (Figure 10.2).

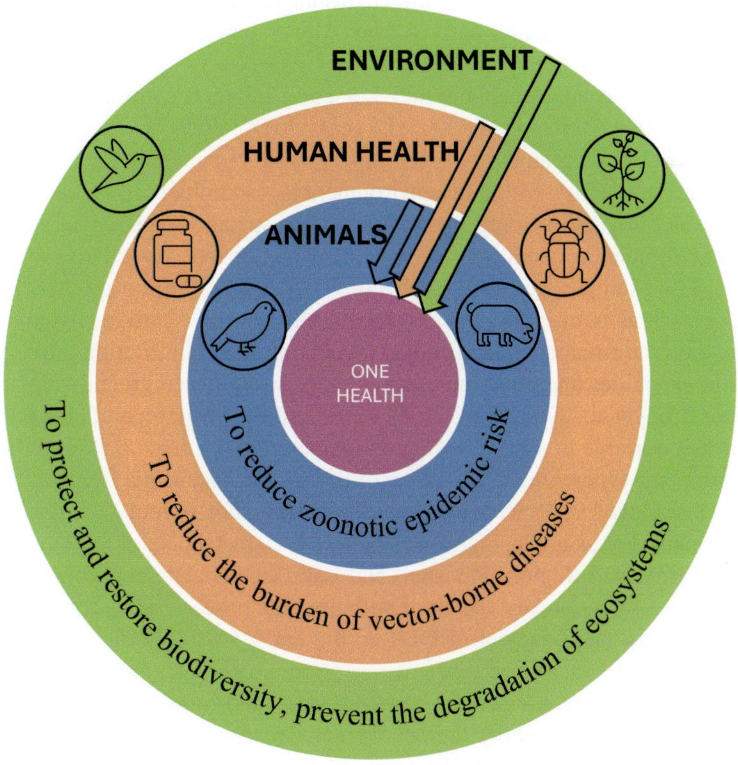

FIGURE 10.2 The three OH targets and specific actions according to the OHJPA.

Despite the wide range of actions reported in the OHJPA, no specific references are available for POPs or ECs. Nonetheless, ECs' effects on humans, animals and the environment are interrelated. For this reason, the present research provides potential OH-based interventions to address POPs, building on the actions suggested by the OHJPA and the PB.

The aim of this study was to investigate and propose potential interventions to mitigate POPs' impact on humans, animals and the environment. To do this, the main OH action plans were discussed, POPs interaction with and effects on the OH components were investigated, and then a strategy made up of risk assessment, control and remediation actions was broken down. Finally, this study sets a starting pathway to tackle POPs, paving the way for future improvements by involving expertise from different disciplines, thus perfectly aligning with the OH approach.

10.2 IMPACT OF POPS ON ENVIRONMENT AND HUMAN HEALTH

10.2.1 Environmental Implications

POPs exhibit persistence, bioaccumulation and high mobility, leading to long-lasting environmental footprints that endanger ecological balance. Their resistance to degradation enables them to persist in the environment for long periods. For instance, HCB and dieldrin (compound derived from Aldrin degradation) are extremely persistent, showing a half-life up to ~ 9 and 7 years respectively, while for PFHxS, PFOS and PFOA was estimated a half-life of 5.3, 3.4 and 2.7 years respectively (Australian Government, 2024; Barber et al., 2005; Li et al., 2018). POPs are transported around the world mainly by ocean currents and atmospheric circulation (Gaur et al., 2022). The distribution of POPs within the environment is also influenced by factors such as volatility, solubility and adsorption capability of the specific compound. Furthermore, several environmental conditions may indirectly contribute to a large-scale POPs diffusion. For example, key factors affecting PFAS fate and transport include soil properties such as permeability, surface charge and organic carbon content; aquifer features; oxidation-reduction conditions; precipitation and infiltration rates; groundwater flow velocities and directions; and interactions between groundwater and surface water (Cara et al., 2022; ITRC, 2022). Furthermore, the effects of climate change add a further variable in understanding the transport and spread of POPs. The rising temperatures may increase volatilization of these compounds and release previously trapped and archived POPs with consequent contamination of remote regions, such as mountain lakes. For example, high levels of pp'DDE (metabolite of DDT) were found in Lesser Himalayan Region (Pakistan) and Lake Iseo (Italy), implying that glacier retreat represents a secondary source of pollution that poses a risk to aquatic life even many years after the phase-out of DDT in the early 1970s (Bettinetti et al., 2011; Riaz et al., 2021). Natural ecosystems are often exposed to complex mixtures of these substances, where interactions between compounds may produce synergistic or antagonistic effects that amplify their ecological consequences (Hansen et al., 2019). Additionally, plastics and microplastics can serve as carriers for other contaminants, facilitating their accumulation in aquatic organisms (Rafa et al., 2024; Wang et al.,

2024). These dynamics shape exposure pathways and determine the extent of POPs environmental and ecological impact. Consequently, evaluating the behaviour of POPs across atmospheric, terrestrial and aquatic systems is essential for assessing risks for humans and animals. Indeed, POPs can accumulate in living organisms and pass from one species to the next through the food chain, reaching toxic concentrations and posing serious risks to biodiversity. Species at higher trophic levels, such as eagles, polar bears and marine mammals, show severe effects, including impaired reproduction and susceptibility to infection due to vitamin and thyroid deficiencies linked to contaminated diets. Additionally, certain POPs like DDT have caused eggshell thinning in birds of prey, drastically reducing their populations (Fitzgerald and Wikoff, 2014). In mammals, endocrine disruption has been linked to reproductive and developmental issues, impacting biodiversity and ecosystem stability (Yadav et al., 2015) (Figure 10.3).

A recent study showed concentrations of ΣHBCDD up to 179 ng/g lipid weight in insects and their predators in a former e-waste recycling site in South China, highlighting the biomagnification of α-HBCDD in terrestrial birds (Liu et al., 2020). Very high concentration of PFAS (up to 7010 ng/g wet weight), especially n-PFOS and perfluorooctanesulfonamide (PFOSA), were found in the livers of striped dolphins living in the North-Western Mediterranean Sea (Garcia-Garin et al., 2023). Very short-, short-, medium- and long-chain chlorinated paraffins were found to be magnified in red-backed rat snakes with maximum biomagnification factors ranging from 2.8 to 4.5 (Du et al., 2020). Although PCBs, PCDDs and PCDFs tend to bioaccumulate in the adipose tissue of animals, they were found in crops, fruit and vegetables as a result of transfer from soil to plants and/or suspension and subsequent deposition on leaves and crops (Grassi et al., 2010; Hülster and Marschner, 1993). Grassi and co-authors found high levels of PCBs in rosemary compared to other crops, suggesting that this plant can be used as a bioindicator. The literature on the effects of POPs in the environment underlines that human exposure to these dangerous substances is inevitable due to the intake of contaminated water, air and food.

10.2.2 Human Health Impact

POPs can enter the human body through multiple exposure pathways: consuming contaminated food or water, inhaling polluted air, as well as dermal contact. Extensive research has demonstrated that exposure to air pollutants (e.g. PFAS, PAHs, PBDEs) can lead to dysfunctions of the cardiovascular and immune systems (Wang et al., 2024). Moreover, approximately 90–98% of human exposure to dioxins and PCBs comes from dietary sources, primarily foods of animal origin, suggesting animal feed plays a significant role in introducing these compounds into the food supply (Malisch and Kotz, 2014). The presence of several PFAS, organochlorine pesticides (OCPs), and other POPs in fishes for human consumption have been documented by several studies (Chang, 2018; Haukås et al., 2007; Li et al., 2024). A well-documented example of POPs human exposure is the contamination of the Great Lakes region in North America, where years of industrial activity resulted in high levels of PCBs and dioxins in fish, leading to health advisories for local communities (US EPA, 2009).

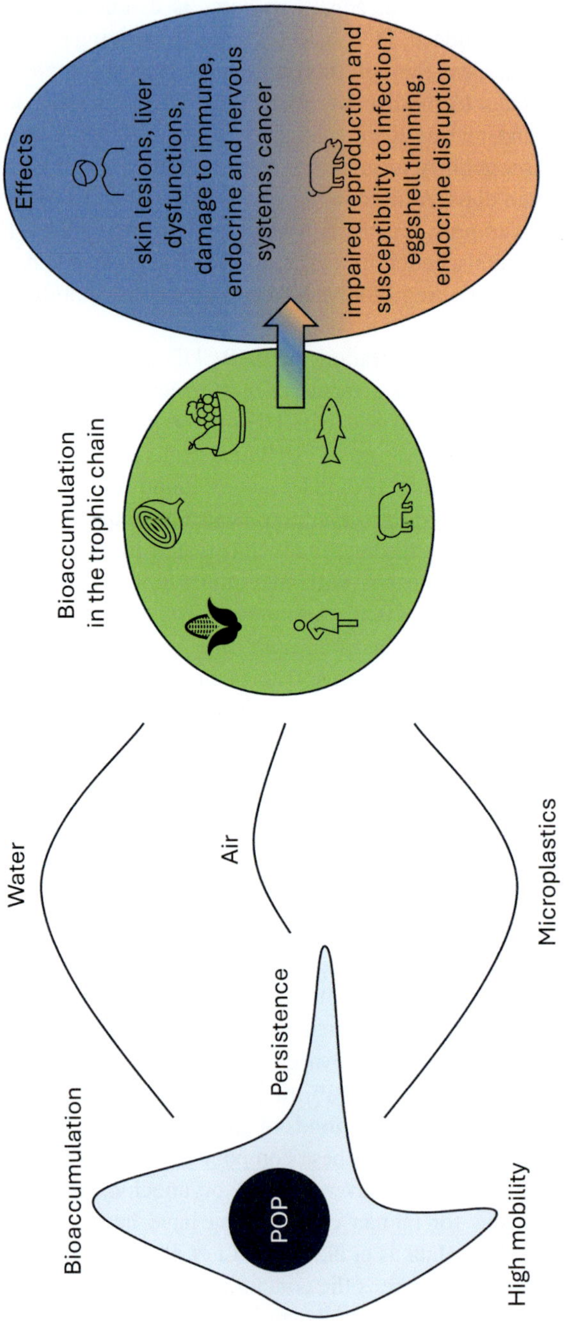

FIGURE 10.3 POP characteristics, pathways, bioaccumulation and effects.

According to the WHO, short-term exposure may result in skin lesions and liver disfunctions; long-term exposure is linked to damages to the immune, endocrine and nervous systems (WHO, 2023). Numerous studies have emphasized the widespread presence of POPs (such as short- and long-chain PFAS, OCPs, and PCBs) in drinking water supplies, posing significant risks to the health of both humans and animal species (Fujii et al., 2007; Hu et al., 2016; Vasseghian et al., 2021a). Since PFAS are not metabolized, once ingested, they readily enter the enterohepatic circulation and accumulate in the liver. Several studies showed PFAS exposure linked to onset of liver, kidney, testicular and prostate cancer (Brunn et al., 2023; Cao et al., 2022). Indeed, the International Agency for Research on Cancer (IARC) classified POPs into Group-1, that is, cancer-causing substances (IARC, 2024). Furthermore, human exposure to POPs may begin during pregnancy, causing potential adverse health effects in the foetus. Indeed, several studies reported POPs levels in breast milk, maternal blood, placenta and cord blood (Aerts et al., 2019; Jeong et al., 2018; Müller et al., 2019). Due to their effects on the thyroid and hormone systems of pregnant women, as well as their impacts on organ development and birth outcomes, PFAS are regarded as endocrine disruption compounds (EDCs) and immunological suppressors (Rickard et al., 2022). Table 10.1 summarizes the exposure hazards associated with POPs.

Given the environmental as well as the human and animal implications related to POPs exposure, adequate strategies to minimize POPs formation, spread and interaction need to be adopted.

TABLE 10.1
Brief list of POPs' possible hazards

POP	Possible Hazards
PAHs	Mutagenicity and carcinogenicity, DNA damage, oxidative stress, impaired male fertility, respiratory diseases, cognitive dysfunction among children and cancer (breast cancer)
Pesticides	Neurological symptoms, endocrine disruption, infertility and foetal malformation, diabetes, cancer (breast cancer, testicular, prostate and kidney cancer), reproductive problems, cardiovascular problems, high blood pressure, glucose intolerance and obesity
Dioxins/furans	Language delay, disturbances in mental and motor development, cancer, diabetes, endocrine disruption, high blood pressure, glucose intolerance and cardiovascular problems
PCBs	Endocrine disruption, neurological disorders, liver injury, diabetes, cancer (breast, prostate, testicular, kidney, ovarian and uterine cancers), cardiovascular problems and obesity
PBDE, HBCD, PCDE, PCN	Reproductive problems, cancer(testicular), diabetes, obesity and cardiovascular problems, Endocrine disruption, reproductive problems and behavioural effects, cancers
PFOS and PFOA	Breast cancer

Source: Guo et al. (2019), CC-BY 4.0 license.

10.3 ONE HEALTH-BASED INTERVENTIONS TO ADDRESS POPS

Preventing, reducing and avoiding the emission of POPs into the environment and in turn in the food chain allow mitigating the risk of contamination through the environment, humans and animals. The OHJPA in actions 1 and 6 suggests adopting upstream prevention measures, protecting and restoring biodiversity preventing degradation, while the PB advocates the need for surveillance, risk assessment and control as OH-based actions. Based on these recommendations, three OH-based strategies have been developed, that is, control and prevention techniques, risk assessment and remediation technologies (Figure 10.4).

10.3.1 POPs Risk Assessment

The risk assessment is a powerful tool that allows identifying pollutants' effects on the ecosystem and human health, supporting the identification of potential polluting sources, pollutant types, main exposition pathways and hit areas. Therefore, control and remediation strategies can be accurately and more precisely adopted for each specific case study.

Ajay et al. (2022) conducted a risk assessment on the exposure of the Indian population to PCDD/Fs and dioxin-like PCBs (dl-PCBs) due to waste burning. The authors investigated the risk associated with the burning source type, that is, a landfill located 7km away from the main city and roadside dumping sites located along the main streets; the age of the exposed citizen, that is, adults or children; and the type of emitted contaminant, that is, PCDD/Fs or dl-PCBs, as well as the exposure pathway, that is, inhalation or dermal contact. The parameters taken into account for the risk assessment were:

- the Hazard Quotient (HQ), that is, the ratio of daily exposure dose ($mg_{pollutant}/kg_{body\_weight}$/day) to reference dose, that is, the dose above which illness can occur, for the non-carcinogenic risk assessment. The threshold value of HQ is 1, where HQ < 1 indicates lower exposure and is considered safe;
- the Hazard Index (HI), that is, the cumulative effect of the HQs arising from various chemicals through different exposure pathways;
- the Incremental Life Cancer Risk (ILCR), that is, the cancer risk for the whole life exposure.

According to this study, the HQ for inhalation is always 10–100 times greater than that of dermal contact. Moreover, based on the HI values, children are the most susceptible to the risk compared to adults; roadside burning dumping sites pose higher risk than municipal solid waste burning sites; and the probability of non-carcinogenic risk arising from PCDD/Fs exposure during open burning is much higher than dl-PCBs. The ILCR trends of carcinogenic risks arising from exposure to PCDDs were found to be 10–100 times higher than dl-PCBs. Cumulative values were higher than 10^{-4} and between 10^{-6} and 10^{-4} in the roadside burning dumpsites, while below the action level, that is, 10^{-6} for the landfill sites. This indicates a moderate to unacceptable level of cancer risk at all the roadside investigated sites.

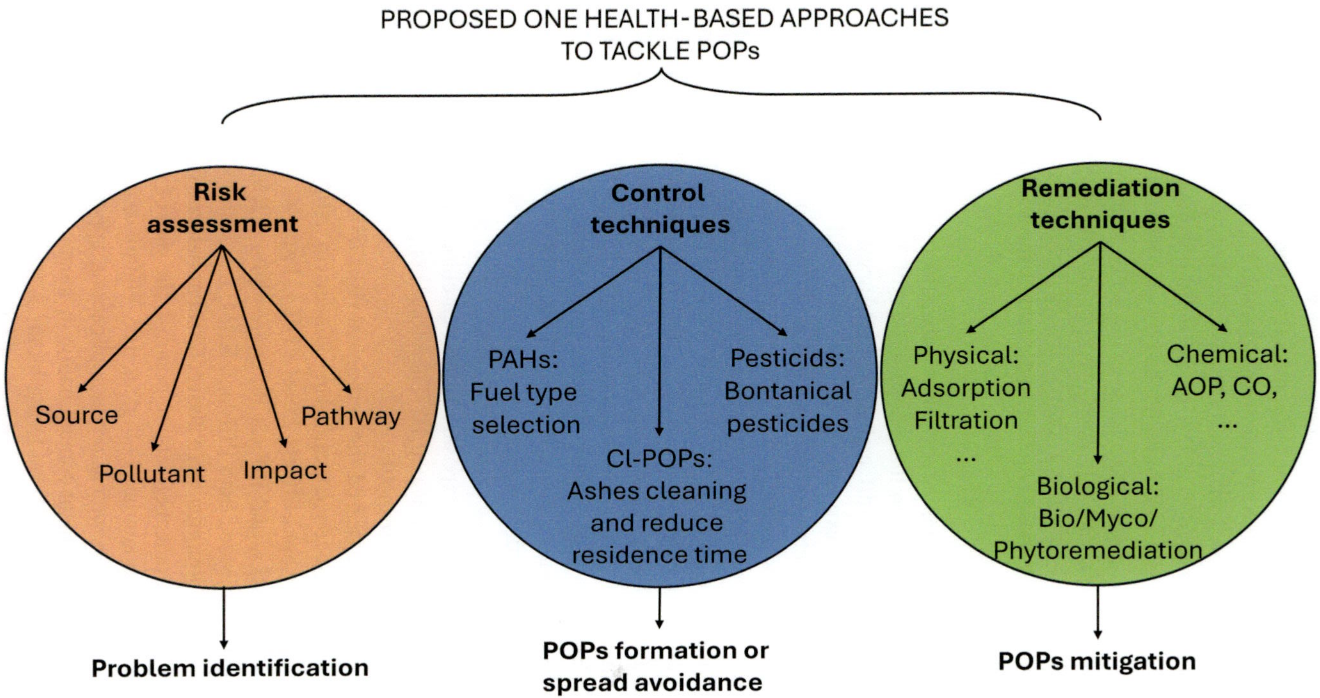

FIGURE 10.4 The proposed OH-based approaches to address POPs. The risk assessment allows the problem identification and the control techniques reduce POP emission and spread, while the remediation techniques improve the environmental quality of a POP polluted area.

Vasseghian et al. (2021) conducted a health risk assessment of POPs concentrations in water resources taking into account data collected from papers published between 1970 and 2020. A health risk assessment model was applied to evaluate both carcinogenic (R) and non-carcinogenic (HQ) risks from contaminants in drinking water, using a Monte Carlo approach to assess uncertainty. The highest carcinogenic risk was observed for β-HCH in Turkey and China, followed by α-HCH in Mexico. The Monte Carlo analysis showed that 75% of the child population had a carcinogenic risk below $1.07*10^{-6}$, while 95% of adults had a risk below $7.35*10^{-6}$, and was thus very close to the USEPA acceptable carcinogenic limit, that is, 10^{-6}. Regarding non-carcinogenic risk, the highest HQ values were identified for Aldrin in all analysed countries, followed by Dieldrin in Turkey and γ-HCH in Mexico. However, acceptable HQ values, meaning those below 1, were found for α-ES in India, Turkey and Mexico, as well as for α-HCH in China, Turkey and Mexico, and β-ES in India and Turkey.

Linares et al. (2010) compared the adult population's environmental and dietary exposure to POPs and metals in the industrial area of Tarragona County. Air inhalation, dermal absorption of soil and dust, and ingestion of soil and dust were considered environmental exposure pathways. Air, soil and food samples were collected for the study. Dietary intake emerged as the dominant exposure route, contributing at least 93% of total intake and exceeding 97% for most pollutants. Dermal absorption was the primary pathway for PCDD/Fs and PCBs, whereas air inhalation significantly contributed to PCNs (55%) and PAHs (51%). Soil and dust ingestion were the main routes of exposure to most metals, except for mercury. Specific food groups played distinct roles in pollutant exposure. Based on the exposure of all the foodstuff investigated, fish and seafood accounted for 25% of PCDD/Fs and 74% of PCBs, while cereals were the largest contributors to PCNs (32%) and PAHs (45%). Meat was the primary source of carcinogenic PAHs (34%). Water consumption had a negligible impact, contributing an average of just 3% to overall chemical intake. Non-carcinogenic risks were below critical thresholds, with no contaminant exceeding a HQ of 1. The HI following exposure to non-carcinogenic pollutants, which was calculated as the sum of HQs, was 0.10. Therefore, none of the POPs exceeded the recommended cancer risk threshold.

Hidayati et al. (2021) assessed the ecological risks of organochlorine pesticides (OCPs) by comparing their concentrations in sediments to sediment-quality guidelines (SQGs), including Threshold-Effects Level (TEL), Probable-Effects Level (PEL), Effects Range-Low (ERL), and Effects Range-Medium (ERM). The sum of chlordane concentrations exceeded both PEL and ERM, indicating a potential ecological risk, while DDT concentrations were above TEL but below PEL, suggesting a possible risk of adverse biological effects. In contrast, the sum of seven PCBs remained below the threshold for potential ecological risk. The intensive use of pesticides in paddy and onion farming, as well as urban applications, appears to contribute to OCP contamination. Surface sediment contamination in aquaculture systems by OCPs and PCBs could pose adverse effects on living biota, as indicated by SQG assessments. The study also analysed the accumulation of OCPs and PCBs in shrimp muscle, detecting values of 19 mg/kg dw for SDDTs and 6.8 mg/kg dw for SHCHs in Brebes.

The highest accumulation of S7 PCBs was recorded in shrimp muscle at 19.8 mg/kg dw in Tegal and 19.3 mg/kg dw in Pemalang. Despite the detected accumulation, the levels of POPs in shrimp muscle remained below the maximum residue limits established by the US Food and Drug Administration.

In a study conducted in the Yuvacık Lake (Turkey), Güzel et al. (2022) investigated the PAH, PCB and OCP concentrations found in sediments with the SGS threshold limits. They found that none of the investigated pollutants were responsible for potential ecological risks, except for DDT with a concentration above TEL but below PEL.

Zhang et al. (2013) conducted a study to assess the quality of water, soil and sediment in the Pearl River Delta (PRD) in China. Their analysis focused on the presence of DDTs and HCHs in different environmental compartments, particularly in freshwater ponds and fishing harbours. They found that most sediments contained DDTs at levels exceeding the ERL, with a significant number of fishing harbour sediments surpassing the ERM. This suggests a potentially severe ecological risk to the aquatic environment, particularly in fishing harbours where DDT contamination is most pronounced. In contrast, no HCH concentrations exceeded the sediment quality guidelines in freshwater ponds. Combined with water quality assessments, this indicates that HCHs in freshwater ponds pose no significant environmental risk.

As shown, the risk assessment allows tracing back the contamination source and narrowing down the array of pollutants posing a risk for human and ecosystems. Therefore, such a tool favours an adequate and precise selection of actions and resources to tackle the identified issue, in order to minimize direct and indirect effects on the OH components.

10.3.2 POPs Prevention and Control Methods

Prevention tops the five steps waste hierarchy according to the EU Waste Framework Directive, as the most preferred approach to avoid or minimize endangering human health and harming the environment. From this perspective, control strategies can be adopted in order to prevent or minimize the formation of POPs and/or their spread into the environment, where they can pose a risk for all the OH components. Following are a few examples of control strategies and innovative approaches.

In a review study conducted by Das et al. (2023), PAHs formation and accumulation on meat during processing as well as control and prevention strategies have been addressed in-depth. PAHs are produced in processed meat during thermo-processing techniques such as smoking, grilling, barbecuing, roasting and frying. These methods involve high temperatures and incomplete combustion of fuels, which generate volatile particles that deposit on the meat. A variety of strategies can be employed to mitigate PAH formation during meat processing, focusing on modifications to cooking techniques, fuel types and pretreatment methods. Cooking methods that minimize direct exposure to high heat, such as steaming, boiling and microwaving, are highly effective at reducing PAH levels. Shallow-pan frying and hot air frying are safer alternatives to deep-frying. Indirect heating methods, such as using electric or gas ovens instead of charcoal grills, also contribute to lower PAH concentrations. Adjusting the cooking set-up to maintain a safe distance of at least

25 cm between the meat and the heat source can significantly reduce PAH accumulation. The choice of fuel is crucial in PAH reduction. 'Flameless and smokeless' charcoal, as well as alternatives like coconut shells and oak wood, produce fewer PAHs than traditional charcoal or wood. Liquid smoking is a safer alternative to conventional smoking, as it involves fractionation and purification processes that limit PAH contamination. Pretreatment of meat with marinades containing antioxidants or acidic ingredients can dramatically reduce PAH formation. For example, marinades made with lemon or tamarind juice, tomato juice, garlic, onion and spices (such as cumin, coriander and black pepper) have been shown to lower PAH levels significantly due to their antioxidative properties. Capsaicin, a biologically active compound found in jalapeños, acts as a free radical scavenger, preventing reactions that form PAHs and enhancing both the safety and the shelf life of grilled meat. Similarly, marinades using beer (e.g. Pilsner or dark beer) and vinegar exhibit strong antioxidative effects, further diminishing PAH concentrations in meat products. Biological approaches offer another promising avenue for PAH reduction. The use of microorganisms such as bacteria, algae and fungi capable of metabolizing PAHs can help detoxify contaminated meat products. These microbes break down PAHs into less harmful hydrophilic metabolites through the activation of antioxidative enzymes like oxidase, manganese peroxidase and lipase. Additionally, these microorganisms bind heat-generated carcinogens to their cell walls and peptidoglycans, reducing their toxicity. Physical barriers can also reduce PAH contamination. Casings, particularly cellulose-based ones, serve as protective layers that prevent smoke particles from directly contacting the meat. By employing a combination of these strategies – such as adjusting cooking methods, using safer fuels, incorporating protective barriers and utilizing antioxidative marinades – producers and consumers can effectively minimize PAH accumulation in thermally processed meat, ensuring a safer and healthier product.

PCDD/F, PCBs and PAHs are involuntary combustion products. Controlling and minimizing the formation and emission of such compounds has been addressed by the European Commission in the Best Available Techniques Conclusions for specific IPPC activities. Overall, among all the IPPC activities surveyed, the BATs for such POPs focus on minimizing the formation as well as the trapping of these compounds into a substrate, reducing the pollutants spreading into the environment. Among the former techniques, the BATC for 'Industrial emissions for iron and steel production' suggest the avoidance of raw materials containing PCDD/F and PCB or their precursors as well as their suppression of formation by adding nitrogen compounds. Despite the accurate choice of the treated material, more preventative techniques might need to be put into place. For instance, post-combustion allows achieving a complete oxidation of the emitted compounds, and rapid quenching (cooling) of the off gas below 250°C through wet scrubbers or cooling towers inhibits the de novo process responsible for the formation of the above-mentioned POPs. Similarly, the "Waste incineration" BATCs suggest the optimization of the incineration process, the control of the waste feed and the cleaning of the boiler to reduce ash residence time and hence the formation of POPs. Finally, the "Large volume organic chemicals" BATCs recommend flue gas cooling. Regarding the abatement of already formed

POPs, the BATCs for the above-mentioned IPPC activities rely mainly on intercepting the flue gas with one or a combination of units. Among these, the most common are the adsorbents (e.g. activated carbon or halogenated activated carbon for large combustion plants; activated carbon, pulverised activated lignite coke or mixtures of these with lime in iron and steel production plants; activated carbon in waste incineration plants and large volume organic chemicals plants), electrostatic precipitators, bag filters or Selective catalytic reduction (SCR).

Cheruiyot et al. (2020) investigated the effects of control strategies in the reduction of Cl-POPs (e.g. PCBs, PCDDs, PCDEs) formation from start-up processes of municipal solid waste incinerators (MSWIs). Cleaning the ash accumulated beneath the furnace beds of the combustion chambers and deposited on the walls of the superheaters and economizers, together with shortening the residence time of the flue gas in the combustion chambers during the optimal Cl-POP formation window (250–450 C), allowed achieving a 77.6–86.8% Cl-POP formation reduction. Moreover, advancing the injection times of the activated carbon and lime slurry allowed an additional 1.9–3.6% of Cl-POP formation reduction. Yang et al. (2022) employed similar control strategies to mitigate the formation and emission of brominated persistent organic pollutants (Br-POPs). Their findings revealed that cleaning the accumulated ash was the most effective measure, contributing to the largest reductions in PBDD/F and PBDE emissions, accounting for 72.8% and 51.3% of the total reductions, respectively. Additionally, combining ash cleaning with a reduction in the residence time of flue gas within the Br-POP formation temperature window further inhibited Br-POP formation. This combined approach, while less dominant, still accounted for 18.6% and 31.1% of the total reductions in PBDD/F and PBDE emissions, respectively, and was particularly effective in reducing PBB emissions, achieving a substantial reduction of 67.5%. In contrast, the contribution of the post-combustion strategy to emission reductions was significantly smaller. This approach accounted for only 8.6% and 17.6% of the total reductions in PBDD/F and PBDE emissions, respectively, and approximately one-third of the reductions in PBB emissions.

Botanical pesticides can be an alternative to the synthetic ones. As reported by Luiz De Oliveira et al. (2018), botanical pesticides, derived from plant extracts or isolated phytochemicals, are considered environmentally friendly alternatives to synthetic pesticides due to their biodegradability and lower potential for long-term environmental and health impacts. According to Lengai et al. (2020), botanical pesticides work as bacteria, fungi, nematodes, viruses and insect pests killers through many different mode of actions. The active ingredients in botanical pesticides, particularly the distinctive structural features of their secondary metabolites – such as alkaloids, essential oils (including terpenes), flavonoids, phenolics, phytosterols, polyketides and resins – possess the inherent ability to exert antibacterial, antifungal, herbicidal and insecticidal effects (Acheuk et al., 2022). Examples of botanical pesticides include essential oils and plant-derived active compounds: neem oil, azadirachtin, garlic oil, thyme essential oil, clove oil, jojoba oil, limonene, peppermint, cinnamon and sesame oils. Nanoparticle coatings, produced from materials like zein, alginate, natural gums (e.g., gum arabic), and chitosan, are innovative systems used to enhance the efficiency of botanical pesticides. These coatings improve bio-efficacy, protect active agents from external degradation (e.g. UV light, humidity), reduce volatility and provide slow-release mechanisms (Luiz De Oliveira et al., 2018). However, their widespread use is limited by high costs associated with raw materials, extraction, formulation and regulatory compliance, as well as variability in the quality and concentration of active ingredients influenced by environmental and geographical factors (Luiz De Oliveira et al., 2018). Moreover, producing botanical pesticides would require readily available raw materials, which are mainly used for food, ornamental or medicine purposes, and large cultivated areas, as well as specific storage and processing facilities (Lengai et al., 2020). Following is a list (Table 10.2) of botanical pesticides highlighting the biological effects and mechanisms of actions.

The examples reported previously, although non-exhaustive, highlight the wide range of approaches that can be adopted to minimize the formation and spread of POPs. For instance, controlling the formation of POPs during combustion reduces the amount of POPs transported by the wind that can deposit on the soil and into the water, in turn building up in plants, animals and humans. Likewise, turning agriculture towards the use of biological pesticides might allow minimizing the domino effect of pollutants building up in the trophic chain, thus reducing the lifetime exposure.

10.3.3 POPs Remediation Technologies

The remediation of POPs from environmental matrices demands a comprehensive and multifaceted approach due to their resistance to degradation, very low concentrations and potential to cause widespread ecological and health damage. Water treatment, soil remediation and gas emission treatment are essential processes to ensure environmental safety and public health. Each of these processes utilizes a variety of techniques tailored to address specific contaminants. However, POPs pose significant challenges due to their recalcitrant nature. Their chemical structure typically includes chlorine atoms (as in PCBs and dioxins), which make them resistant to chemical degradation, photolysis and biodegradation. For instance, the Carbon-Fluorine bond present in PFAS compounds is considered the strongest covalent bond in chemistry, giving PFAS high thermal and chemical stability. These kinds of properties mean that standard treatment technologies are generally not sufficient for effectively removing POPs from environmental media. In water treatment, the hydrophobic nature of POPs allows them to bind to particles rather than remain in the aqueous phase, making them difficult to remove through conventional processes, such as sand filtration, sedimentation and coagulation-flocculation (Nguyen and Le Tran, 2024). Similarly, in soil, the strong adsorption of POPs to soil particles and their slow rate of biodegradation limit the effectiveness of standard remediation technologies. Another limitation is that POPs encompass a wide range of compounds having different physical-chemical properties. Then, the technology must be able to guarantee a sufficiently high performance for the highest possible array of compounds which can present very different concentrations, behaviour and removal mechanisms. Therefore, POPs remediation requires specialized approaches, which often involve advanced treatment methods. Various remediation technologies have been used to address the contamination caused by POPs, based on separation/destruction methods.

TABLE 10.2
Some commercial botanical pesticides, bioactive compounds, biological effects and mechanism of action

Product	Botanical Name	Main Bioactive Compound(s)	Biological Effects	Target Site	Mechanism of Action
Capsicum oleoresin	Capsicum spp.	Capsaicin	Repellent, Fungicide, Nematicide, Bactericide	Nerve	Induced metabolism, impaired cell membrane, and nervous system. Acts as a physical repellent.
Cinnamon essential oil	Cinnamomum zeylanicum	Cinnamaldehyde	Insecticide, Herbicide	Octopaminergic system	Increase the level of intracellular messenger and effectively inhibit cyclic AMP of abdominal epidermal tissue.
Clove essential oil	Syzygium aromaticum	Eugenol	Insecticide, Herbicide	Octopaminergic system	Increase the level of intracellular messenger and effectively inhibit cyclic AMP of abdominal epidermal tissue.
Jojoba essential oil	Simmondsia californica Nutt., S. chinensis	Straight-chain wax esters	Fungicide, Insecticide	Octopaminergic system	Increase the level of intracellular messenger and effectively inhibit cyclic AMP of abdominal epidermal tissue.
Karanjin	Derris indica (Lam.) Bennet	Karanjin	Insecticide, Acaricide		Serves as feeding restraint, repellent, reduced growth, oviposition suppressor, and low or no fertilization.
Lemongrass essential oil	Cymbopogon nardus, C. citratus, Cymbopogon flexuosus	Citronellal, Citral	Insecticide, Herbicide	Octopaminergic system	Increase the level of intracellular messenger and effectively inhibit cyclic AMP of abdominal epidermal tissue.
Neem (neem oil, medium polarity extracts)	A. indica	Azadirachtin, Dihydroazadirachtin, Triterpenoids (Nimbin, Salannin	Insecticide, Acaricide, Fungicide	Endocrine System	Inhibit Prothoracicotropic hormone (PTTH); distort phagostimulant disruptor by cholinergic transmission

(continued)

TABLE 10.2 (Continued)
Some commercial botanical pesticides, bioactive compounds, biological effects and mechanism of action

Product	Botanical Name	Main Bioactive Compound(s)	Biological Effects	Target Site	Mechanism of Action
Nicotine	Nicotiana spp.	(S)-isomer, (RS)-isomers, and (S)-isomer of nicotine sulphate.	Insecticide	Nervous system	It competes with the neurotransmitter by attaching to acetylcholine receptors (nAChRs) at neuron synapses, producing unregulated nerve firing. The disturbance of normal nerve impulse performance caused physiological system malfunctions of the neurons.
Rotenone	Derris spp., Lonchocarpus spp., and Tephrosia spp.	Rotenone, Deguelin, (isoflavonoids)	Insecticide, Acaricide	Mitochondria	Cell respiratory enzyme inhibitor disrupts cellular metabolism, reduces ATP output. Nerve and muscle cell malfunctions lead to low feeding rates.
Ryania	Ryania spp. (Ryania speciose Vahl)	Ryanodine, Ryania, 9,21didehydroryanodine (alkaloids)	Insecticide	Muscles	Activation of sarcoplasmic reticulum. Affect calcium development and cause improper function of muscles.
Sabadilla	Schoenocaulon spp. (S. officinale)	Mixture of alkaloids (cevadine, veratridine)	Insecticide	Nerve	Obstruct the movement of neurons and potassium ion in nerve axons.
Thyme essential oil	Thymus vulgaris L. Thymus spp.	Thymol, Carvacrol	Insecticide, Fungicide, Herbicide	Octopaminergic system	Prevent octopamine receptors via tyramine receptors cascade.

Source: Ngegba et al. (2022), CC-BY 4.0 license.

A general and non-exhaustive overview of both well-established and emerging technologies for POPs remediation is presented in Table 10.3.

In general, conventional physical methods involved in water treatment plants result in modest removal of recalcitrant compounds such as POPs. Coagulation-flocculation techniques achieve limited removal of POPs, including PAHs, aldrin, dieldrin and lindane (Nguyen and Le Tran, 2024). Nonetheless, the same review article emphasizes the importance of integrating advanced treatments, such as adsorption processes on granular/powdered activated carbon to enhance the removal efficiency of PAHs and PFAS. However, even advanced physical methods exhibit clear limitations. Although they offer good removal efficiencies, processes based on adsorption and membrane filtration merely transfer the compound from the aqueous or the gas phase to the adsorbent (or retentate) without degrading the contaminant. As a result, they pose potential risks of secondary contamination and lead to additional costs (Sun et al., 2020). Nonetheless, many strategies can be explored to overcome the limitations of these technologies. For instance, bio-based adsorbents (e.g. zeolites, biochar), as well as biodegradable reactive media for permeable reactive barriers, may be considered to reduce carbon footprint and enhance environmental compatibility (Bobde et al., 2023). The use of innovative regeneration methods could be taken into account for both improving adsorbent lifetime and removing POPs (Gagliano et al., 2021). Furthermore, adopting renewable energy sources and optimizing electrode materials to reduce power consumption may improve the environmental sustainability of electro-based methods (Aryanti et al., 2024). However, more research is needed to extend the application of innovative technologies, such as sonolysis and non-thermal plasma treatments, in order to increase their economic and environmental feasibility on a real scale (Sidnell et al., 2022) (Figure 10.5).

Several chemical methods have been developed for the degradation of POPs into less hazardous substances (Table 10.2). Such technologies involve the addition of chemicals (e.g. sodium hypochlorite, hydrogen peroxide); energy sources (e.g. UV irradiation) for activating chemical reactions; and catalysts (e.g. titanium dioxide, nanomaterials) for improving degradation of target contaminants. Advanced oxidation processes (AOPs) are among the most effective and established methods (e.g. Fenton's process, UV- and ozone-based AOPs) for degrading persistent contaminants in water by generating highly reactive and non-selective species, such as hydroxyl radical •OH (1.8-2.7 eV) and sulphate radical $SO_4^{\bullet-}$ (2.5-3.1 eV) (Dong et al., 2022; Marino et al., 2024; Sil et al., 2020). Advanced reduction processes (ARPs) are promising alternatives for the removal of oxidation-resistant contaminants, such as PFAS compounds (Chen et al., 2022; Mukherjee et al., 2023). The addition of chemicals to the treatment process presents notable disadvantages, including the associated costs and formation of toxic by-products. On the other hand, integrating innovative systems for the online control of the processes may contribute to reducing the environmental impact of these techniques by optimizing the use of chemical reagents and minimizing as a result the by-products formation (Marino et al., 2025a, 2025b). Despite technologies like electrokinetic remediation being regarded as environmentally friendly solutions for controlling POPs by several mechanisms (e.g. electroosmosis,

TABLE 10.3
Overview of established and emerging technologies reported in literature for POPs remediation in different environmental matrices

Technology	Matrix	Advantages	Disadvantages	References
		Physical Methods		
Adsorption on activated carbon	Water, Air, Soil	Well-established; high efficiency; simple design and operation	Requires regeneration or disposal of spent carbon; challenges for short-chain PFAS	(Chi et al., 2006; Gagliano et al., 2020; Usman et al., 2024)
Adsorption on Biochar	Water, Soil	Sustainable; cost-effective	Lower efficiency; it could release PAHs and PCDDs	(Ahmaruzzaman, 2021; El-Naggar et al., 2019; Nguyen et al., 2023)
Cold Atmospheric Pressure Plasma (CAPP)	Water, Soil	High efficiency; chemical-free	High energy demand; control challenges	(Aggelopoulos, 2022; Cyganowski et al., 2024; Tomei et al., 2023)
Electrodialysis	Water, Soil	Inexpensive; non-disruptive	Limited scalability	(Ahmed et al., 2020; Sörengård et al., 2019)
Electrostatic Precipitation	Air	Targets particulate-bound POPs; low operational cost	Ineffective for gaseous POPs; suitable only for large volumes	(Meij and te Winkel, 2007)
Electrocoagulation flotation (ECF)	Water	Effective for particle-absorbed POPs	Limited effectiveness and scalability; high energy demand; electrode consumption	(Titchou et al., 2021)
Membrane Filtration	Water	Very high removal efficiency	Fouling, high costs; need to manage the retentate	(Kafle et al., 2024; Roy et al., 2020)
Permeable Reactive Barriers (PRBs)	Water, Soil	Prevents contaminant spread	Just isolates POPs; need to combine other technologies for removal	(Abbas et al., 2021; Ganbat et al., 2023)

Soil washing	Soil	Cost-effective, controlled system; concentrates POPs into a smaller phase	Need to treat/dispose the liquid phase	(Liu et al., 2022; Trellu et al., 2016)
Sonolysis	Water	High efficiency; possibility of POPs destruction in ex situ applications;	High energy consumption; limited scale-up for Large applications; control challenges	(Kulkarni et al., 2022; Ryu et al., 2021)
		Chemical Methods		
Advanced Oxidation Processes (AOPs)	Water, Air	Possibility of complete mineralization; effective for a wide range of POPs	By-product formation; high energy/ chemical demand	(Demeestere et al., 2007; Liu et al., 2023, 2019; Oturan and Aaron, 2014; Yang et al., 2019)
Advanced Reduction Processes (ARPs)	Water	Degradation of highly persistent compounds (e.g. PFAS)	Selective; limited research	(Tenorio et al., 2020; Yang et al., 2020)
Catalytic Oxidation (CO)	Water, Air	Degradation of highly recalcitrant POPs	Catalyst degradation	(Liu et al., 2021; Ruan et al., 2021)
Electrochemical Oxidation (EO)	Water, soil	High efficiency; on-site; controlled process	Electrode degradation; high-energy demand	(Brillas, 2021; Qiao and Xiong, 2021)
Electrokinetic Remediation (ER)	Soil	Suitable for in situ applications; cost-effective	Low efficiency for strongly adsorbed POPs; high-energy demand; need to combine other technologies	(Cameselle and Gouveia, 2018; Chen et al., 2021; Ganbat et al., 2023)
Supercritical Water Oxidation (SCWO)	Water, Soil	Complete and rapid degradation of POPs	High operational costs; complex system; corrosion	(Aviezer et al., 2020; Kronholm et al., 2002; Scheitlin et al., 2023)
		Biological Methods		
Bioremediation	Water, Soil	Natural and eco-friendly; cost-effective; synergistic effects	Slow and dependent on environmental conditions; recalcitrant POPs	(Gaur et al., 2018; Guo et al., 2024; Mokrani et al., 2024)

(continued)

TABLE 10.3 (Continued)
Overview of established and emerging technologies reported in literature for POPs remediation in different environmental matrices

Technology	Matrix	Advantages	Disadvantages	References
Mycoremediation	Water, sewage sludge, soil	Eco-friendly, cost-effective, sustainable;	Slow and dependent on environmental conditions; limited removals	(Stiebeling and Labes, 2022; Yadav et al., 2021)
Phytoremediation	Water, Soil	Sustainable; cost-effective; reduces soil erosion; aesthetic, restoring ecosystems	Slow and dependent on environmental conditions; selective; require plants disposal	(Lekshmi et al., 2024; Malik et al., 2022; Marzi et al., 2025)
Thermal Methods				
Incineration (750-1200°C)	Solid waste fly ash	High efficiency; large scale applications	High-energy demand; need to control toxic emissions	(Li et al., 2023)
Pyrolysis (450-800°C)	Soil, sewage sludge	Destruction of POPs; energy recovery	High operational costs; gas emissions to be treated	(Buss, 2021)
Thermal desorption (300-800°C)	Soil	Effective for VOCs; reduced energy consumption	Not complete destruction; less effective on non-volatile POPs	(Chen et al., 2024; Olasupo et al., 2024)
Plasma arc technology (3.000-10.000 °C)	Water	Destruction of recalcitrant POPs (PFAS)	Very high costs; hard to scale up	(Singh et al., 2019; Verma et al., 2023; Zhang et al., 2024)

electromigration), these techniques are hard to scale-up due to high energy demand (Wang et al., 2024) (Figure 10.5).

Biological methods for the remediation of POPs are gaining attention due to their sustainability and minimal ecological impact as they rely on natural processes that do not involve the use of chemicals and external energy sources. Moreover, although biological methods require more time, they are much less expensive than chemical and physical methods. Biological methods utilize microorganisms, enzymes, algae, fungi and plants to degrade, transform or immobilize POPs. For instance, PCBs can be successfully biodegraded by microorganisms and under both anaerobic and aerobic conditions through anaerobic reductive dichlorination and aerobic breakdown of the biphenyl structure, respectively (Guo et al., 2024). Microbial activity and enzyme action have shown promise in degrading PAHs, atrazine, lindane and DDT as well (Devendrapandi et al., 2024). Furthermore, algae and microalgae play an important role in the bioremediation of POPs through bioaccumulation and biodegradation (El-Maradny et al., 2022; Kaleem et al., 2023; Sharma et al., 2024). Additionally,

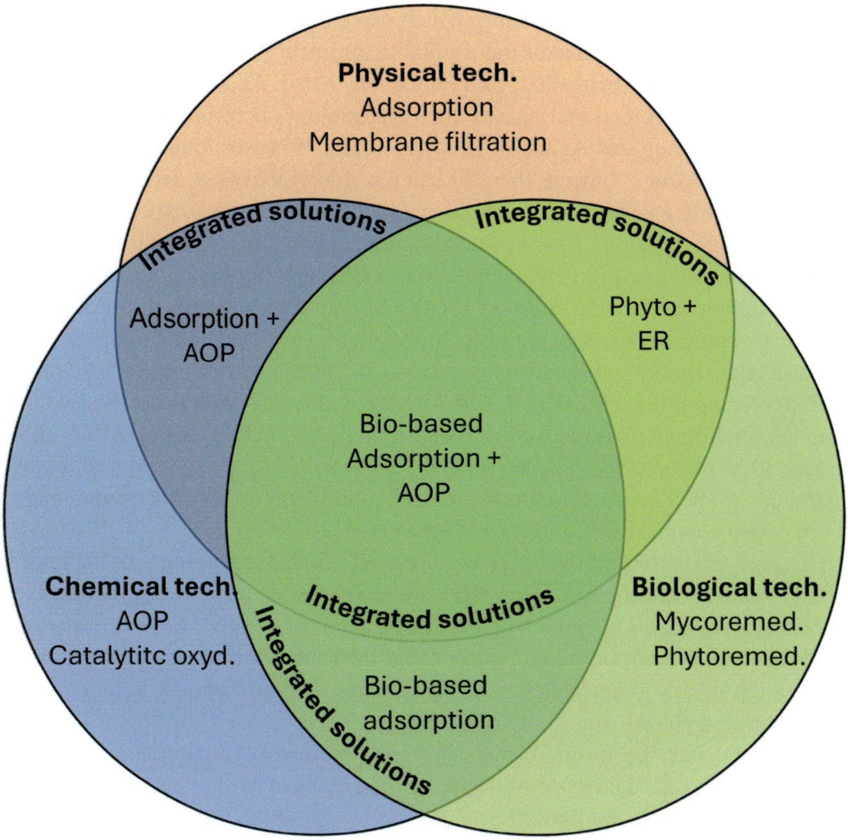

FIGURE 10.5 Synthesis of remediation technique combinations. By combining multiple remediation techniques, a more efficient and broad range of POP removal can be achieved.

mycoremediation has proven to be effective in removing POPs, such as PCBs, from both liquid and soil media, primarily through biodegradation and the adsorption of contaminants onto fungal biomass (Kaleem et al., 2023; Yadav et al., 2021). Phytoremediation is another viable alternative for POPs remediation. A recent study successfully reported the hemp phytoextraction of several PFAS and consequent degradation in harvested plants through hydrothermal liquefaction, highlighting that these techniques can also be effective for highly recalcitrant compounds (Nason et al., 2024) (Figure 10.5).

Biological methods present clear advantages for a sustainable approach to POPs remediation. Compared to chemical methods that often require high energy inputs and generate toxic by-products, bioremediation can completely degrade POPs at much lower costs, helping in reducing the carbon footprint (Truu et al., 2015). Furthermore, synergistic effects through plant-microbial interactions may be promoted to enhance the biodegradation process and achieve higher removal efficiencies (Guo et al., 2024). In addition to its environmental benefits, simplicity and flexibility, phytoremediation in constructed wetlands, green roofs and green walls offers numerous other advantages (Lekshmi et al., 2024; Vijayaraghavan et al., 2019). Indeed, these applications can be designed to be aesthetically pleasing, which can enhance the visual appeal of urban and rural landscapes. This makes them suitable for integration into parks, schools and residential areas (Rane et al., 2023; Zamora et al., 2019). Additionally, these systems can also provide recreational spaces for communities, contributing to the overall quality of life (Wang and Aghajani Delavar, 2023). However, biological techniques have some limitations. Indeed, the efficiency of bioremediation depends on several factors, including environmental conditions such as temperature, pH, nutrient availability and the presence of heavy metals and/or competing microbial species. On the other hand, addressing POPs contamination through biological methods appears as a viable and sustainable solution for the future, aligning with the principles of the One Health framework, which recognizes the interconnection of human health, animal health, and environment. However, in the technical-economic evaluation of a real application, it is essential to consider the additional costs associated with the extended time frame required for long-term remediation (15+ years) and the value of the land to be restored (Wang and Aghajani Delavar, 2023). Furthermore, in cases of urgent remediation needs, bioremediation techniques are not a viable option, making alternative treatments more attractive (Figure 10.5).

This analysis highlights that selecting the most suitable technology for the removal of POPs is rarely a straightforward decision. In general, interest in destruction-based methods has increased as target POPs can be totally removed. Indeed, separation-based methods present additional costs to be incurred, such as those related to the regeneration of adsorbent material or membranes or the treatment and/or disposal of concentrated phases (liquid or solid). However, the selection of appropriate technology is influenced by several factors, including regulatory requirements (e.g. PFAS limits in the new EU Directive on drinking water), financial budget, the urgency of contamination events, and numerous site-specific conditions. A risk-based approach is particularly valuable in this context, as it helps to identify the most directly affected targets while minimizing risks to both ecosystems and human health. Additionally, a comprehensive cost-benefit analysis encompassing not only economic considerations but also environmental and societal benefits is essential for making informed decisions. In line with the One Health approach, the selection of the 'best available technology' (BAT) should prioritize methods that achieve effective pollutant removal while maintaining the lowest possible environmental impact. Often, employing combined approaches through the integration of multiple treatment techniques, such as adsorption and AOPs, is a viable solution in order to leverage synergistic effects and enhance overall efficiency (Figure 10.5). Further research is strongly recommended to integrate new technologies that enable real-time process monitoring and efficiency assessment over time. In this regard, greater efforts could be directed towards exploring the potential role of artificial intelligence in enhancing data management and optimization. Therefore, an overarching, multidisciplinary holistic perspective is crucial: remediation efforts should not only be efficient but should also align with sustainability principles to minimize secondary environmental burdens. Furthermore, the One Health approach may play a key role in emerging sectors, such as wastewater reuse, where sustainable and integrated treatment solutions are needed to ensure safe resource recovery, high water quality and strict hygiene standards. Future advancements in POPs removal technologies should emphasize energy efficiency, resource optimization, and the development of eco-friendly materials to further enhance sustainability. By adopting this multidisciplinary perspective, remediation strategies can contribute to long-term environmental protection while supporting circular economy principles.

10.4 CONCLUSIONS

The scope of this chapter was to identify some potential One Health approaches to address POPs' effects on human, animal and environmental health. According to the One Health Action Plan, the risk assessment as well as control and remediation are among the most adequate actions for tackling POPs. Indeed, the risk assessment allows defining POP sources, pollutant types, hit areas and exposition pathways, hence, narrowing the size of the interventions. Moreover, such data support the identification of POPs' direct and indirect effects on human and animal health as well as on the environment. Control and remediation measures are implemented to prevent the spread and removal of POPs from the environment, in order to minimize the exposure to such pollutants as well as the surging of diseases and the impoverishing of the environment. The examples reported in this chapter highlight POPs' wide area of influence in terms of sources, environmental matrixes, targets and pathways. Therefore, such an issue requires to be addressed from multiple perspectives. To achieve this, scientists, practitioners and policymakers need to collaborate, transcending disciplinary, sectoral and ideological boundaries, and work together to fulfil the One Health vision.

REFERENCES

Abbas, T., Wadhawan, T., Khan, A., McEvoy, J., Khan, E., 2021. Iron turning waste: Low cost and sustainable permeable reactive barrier media for remediating dieldrin, endrin,

DDT and lindane in groundwater. *Environmental Pollution* 289, 117825. https://doi.org/10.1016/j.envpol.2021.117825

Acheuk, F., Basiouni, S., Shehata, A.A., Dick, K., Hajri, H., Lasram, S., Yilmaz, M., Emekci, M., Tsiamis, G., Spona-Friedl, M., May-Simera, H., Eisenreich, W., Ntougias, S., 2022. Status and prospects of botanical biopesticides in Europe and Mediterranean countries. *Biomolecules* 12, 311. https://doi.org/10.3390/biom12020311

Aerts, R., Van Overmeire, I., Colles, A., Andjelković, M., Malarvannan, G., Poma, G., Den Hond, E., Van de Mieroop, E., Dewolf, M.-C., Charlet, F., Van Nieuwenhuyse, A., Van Loco, J., Covaci, A., 2019. Determinants of persistent organic pollutant (POP) concentrations in human breast milk of a cross-sectional sample of primiparous mothers in Belgium. *Environment International* 131, 104979. https://doi.org/10.1016/j.envint.2019.104979

Aggelopoulos, C.A., 2022. Recent advances of cold plasma technology for water and soil remediation: A critical review. *Chemical Engineering Journal* 428, 131657. https://doi.org/10.1016/j.cej.2021.131657

Ahmaruzzaman, M., 2021. Biochar based nanocomposites for photocatalytic degradation of emerging organic pollutants from water and wastewater. *Materials Research Bulletin* 140, 111262. https://doi.org/10.1016/j.materresbull.2021.111262

Ahmed, M.B., Johir, M.A.H., McLaughlan, R., Nguyen, L.N., Xu, B., Nghiem, L.D., 2020. Per- and polyfluoroalkyl substances in soil and sediments: Occurrence, fate, remediation and future outlook. *Science of The Total Environment* 748, 141251. https://doi.org/10.1016/j.scitotenv.2020.141251

Ajay, S.V., Kirankumar, P.S., Varghese, A., Prathish, K.P., 2022. Assessment of dioxin-like POP's emissions and human exposure risk from open burning of municipal solid wastes in streets and dumpyard fire breakouts. *Expo Health* 14, 763–778. https://doi.org/10.1007/s12403-021-00450-4

Aryanti, P.T.P., Nugroho, F.A., Phalakornkule, C., Kadier, A., 2024. Energy efficiency in electrocoagulation processes for sustainable water and wastewater treatment. *Journal of Environmental Chemical Engineering* 12, 114124. https://doi.org/10.1016/j.jece.2024.114124

Australian Government, 2024. *Benzene hexachloro – (HCB) – DCCEEW* [WWW Document]. www.dcceew.gov.au/environment/protection/npi/substances/fact-sheets/benzene-hexachloro-hcb (accessed 12.5.24).

Aviezer, Y., Birnhack, L., Lahav, O., 2020. Supercritical water oxidation of recalcitrant organic pollutants present in retentates from RO desalination of secondary effluents. *Desalination and Water Treatment* 183, 7–22. https://doi.org/10.5004/dwt.2020.25377

Barber, J.L., Sweetman, A.J., van Wijk, D., Jones, K.C., 2005. Hexachlorobenzene in the global environment: Emissions, levels, distribution, trends and processes. *Science of The Total Environment* 349, 1–44. https://doi.org/10.1016/j.scitotenv.2005.03.014

Bettinetti, R., Galassi, S., Guilizzoni, P., Quadroni, S., 2011. Sediment analysis to support the recent glacial origin of DDT pollution in Lake Iseo (Northern Italy). *Chemosphere, Special Issue in Honour and Remembrance of Dr. Larry Needham'* 85, 163–169. https://doi.org/10.1016/j.chemosphere.2011.06.037

Bobde, V.P., Sharma, K., Kumar, A., Pandey, R.K., Wadhwa, S.J., 2023. Recent advances on sustainable removal of emerging contaminants from water by bio-based adsorbents. *New Journal of Chemistry* 47, 17626–17644. https://doi.org/10.1039/D3NJ03731B

Brillas, E., 2021. Recent development of electrochemical advanced oxidation of herbicides. A review on its application to wastewater treatment and soil remediation. *Journal of Cleaner Production* 290, 125841. https://doi.org/10.1016/j.jclepro.2021.125841

Brunn, H., Arnold, G., Körner, W., Rippen, G., Steinhäuser, K.G., Valentin, I., 2023. PFAS: Forever chemicals—persistent, bioaccumulative and mobile. Reviewing the status and the need for their phase out and remediation of contaminated sites. *Environmental Sciences Europe* 35, 1–50. https://doi.org/10.1186/S12302-023-00721-8

Buss, W., 2021. Pyrolysis solves the issue of organic contaminants in sewage sludge while retaining carbon—making the ccase for sewage sludge treatment via pyrolysis. *ACS Sustainable Chemical Engineering* 9, 10048–10053. https://doi.org/10.1021/acssuschemeng.1c03651

Cameselle, C., Gouveia, S., 2018. Electrokinetic remediation for the removal of organic contaminants in soils. *Current Opinion in Electrochemistry, Environmental Electrochemistry – Solar Cells* 11, 41–47. https://doi.org/10.1016/j.coelec.2018.07.005

Cao, H., Zhou, Z., Hu, Z., Wei, C., Li, J., Wang, L., Liu, G., Zhang, J., Wang, Y., Wang, T., Liang, Y., 2022. Effect of enterohepatic circulation on the accumulation of per- and polyfluoroalkyl substances: Evidence from experimental and computational studies. *Environmental Science Technology* 56, 3214–3224. https://doi.org/10.1021/acs.est.1c07176

Cara, B., Lies, T., Thimo, G., Robin, L., Lieven, B., 2022. Bioaccumulation and trophic transfer of perfluorinated alkyl substances (PFAS) in marine biota from the Belgian North Sea: Distribution and human health risk implications. *Environmental Pollution* 311, 119907. https://doi.org/10.1016/j.envpol.2022.119907

Chang, G.-R., 2018. Persistent organochlorine pesticides in aquatic environments and fishes in Taiwan and their risk assessment. *Environmental Science Pollution Research* 25, 7699–7708. https://doi.org/10.1007/s11356-017-1110-z

Chen, Y., Zhi, D., Zhou, Y., Huang, A., Wu, S., Yao, B., Tang, Y., Sun, C., 2021. Electrokinetic techniques, their enhancement techniques and composite techniques with other processes for persistent organic pollutants remediation in soil: A review. *Journal of Industrial and Engineering Chemistry* 97, 163–172. https://doi.org/10.1016/j.jiec.2021.03.009

Chen, Z., Wang, X., Dong, R., Zhang, Y., Jin, X., Gu, C., 2022. Challenging the contamination of per-and polyfluoroalkyl substances in water: Advanced oxidation or reduction? *Environmental Functional Materials* 1(3), 325–337. https://doi.org/10.1016/j.efmat.2022.11.002

Chen, Zefeng, Chen, Zhiguo, Li, Y., Zhang, R., Liu, Y., Hui, A., Cao, W., Liu, J., Bai, H., Song, J., 2024. A review on remediation of chlorinated organic contaminants in soils by thermal desorption. *Journal of Industrial and Engineering Chemistry* 133, 112–121. https://doi.org/10.1016/j.jiec.2023.12.022

Cheruiyot, N.K., Yang, H.-H., Wang, L.-C., Lin, C.-C., 2020. Feasible and effective control strategies on extreme emissions of chlorinated persistent organic pollutants during the start-up processes of municipal solid waste incinerators. *Environmental Pollution* 267, 115469. https://doi.org/10.1016/j.envpol.2020.115469

Chi, K.H., Chang, S.H., Huang, C.H., Huang, H.C., Chang, M.B., 2006. Partitioning and removal of dioxin-like congeners in flue gases treated with activated carbon adsorption. *Chemosphere* 64, 1489–1498. https://doi.org/10.1016/j.chemosphere.2005.12.072

Cyganowski, P., Terefinko, D., Motyka-Pomagruk, A., Babinska-Wensierska, W., Khan, M.A., Klis, T., Sledz, W., Lojkowska, E., Jamroz, P., Pohl, P., Caban, M., Magureanu, M., Dzimitrowicz, A., 2024. The potential of cold atmospheric pressure plasmas for the direct degradation of organic pollutants derived from the food production industry. *Molecules* 29, 2910. https://doi.org/10.3390/molecules29122910

Das, A.K., Bhattacharya, D., Das, A., Nath, S., Bandyopadhyay, S., Nanda, P.K., Gagaoua, M., 2023. Current innovative approaches in reducing polycyclic aromatic hydrocarbons

(PAHs) in processed meat and meat products. *Chemical Biological Technology Agricultural* 10, 109. https://doi.org/10.1186/s40538-023-00483-8

Demeestere, K., Dewulf, J., Van Langenhove, H., 2007. Heterogeneous photocatalysis as an advanced oxidation process for the abatement of chlorinated, monocyclic aromatic and sulfurous volatile organic compounds in air: State of the art. *Critical Reviews in Environmental Science and Technology* 37, 489–538. https://doi.org/10.1080/10643380600966467

Devendrapandi, G., Liu, X., Balu, R., Ayyamperumal, R., Valan Arasu, M., Lavanya, M., Minnam Reddy, V.R., Kim, W.K., Karthika, P.C., 2024. Innovative remediation strategies for persistent organic pollutants in soil and water: A comprehensive review. *Environmental Research* 249, 118404. https://doi.org/10.1016/j.envres.2024.118404

Dong, C., Fang, W., Yi, Q., Zhang, J., 2022. A comprehensive review on reactive oxygen species (ROS) in advanced oxidation processes (AOPs). *Chemosphere* 308(Part 1), 136205–136205. ISSN 0045-6535 https://doi.org/10.1016/j.chemosphere.2022.136205

Du, X., Yuan, B., Zhou, Y., de Wit, C.A., Zheng, Z., Yin, G., 2020. Chlorinated paraffins in two snake species from the Yangtze River Delta: Tissue distribution and biomagnification. *Environmental Science & Technology* 54, 2753–2762. https://doi.org/10.1021/acs.est.9b06467

El-Maradny, A., Salem, D.M.S.A., Mohamed, L.A., 2022. Utilization of algae for bioextraction of POPs in highly polluted coastal area, eastern Alexandria, Mediterranean Sea. *International Journal of Environmental Science Technology* 19, 3975–3988. https://doi.org/10.1007/s13762-021-03470-4

El-Naggar, A., El-Naggar, A.H., Shaheen, S.M., Sarkar, B., Chang, S.X., Tsang, D.C.W., Rinklebe, J., Ok, Y.S., 2019. Biochar composition-dependent impacts on soil nutrient release, carbon mineralization, and potential environmental risk: A review. *Journal of Environmental Management* 241, 458–467. https://doi.org/10.1016/j.jenvman.2019.02.044

El-Shahawi, M.S., Hamza, A., Bashammakh, A.S., Al-Saggaf, W.T., 2010. An overview on the accumulation, distribution, transformations, toxicity and analytical methods for the monitoring of persistent organic pollutants. *Talanta* 80, 1587–1597. https://doi.org/10.1016/j.talanta.2009.09.055

Fitzgerald, L., Wikoff, D.S., 2014. Persistent organic pollutants, in: Wexler, P. (Ed.), *Encyclopedia of Toxicology* (3rd Edition). Academic Press, pp. 820–825. https://doi.org/10.1016/B978-0-12-386454-3.00211-6

Fujii, S., Polprasert, C., Tanaka, S., Hong Lien, N.P., Qiu, Y., 2007. New POPs in the water environment: Distribution, bioaccumulation and treatment of perfluorinated compounds – A review paper. *Journal of Water Supply: Research and Technology-Aqua* 56, 313–326. https://doi.org/10.2166/aqua.2007.005

Gagliano, E., Falciglia, P.P., Zaker, Y., Karanfil, T., Roccaro, P., 2021. Microwave regeneration of granular activated carbon saturated with PFAS. *Water Research* 198, 117121. https://doi.org/10.1016/j.watres.2021.117121

Gagliano, E., Sgroi, M., Falciglia, P.P., Vagliasindi, F.G.A., Roccaro, P., 2020. Removal of poly- and perfluoroalkyl substances (PFAS) from water by adsorption: Role of PFAS chain length, effect of organic matter and challenges in adsorbent regeneration. *Water Research* 171, 115381. https://doi.org/10.1016/j.watres.2019.115381

Ganbat, N., Hamdi, F.M., Ibrar, I., Altaee, A., Alsaka, L., Samal, A.K., Zhou, J., Hawari, A.H., 2023. Iron slag permeable reactive barrier for PFOA removal by the electrokinetic process. *Journal of Hazardous Materials* 460, 132360. https://doi.org/10.1016/j.jhazmat.2023.132360

Garcia-Garin, O., Borrell, A., Colomer-Vidal, P., Vighi, M., Trilla-Prieto, N., Aguilar, A., Gazo, M., Jiménez, B., 2023. Biomagnification and temporal trends (1990–2021) of perfluoroalkyl substances in striped dolphins (*Stenella coeruleoalba*) from the NW Mediterranean sea. *Environmental Pollution* 339, 122738. https://doi.org/10.1016/j.envpol.2023.122738

Gaur, N., Dutta, D., Jaiswal, A., Dubey, R., Kamboj, D.V., Gaur, N., Dutta, D., Jaiswal, A., Dubey, R., Kamboj, D.V., 2022. Role and effect of persistent organic pollutants to our environment and wildlife, in: *Persistent Organic Pollutants (POPs) – Monitoring, Impact and Treatment*. IntechOpen. https://doi.org/10.5772/intechopen.101617

Gaur, N., Narasimhulu, K., Y, P., 2018. Recent advances in the bio-remediation of persistent organic pollutants and its effect on environment. *Journal of Cleaner Production* 198, 1602–1631. https://doi.org/10.1016/j.jclepro.2018.07.076

Grassi, P., Fattore, E., Generoso, C., Fanelli, R., Arvati, M., Zuccato, E., 2010. Polychlorobiphenyls (PCBs), polychlorinated dibenzo-*p*-dioxins (PCDDs) and dibenzofurans (PCDFs) in fruit and vegetables from an industrial area in northern Italy. *Chemosphere* 79, 292–298. https://doi.org/10.1016/j.chemosphere.2010.01.028

Guo, W., Pan, B., Sakkiah, S., Yavas, G., Ge, W., Zou, W., Tong, W., Hong, H., 2019. Persistent organic pollutants in food: Contamination sources, health effects and detection methods. *International Journal of Environmental Research and Public Health* 16, 4361. https://doi.org/10.3390/ijerph16224361

Guo, W., Ren, H., Jin, Y., Chai, Z., Liu, B., 2024. The bioremediation of the typical persistent organic pollutants (POPs) by microalgae-bacteria consortia: A systematic review. *Chemosphere* 355, 141852. https://doi.org/10.1016/j.chemosphere.2024.141852

Güzel, B., Canlı, O., Çelebi, A., 2022. Characterization, source and risk assessments of sediment contaminants (PCDD/Fs, DL-PCBs, PAHs, PCBs, OCPs, metals) in the urban water supply area. *Applied Geochemistry* 143, 105394. https://doi.org/10.1016/j.apgeochem.2022.105394

Hansen, K.E.A., Johanson, S.M., Steppeler, C., Sødring, M., Østby, G.C., Berntsen, H.F., Zimmer, K.E., Aleksandersen, M., Paulsen, J.E., Ropstad, E., 2019. A mixture of Persistent Organic Pollutants (POPs) and Azoxymethane (AOM) show potential synergistic effects on intestinal tumorigenesis in the A/J Min/+ mouse model. *Chemosphere* 214, 534–542. https://doi.org/10.1016/j.chemosphere.2018.09.126

Haukås, M., Berger, U., Hop, H., Gulliksen, B., Gabrielsen, G.W., 2007. Bioaccumulation of per- and polyfluorinated alkyl substances (PFAS) in selected species from the Barents Sea food web. *Environmental Pollution* 148, 360–371. https://doi.org/10.1016/j.envpol.2006.09.021

Hidayati, N.V., Asia, L., Khabouchi, I., Torre, F., Widowati, I., Sabdono, A., Doumenq, P., Syakti, A.D., 2021. Ecological risk assessment of persistent organic pollutants (POPs) in surface sediments from aquaculture system. *Chemosphere* 263, 128372. https://doi.org/10.1016/j.chemosphere.2020.128372

Hu, X.C., Andrews, D.Q., Lindstrom, A.B., Bruton, T.A., Schaider, L.A., Grandjean, P., Lohmann, R., Carignan, C.C., Blum, A., Balan, S.A., Higgins, C.P., Sunderland, E.M., 2016. Detection of poly- and perfluoroalkyl substances (PFASs) in U.S. drinking water linked to industrial sites, military fire training areas, and wastewater treatment plants. *Environmental Science and Technology Letters* 3, 344–350. https://doi.org/10.1021/acs.estlett.6b00260

Hülster, A., Marschner, H., 1993. Transfer of PCDD/PCDF from contaminated soils to food and fodder crop plants. *Chemosphere, Proceedings of the Twelfth International Symposium* 27, 439–446. https://doi.org/10.1016/0045-6535(93)90324-X

Hung, H., Halsall, C., Ball, H., Bidleman, T., Dachs, J., Silva, A.D., Hermanson, M., Kallenborn, R., Muir, D., Sühring, R., Wang, X., Wilson, S., 2022. Climate change influence on the levels and trends of persistent organic pollutants (POPs) and chemicals of emerging Arctic concern (CEACs) in the Arctic physical environment – A review. *Environmental Science: Processes & Impacts* 24, 1577–1615. https://doi.org/10.1039/D1EM00485A

IARC, 2024. *List of Classifications* [WWW Document]. https://monographs.iarc.who.int/list-of-classifications (accessed 1.29.25).

ITRC, 2022. Per-and Polyfluoroalkyl Substances (PFAS) technical and regulatory guidance.

Jeong, Y., Lee, S., Kim, Sunmi, Park, J., Kim, H.-J., Choi, G., Choi, S., Kim, Sungjoo, Kim, S.Y., Kim, Sungkyoon, Choi, K., Moon, H.-B., 2018. Placental transfer of persistent organic pollutants and feasibility using the placenta as a non-invasive biomonitoring matrix. *Science of The Total Environment* 612, 1498–1505. https://doi.org/10.1016/j.scitotenv.2017.07.054

Kafle, S.R., Adhikari, S., Shrestha, R., Ban, S., Khatiwada, G., Gaire, P., Tuladhar, N., Jiang, G., Tiwari, A., 2024. Advancement of membrane separation technology for organic pollutant removal. *Water Science and Technology* 89, 2290–2310. https://doi.org/10.2166/wst.2024.117

Kaleem, M., Mumtaz, A.S., Hashmi, M.Z., Saeed, A., Inam, F., Waqar, R., Jabeen, A., 2023. Myco- and phyco-remediation of polychlorinated biphenyls in the environment: A review. *Environmental Science Pollution Research* 30, 13994–14007. https://doi.org/10.1007/s11356-022-24902-9

Kronholm, J., Kalpala, J., Hartonen, K., Riekkola, M.-L., 2002. Pressurized hot water extraction coupled with supercritical water oxidation in remediation of sand and soil containing PAHs. *Journal of Supercritical Fluids* 23, 123–134. https://doi.org/10.1016/S0896-8446(02)00008-6

Kulkarni, P.R., Richardson, S.D., Nzeribe, B.N., Adamson, D.T., Kalra, S.S., Mahendra, S., Blotevogel, J., Hanson, A., Dooley, G., Maraviov, S., Popovic, J., 2022. Field demonstration of a sonolysis reactor for treatment of PFAS-contaminated groundwater. *Journal of Environmental Engineering* 148, 06022005. https://doi.org/10.1061/(ASCE)EE.1943-7870.0002064

Kumar, R., Qureshi, M., Vishwakarma, D.K., Al-Ansari, N., Kuriqi, A., Elbeltagi, A., Saraswat, A., 2022. A review on emerging water contaminants and the application of sustainable removal technologies. *Case Studies in Chemical and Environmental Engineering* 6, 100219. ISSN 2666-0164. https://doi.org/10.1016/J.CSCEE.2022.100219

Lekshmi, S., Ardra Lekshmi, A., Achuthavarier, D., Smitha Chandran, S., 2024. Critical analysis of sustainable ways of removing insidious pollutants from the environment through phytoremediation techniques. *Chemistry and Ecology* 40, 892–917. https://doi.org/10.1080/02757540.2024.2365164

Lengai, G.M.W., Muthomi, J.W., Mbega, E.R., 2020. Phytochemical activity and role of botanical pesticides in pest management for sustainable agricultural crop production. *Scientific African* 7, e00239. https://doi.org/10.1016/j.sciaf.2019.e00239

Li, W., Yan, D., Li, L., Wen, Z., Liu, M., Lu, S., Huang, Q., 2023. Review of thermal treatments for the degradation of dioxins in municipal solid waste incineration fly ash: Proposing a suitable method for large-scale processing. *Science of The Total Environment* 875, 162565. https://doi.org/10.1016/j.scitotenv.2023.162565

Li, Y., Baumert, B.O., Costello, E., Chen, J.C., Rock, S., Stratakis, N., Goodrich, J.A., Zhao, Y., Eckel, S.P., Walker, D.I., Valvi, D., La Merrill, M.A., McConnell, R., Cortessis, V.K., Aung, M., Wu, H., Baccarelli, A., Conti, D., Chatzi, L., 2024. Per- and polyfluoroalkyl substances, polychlorinated biphenyls, organochlorine pesticides, and polybrominated diphenyl ethers and dysregulation of MicroRNA expression in humans and animals—A systematic review. *Environmental Research* 244, 117832. https://doi.org/10.1016/j.envres.2023.117832

Li, Y., Fletcher, T., Mucs, D., Scott, K., Lindh, C.H., Tallving, P., Jakobsson, K., 2018. Half-lives of PFOS, PFHxS and PFOA after end of exposure to contaminated drinking water. *Occupational and Environmental Medicine* 75, 46–51. https://doi.org/10.1136/oemed-2017-104651

Linares, V., Perelló, G., Nadal, M., Gómez-Catalán, J., Llobet, J.M., Domingo, J.L., 2010. Environmental versus dietary exposure to POPs and metals: A probabilistic assessment of human health risks. *Journal of Environmental Monitoring* 12, 681–688. https://doi.org/10.1039/B914962G

Liu, H., Li, X., Dai, Q., Zhao, H., Chai, G., Guo, Yanglong, Guo, Yun, Wang, L., Zhan, W., 2021. Catalytic oxidation of chlorinated volatile organic compounds over Mn-Ti composite oxides catalysts: Elucidating the influence of surface acidity. *Applied Catalysis B: Environmental* 282, 119577. https://doi.org/10.1016/j.apcatb.2020.119577

Liu, J., Zhao, L., Liu, Q., Li, J., Qiao, Z., Sun, P., Yang, Y., 2022. A critical review on soil washing during soil remediation for heavy metals and organic pollutants. *International Journal of Environmental Science and Technology* 19, 601–624. https://doi.org/10.1007/s13762-021-03144-1

Liu, Y., Luo, X., Zeng, Y., Deng, M., Tu, W., Wu, Y., Mai, B., 2020. Bioaccumulation and biomagnification of hexabromocyclododecane (HBCDD) in insect-dominated food webs from a former e-waste recycling site in South China. *Chemosphere* 240, 124813. https://doi.org/10.1016/j.chemosphere.2019.124813

Liu, Z., Chys, M., Yang, Y., Demeestere, K., Van Hulle, S., 2019. Oxidation of trace organic contaminants (TrOCs) in wastewater effluent with different ozone-based AOPs: Comparison of ozone exposure and • OH formation. *Industrial & Engineering Chemistry Research* 58(20), 8896–8902. https://doi.org/10.1021/acs.iecr.9b00293

Liu, Z., Ren, X., Duan, X., Sarmah, A.K., Zhao, X., 2023. Remediation of environmentally persistent organic pollutants (POPs) by persulfates oxidation system (PS): A review. *Science of The Total Environment* 863, 160818. https://doi.org/10.1016/j.scitotenv.2022.160818

Lorenzo, M., Campo, J., Picó, Y., 2018. Analytical challenges to determine emerging persistent organic pollutants in aquatic ecosystems. *TrAC – Trends in Analytical Chemistry* 103, 137–155. https://doi.org/10.1016/j.trac.2018.04.003

Luiz De Oliveira, J., Ramos Campos, E.V., Fraceto, L.F., 2018. Recent developments and challenges for nanoscale formulation of botanical pesticides for use in sustainable agriculture. *Journal of Agricultural Food Chemistry* 66, 8898–8913. https://doi.org/10.1021/acs.jafc.8b03183

Malik, B., Pirzadah, T.B., Hakeem, K.R., 2022. Chapter 20 – Phytoremediation of persistent organic pollutants (POPs), in: Bhat, R.A., Tonelli, F.M.P., Dar, G.H., Hakeem, K. (Eds.), *Phytoremediation*. Academic Press, pp. 415–436. https://doi.org/10.1016/B978-0-323-89874-4.00010-8

Malisch, R., Kotz, A., 2014. Dioxins and PCBs in feed and food — Review from European perspective. *Science of The Total Environment, Halogenated Persistent Organic Pollutants (Dioxin2013, Daegu/Korea)* 491–492, 2–10. https://doi.org/10.1016/j.scitotenv.2014.03.022

Marican, A., Durán-Lara, E.F., 2018. A review on pesticide removal through different processes. *Environmental Science Pollution Research* 25, 2051–2064. https://doi.org/10.1007/s11356-017-0796-2

Marino, L., Gagliano, E., Santoro, D., Roccaro, P., 2025a. Fluorescence sensor enabled control of contaminants of emerging concern in reclaimed wastewater using ozone-based treatment processes. *Water Research* 268, 122616. https://doi.org/10.1016/j.watres.2024.122616

Marino, L., Todesco, R., Gagliano, E., Santoro, D., Roccaro, P., 2025b. Real-time wastewater quality monitoring by fluorescence sensors: Validation for COD and CEC monitoring and implication for carbon footprint reduction. *Science of The Total Environment* 963, 178464. https://doi.org/10.1016/j.scitotenv.2025.178464

Marino, L., Gagliano, E., Santoro, D., Roccaro, P., 2024. Control of contaminants of emerging concern (CEC) during pilot scale advanced oxidation processes (AOPs) for water reuse, in: Mannina, G., Cosenza, A., Mineo, A. (Eds.), *Resource Recovery from Wastewater Treatment*. Springer Nature Switzerland, pp. 348–352. https://doi.org/10.1007/978-3-031-63353-9_60

Marzi, D., Valente, F., Luche, S., Caissutti, C., Sabia, A., Capitani, I., Capobianco, G., Serranti, S., Masi, A., Panozzo, A., Ricci, A., Bolla, P.K., Vamerali, T., Brunetti, P., Visioli, G., 2025. Phytoremediation of perfluoroalkyl and polyfluoroalkyl substances (PFAS): Insights on plant uptake, omics analysis, contaminant detection and biomass disposal. *Science of The Total Environment* 959, 178323. https://doi.org/10.1016/j.scitotenv.2024.178323

Meij, R., te Winkel, H., 2007. The emissions of heavy metals and persistent organic pollutants from modern coal-fired power stations. *Atmospheric Environment* 41, 9262–9272. https://doi.org/10.1016/j.atmosenv.2007.04.042

Mojiri, A., Zhou, J.L., Ohashi, A., Ozaki, N., Kindaichi, T., 2019. Comprehensive review of polycyclic aromatic hydrocarbons in water sources, their effects and treatments. *Science of The Total Environment* 696, 133971. https://doi.org/10.1016/j.scitotenv.2019.133971

Mokrani, S., Houali, K., Yadav, K.K., Arabi, A.I.A., Eltayeb, L.B., AwjanAlreshidi, M., Benguerba, Y., Cabral-Pinto, M.M.S., Nabti, E., 2024. Bioremediation techniques for soil organic pollution: Mechanisms, microorganisms, and technologies – A comprehensive review. *Ecological Engineering* 207, 107338. https://doi.org/10.1016/j.ecoleng.2024.107338

Mukherjee, P., Sathiyan, K., Zidki, T., Nadagouda, M.N., Sharma, V.K., 2023. Can ultraviolet-assisted advanced reduction processes effectively destroy per- and polyfluoroalkyl substances in real water matrices? *Current Opinion in Chemical Engineering* 42, 100971. https://doi.org/10.1016/j.coche.2023.100971

Müller, M.H.B., Polder, A., Brynildsrud, O.B., Grønnestad, R., Karimi, M., Lie, E., Manyilizu, W.B., Mdegela, R.H., Mokiti, F., Murtadha, M., Nonga, H.E., Skaare, J.U., Solhaug, A., Lyche, J.L., 2019. Prenatal exposure to persistent organic pollutants in Northern Tanzania and their distribution between breast milk, maternal blood, placenta and cord blood. *Environmental Research* 170, 433–442. https://doi.org/10.1016/j.envres.2018.12.026

Nason, S.L., Thomas, S., Stanley, C., Silliboy, R., Blumenthal, M., Zhang, W., Liang, Y., Jones, J.P., Zuverza-Mena, N., White, J.C., Haynes, C.L., Vasiliou, V., Timko, M.P., Berger, B.W., 2024. A comprehensive trial on PFAS remediation: Hemp phytoextraction and PFAS degradation in harvested plants. *Environmental Science: Advances* 3, 304–313. https://doi.org/10.1039/D3VA00340J

Ngegba, P.M., Cui, G., Khalid, M.Z., Zhong, G., 2022. Use of botanical pesticides in agriculture as an alternative to synthetic pesticides. *Agriculture* 12, 600. https://doi.org/10.3390/agriculture12050600

Nguyen, A.T., Le Tran, L., 2024. A review about the occurrence and effectiveness of conventional and advanced treatment technologies of persistent organic pollutants in surface water. *Reviews Environmental Contamination (formerly:Residue Reviews)* 262, 11. https://doi.org/10.1007/s44169-024-00062-4

Nguyen, T.-B., Sherpa, K., Bui, X.-T., Nguyen, V.-T., Vo, T.-D.-H., Ho, H.-T.-T., Chen, C.-W., Dong, C.-D., 2023. Biochar for soil remediation: A comprehensive review of current research on pollutant removal. *Environmental Pollution* 337, 122571. https://doi.org/10.1016/j.envpol.2023.122571

Olasupo, A., Corbin, D.R., Shiflett, M.B., 2024. Trends in low temperature and non-thermal technologies for the degradation of persistent organic pollutants. *Journal of Hazardous Materials* 468, 133830. https://doi.org/10.1016/j.jhazmat.2024.133830

Oturan, M.A., Aaron, J.-J., 2014. Advanced oxidation processes in water/wastewater treatment: Principles and applications. A review. *Critical Reviews in Environmental Science and Technology* 44, 2577–2641. https://doi.org/10.1080/10643389.2013.829765

Qiao, J., Xiong, Y., 2021. Electrochemical oxidation technology: A review of its application in high-efficiency treatment of wastewater containing persistent organic pollutants. *Journal of Water Process Engineering* 44, 102308. https://doi.org/10.1016/j.jwpe.2021.102308

Rafa, N., Ahmed, B., Zohora, F., Bakya, J., Ahmed, S., Ahmed, S.F., Mofijur, M., Chowdhury, A.A., Almomani, F., 2024. Microplastics as carriers of toxic pollutants: Source, transport, and toxicological effects. *Environmental Pollution* 343, 123190. https://doi.org/10.1016/j.envpol.2023.123190

Rane, N.R., Kanojia, A., Patil, S.M., Khandare, R., Kodam, K.M., Jeon, B.-H., 2023. 20 – Constructed wetland system and its engineered designs for the treatment of textile industry effluent, in: Govindwar, S.P., Kurade, M.B., Jeon, B.-H., Pandey, A. (Eds.), *Current Developments in Bioengineering and Biotechnology*. Elsevier, pp. 601–626. https://doi.org/10.1016/B978-0-323-91235-8.00004-8

Riaz, R., de Wit, C.A., Malik, R.N., 2021. Persistent organic pollutants (POPs) in fish species from different lakes of the lesser Himalayan region (LHR), Pakistan: The influence of proximal sources in distribution of POPs. *Science of The Total Environment* 760, 143351. https://doi.org/10.1016/j.scitotenv.2020.143351

Rickard, B.P., Rizvi, I., Fenton, S.E., 2022. Per- and poly-fluoroalkyl substances (PFAS) and female reproductive outcomes: PFAS elimination, endocrine-mediated effects, and disease. *Toxicology* 465, 153031. https://doi.org/10.1016/j.tox.2021.153031

Roy, D., Khosravanipour Mostafazadeh, A., Drogui, P., Tyagi, R.D., 2020. Chapter 12 – Removal of organic micro-pollutants by membrane filtration, in: Varjani, S., Pandey, A., Tyagi, R.D., Ngo, H.H., Larroche, C. (Eds.), *Current Developments in Biotechnology and Bioengineering*. Elsevier, pp. 281–307. https://doi.org/10.1016/B978-0-12-819594-9.00012-7

Ruan, Y., Kong, L., Zhong, Y., Diao, Z., Shih, K., Hou, L., Wang, S., Chen, D., 2021. Review on the synthesis and activity of iron-based catalyst in catalytic oxidation of refractory organic pollutants in wastewater. *Journal of Cleaner Production* 321, 128924. https://doi.org/10.1016/j.jclepro.2021.128924

Ryu, B., Wong, K.T., Choong, C.E., Kim, J.-R., Kim, H., Kim, S.-H., Jeon, B.-H., Yoon, Y., Snyder, S.A., Jang, M., 2021. Degradation synergism between sonolysis and photocatalysis for organic pollutants with different hydrophobicity: A perspective of mechanism and application for high mineralization efficiency. *Journal of Hazardous Materials* 416, 125787. https://doi.org/10.1016/j.jhazmat.2021.125787

Schaffner, U., Heimpel, G.E., Mills, N.J., Muriithi, B.W., Thomas, M.B., Gc, Y.D., Wyckhuys, K.A.G., 2024. Biological control for One Health. *Science of The Total Environment* 951, 175800. https://doi.org/10.1016/j.scitotenv.2024.175800

Scheitlin, C.G., Dasu, K., Rosansky, S., Dejarme, L.E., Siriwardena, D., Thorn, J., Mullins, L., Haggerty, I., Shqau, K., Stowe, J., 2023. Application of supercritical water oxidation to effectively destroy Per- and polyfluoroalkyl substances in aqueous matrices. *ACS EST Water* 3, 2053–2062. https://doi.org/10.1021/acsestwater.2c00548

Sharma, M., Sharma, A., Batra, N., Pareek, R., Patel, S., 2024. Persistent organic pollutants in water and their microalgae based bioremediation. *Biosciences Biotechnology Research Asia* 21, 913–928. https://doi.org/10.13005/bbra/3273

Sidnell, T., Hurst, J., Lee, J., Bussemaker, M., 2022. Towards cost effective scale up of the sonolytic defluorination of perfluorooctane sulfonic acid. Available at SSRN: https://ssrn.com/abstract=4229483 or https://doi.org/10.2139/ssrn.4229483

Sil, A., Barman, M., Bhattacharyya, K., Das, S., Horijan, B., Lohara, S., 2020. Strategies for mitigation of polychlorinated biphenyls (PCBs): A review. *International Journal of Chemical Studies* 8, 1700–1708. https://doi.org/10.22271/chemi.2020.v8.i4q.9857

Singh, R.K., Fernando, S., Baygi, S.F., Multari, N., Thagard, S.M., Holsen, T.M., 2019. Breakdown products from Perfluorinated Alkyl Substances (PFAS) degradation in a plasma-based water treatment process. *Environmental Science Technology* 53, 2731–2738. https://doi.org/10.1021/acs.est.8b07031

Sörengård, M., Niarchos, G., Jensen, P.E., Ahrens, L., 2019. Electrodialytic per- and polyfluoroalkyl substances (PFASs) removal mechanism for contaminated soil. *Chemosphere* 232, 224–231. https://doi.org/10.1016/j.chemosphere.2019.05.088

Stiebeling, D.A., Labes, A., 2022. Mycoremediation of sewage sludge and manure with marine fungi for the removal of organic pollutants. *Frontiers in Marine Science* 9, 946220. https://doi.org/10.3389/fmars.2022.946220

Stockholm Convention, 2023. Printed at United Nations, Geneva, 2319428 (E), July 2024, p. 1500, UNEP/BRS/2023/12.

Sun, B., Li, Q., Zheng, M., Su, G., Lin, S., Wu, M., Li, C., Wang, Q., Tao, Y., Dai, L., Qin, Y., Meng, B., 2020. Recent advances in the removal of persistent organic pollutants (POPs) using multifunctional materials: A review. *Environmental Pollution* 265, 114908. https://doi.org/10.1016/j.envpol.2020.114908

Tenenbaum, D.J., 2004. POPs in polar bears: Organochlorines affect bone density. *Environmental Health Perspectives* 112, A1011–A1011. https://doi.org/10.1289/ehp.112-a1011

Tenorio, R., Liu, J., Xiao, X., Maizel, A., Higgins, C.P., Schaefer, C.E., Strathmann, T.J., 2020. Destruction of Per- and Polyfluoroalkyl Substances (PFASs) in Aqueous Film-Forming Foam (AFFF) with UV-sulfite photoreductive treatment. *Environmental Science Technology* 54, 6957–6967. https://doi.org/10.1021/acs.est.0c00961

Titchou, F.E., Zazou, H., Afanga, H., El Gaayda, J., Akbour, R.A., Hamdani, M., 2021. Removal of Persistent Organic Pollutants (POPs) from water and wastewater by adsorption and electrocoagulation process. *Groundwater for Sustainable Development* 13, 100575. https://doi.org/10.1016/j.gsd.2021.100575

Tomei, G., Saleem, M., Ceriani, E., Pinton, A., Marotta, E., Paradisi, C., 2023. Cold plasma for green advanced reduction/oxidation processes (AROPs) of organic pollutants in water. *Chemistry – A European Journal* 29, e202302090. https://doi.org/10.1002/chem.202302090

Trellu, C., Mousset, E., Pechaud, Y., Huguenot, D., van Hullebusch, E.D., Esposito, G., Oturan, M.A., 2016. Removal of hydrophobic organic pollutants from soil washing/flushing solutions: A critical review. *Journal of Hazardous Materials* 306, 149–174. https://doi.org/10.1016/j.jhazmat.2015.12.008

Truu, J., Truu, M., Espenberg, M., Nõlvak, H., Juhanson, J., 2015. Phytoremediation and plant-assisted bioremediation in soil and treatment wetlands: A review. *Open Biotechnology Journal*, 9(Suppl 1-M9), 85–92. https://doi.org/10.2174/1874070701509010085

US EPA, O., 2009. Persistent organic pollutants: A global issue, a global response [WWW Document]. https://19january2017snapshot.epa.gov/international-cooperation/persistent-organic-pollutants-global-issue-global-response (accessed 12.9.24).

Usman, M., Chaudhary, A., Hanna, K., 2024. Efficient PFAS removal from contaminated soils through combined washing and adsorption in soil effluents. *Journal of Hazardous Materials* 476, 135118. https://doi.org/10.1016/j.jhazmat.2024.135118

Vasseghian, Y., Hosseinzadeh, S., Khataee, A., Dragoi, E.-N., 2021a. The concentration of persistent organic pollutants in water resources: A global systematic review, meta-analysis and probabilistic risk assessment. *Science of The Total Environment* 796, 149000. https://doi.org/10.1016/j.scitotenv.2021.149000

Vasseghian, Y., Hosseinzadeh, S., Khataee, A., Dragoi, E.-N., 2021b. The concentration of persistent organic pollutants in water resources: A global systematic review, meta-analysis and probabilistic risk assessment. *Science of The Total Environment* 796, 149000. https://doi.org/10.1016/j.scitotenv.2021.149000

Verma, S., Lee, T., Sahle-Demessie, E., Ateia, M., Nadagouda, M.N., 2023. Recent advances on PFAS degradation via thermal and nonthermal methods. *Chemical Engineering Journal Advances* 13, 100421. https://doi.org/10.1016/j.ceja.2022.100421

Vijayaraghavan, K., Reddy, D.H.K., Yun, Y.-S., 2019. Improving the quality of runoff from green roofs through synergistic biosorption and phytoremediation techniques: A review. *Sustainable Cities and Society* 46, 101381. https://doi.org/10.1016/j.scs.2018.12.009

Wang, F., Xiang, L., Sze-Yin Leung, K., Elsner, M., Zhang, Y., et al., 2024. Emerging contaminants: A One Health perspective. *The Innovation* 5, 100612. https://doi.org/10.1016/j.xinn.2024.100612

Wang, J., Aghajani Delavar, M., 2023. Techno-economic analysis of phytoremediation: A strategic rethinking. *Science of The Total Environment* 902, 165949. https://doi.org/10.1016/j.scitotenv.2023.165949

WHO, 2023. Dioxins [WWW Document]. www.who.int/news-room/fact-sheets/detail/dioxins-and-their-effects-on-human-health (accessed 12.9.24).

Yadav, I.C., Devi, N.L., Syed, J.H., Cheng, Z., Li, J., Zhang, G., Jones, K.C., 2015. Current status of persistent organic pesticides residues in air, water, and soil, and their possible effect on neighboring countries: A comprehensive review of India. *Science of The Total Environment* 511, 123–137. https://doi.org/10.1016/j.scitotenv.2014.12.041

Yadav, P., Rai, S.N., Mishra, V., Singh, M.P., 2021. Mycoremediation of environmental pollutants: A review with special emphasis on mushrooms. *Environmental Sustainability* 4, 605–618. https://doi.org/10.1007/s42398-021-00197-3

Yang, H.-H., Cheruiyot, N.K., Lin, C., Wang, L.-C., 2022. Control of extreme brominated persistent organic pollutant emissions from start-ups of waste-to-energy incinerators. *Journal of Cleaner Production* 345, 131108. https://doi.org/10.1016/j.jclepro.2022.131108

Yang, L., He, L., Xue, J., Ma, Y., Shi, Y., Wu, L., Zhang, Z., 2020. UV/SO32− based advanced reduction processes of aqueous contaminants: Current status and prospects. *Chemical Engineering Journal* 397, 125412. https://doi.org/10.1016/j.cej.2020.125412

Yang, Q., Ma, Y., Chen, F., Yao, F., Sun, J., Wang, S., Yi, K., Hou, L., Li, X., Wang, D., 2019. Recent advances in photo-activated sulfate radical-advanced oxidation process (SR-AOP) for refractory organic pollutants removal in water. *Chemical Engineering Journal* 378, 122149. https://doi.org/10.1016/j.cej.2019.122149

Yuan, B., McLachlan, M.S., Roos, A.M., Simon, M., Strid, A., de Wit, C.A., 2021. Long-chain chlorinated paraffins have reached the arctic. *Environmental Science Technology Letters* 8, 753–759. https://doi.org/10.1021/acs.estlett.1c00470

Zamora, S., Sandoval, L., Marín-Muñíz, J.L., Fernández-Lambert, G., Hernández-Orduña, M.G., 2019. Impact of ornamental vegetation type and different substrate layers on pollutant removal in constructed wetland mesocosms treating rural community wastewater. *Processes* 7(8), 531. https://doi.org/10.3390/PR7080531

Zhang, H., Zhu, L., Zhang, Y., Héroux, P., Cai, L., Liu, Y., 2024. Removal of per- and polyfluoroalkyl substances from water by plasma treatment: Insights into structural effects and underlying mechanisms. *Water Research* 253, 121316. https://doi.org/10.1016/j.watres.2024.121316

Zhang, K., Wei, Y.-L., Zeng, E.Y., 2013. A review of environmental and human exposure to persistent organic pollutants in the Pearl River Delta, South China. *Science of The Total Environment* 463–464, 1093–1110. https://doi.org/10.1016/j.scitotenv.2012.10.104

11 Scope of Machine Learning and Blockchain Technology in Management of Emerging Contaminants

R. Archana Reddy, G. Shyamala, Ravindran Gobinath, and M. Sai Anjani

11.1 INTRODUCTION TO EMERGING CONTAMINANTS

Natural resource depletion has intensified due to growing population and food needs. Demand for sustainable, high-quality food production is rising. Scholars view the circular economy (CE) as a potential solution (S. Harris et al., 2021; Shikha Ojha et al., 2020; Oliveira et al., 2021). Circular food supply chain transformation includes preventing food losses, increasing resource efficiency, and developing regenerative ecosystems. Multiple CE methodologies are described in 'R'-frameworks (Demestichas & Daskalakis, 2020; Purva Mhatre et al., 2021). The R-framework includes these strategies: Refuse, Rethink, Reduce, Reuse, Repair, Refurbish, Remanufacture, Repurpose, Recycle, and Recover. 'Regenerative' practices in agriculture protect resources. Ecological systems face dangers from emerging contaminants and uncommon substances in the environment, including pharmaceuticals, personal care products (PPCPs), industrial chemicals, insecticides, and nanomaterials. Scientific interest has increased as these compounds exhibit long-lasting properties, accumulation behaviour, and health risks. Human-made processes produce environmental chemicals that release into the environment. The international community scrutinizes the impact of micro-contaminants in water sources.

The four fundamental elements: population numbers, food requirements, circular economy procedures, and emerging pollutants are interconnected in Figure 11.1. Rising population demands more sustainable resource management, increasing food requirements. Circular economy methods valorize and regenerate to reduce environmental deterioration. Strategies include preventive practices like resource reduction and recovering practices like material recovery and recycling. Long-term preservation needs regenerative strategies involving organic production, fair trade,

DOI: 10.1201/9781032623801-11

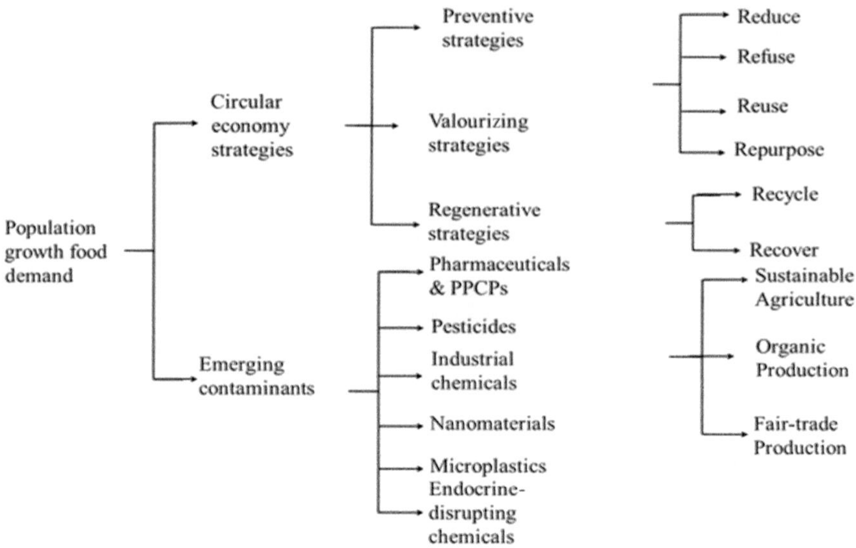

FIGURE 11.1 Circular economy strategies and emerging contaminants.

and sustainable farming. New chemical pollutants, including pesticides, industrial chemicals, nanomaterials, microplastics, pharmaceuticals, and endocrine-disrupting chemicals, have increased with industrial actions and food consumption. These threats necessitate methods for managing pollutants. Sustainable farming using circular economy principles decreases resource usage and protects ecosystems, addressing new pollutant problems. Organizations must link food supply control systems with substance reduction and sustainable development.

Pharmaceuticals and Personal Care Products (PPCPs): include medications, cosmetics, detergents, fragrances, and both prescription and over-the-counter medications. Humans and animals excrete active pharmaceutical ingredients (APIs), entering water bodies through sewage systems. Hormones and antibiotics can disturb aquatic environments and fuel antibiotic resistance.

Pesticides: Soil microorganisms and unintended species can suffer when agricultural chemicals for herbicide, insecticide, and fungicide contamination leak into water sources. Environmental agencies classify neonicotinoids and persistent organic pesticides as endocrine-disrupting chemicals.

Industrial Chemicals: Frictionless substances like PFAS, plasticizing agents, and heat-resistant chemicals are dispersed into the environment by manufacturing operations. PFAS, or timeless chemicals, resist degradation but are linked to negative health impacts.

Nanomaterials: Engineered nanomaterials (ENMs) serve as the foundation for electronics, textiles, and medical applications. Their special properties create unknown environmental issues due to substantial surface area and chemical reactivity.

Microplastics: Studies indicate that small plastic particles infiltrate marine ecosystems through direct product discharges and decomposition of larger plastics. Terrestrial ecosystems also contain similar particles. Microplastics harm aquatic organisms and transmit pollutants.

Endocrine-disrupting chemicals (EDCs): Human and animal hormone systems experience interference from these drugs, causing metabolic, developmental, and reproductive complications. Annually, thousands of tons of phthalates and bisphenol A (BPA) become widespread, as examples of this.

11.2 OVERVIEW OF MACHINE LEARNING IN ENVIRONMENTAL MANAGEMENT

Artificial intelligence (AI) and machine learning (ML) technologies transform environment management systems, optimizing decision-making. ML models analyse large data sets to extract patterns, predicting issues like air pollution, biodiversity loss, climate change, and resource protection. Reichstein et al. (2019) explained ML execution in environmental operations and presents industrial uses. Researchers use support vector machines (SVMs) and deep learning to predict temperature, rainfall, and to identify rare weather patterns (Reichstein et al., 2019). Recurrent neural networks analyse temperature anomalies to detect global warming patterns (Huntingford et al., 2019). ML algorithms enable air quality monitoring and pollutant hotspot detection. Supervised learning models utilize pollution and meteorological data for air pollution forecasting (Baklanov & Zhang, 2020). Cities receive pollution reduction plans due to forecasting systems. Computer learning systems analysing water resource data improve distribution systems. Zounemat-Kermani et al. (2021) explain how SVMs and k-nearest neighbours (k-NN) evaluate waterflow alterations and water quality changes. Applications lead to water supply operations in drought-struck regions.

As shown in Table 11.1, ML applications in environmental management summarize significant findings by ML methods. Techniques like deep learning, support vector machines (SVMs), recurrent neural networks (RNNs), and convolutional neural networks (CNNs) have solved many environmental challenges. ML applications provide accurate forecasting of water flow and quality indicators, contributing to resource sustainability and generating weather-related alerts for decisions. Hybrid CNN use for species recognition and acoustic data evaluation forms an efficient method for protecting biodiversity. Robotic systems with CNN architectures are used for quick waste sorting and autonomic recycling. Decision tree algorithms and ensemble models support precision farming, optimizing crop productivity, irrigation regulation, and pest control. These applications show how ML can redefine data-driven decision systems for environmental sustainability.

ML tracks bird and marine populations using CNNs in camera trap photos, aiding conservation with species location and habitat data. ML automates material identification for recycling. Robotic systems with CNNs enhance waste separation, reducing landfill waste. ML boosts renewable energy efficiency by improving generation and storage. Reinforcement learning and predictive analytics enhance renewable energy integration by predicting solar and wind patterns (Dileep, 2020). ML enhances pest

TABLE 11.1
Applications of machine learning in environmental management

Application Area	ML Techniques Used	Key Outcomes
Climate Modelling and Prediction	Deep learning, Support Vector Machines (SVMs), Recurrent Neural Networks (RNNs)	Improved climate predictions, better understanding of global warming trends
Air Quality Monitoring and Pollution Control	Supervised learning, Random forests, Gradient boosting	Identification of pollution hotspots, targeted policy measures
Water Resource Management	SVMs, k-nearest neighbours (k-NN)	Optimized water resource allocation, sustainable water management
Biodiversity Conservation	Convolutional Neural Networks (CNNs), Acoustic data analysis	Enhanced species identification and population monitoring
Waste Management and Circular Economy	Image recognition, CNN-powered robotic systems	Automated waste sorting, improved recycling rates
Renewable Energy Optimization	Reinforcement learning, Predictive analytics	Optimized renewable energy generation and storage
Agriculture and Precision Farming	Decision tree algorithms, Ensemble models	Improved crop yield, efficient irrigation, and pest management

management, irrigation, and crop productivity in precision agriculture. Decision tree algorithms and ensemble models analyse soil, weather, and crop health data to aid farmers (Liakos et al., 2018). Random forests detect illegal logging in the Amazon using satellite images. CNN-based ML processed Serengeti camera trap pictures, reducing analysis by 75% (Norouzzadeh et al., 2018). Learning Technique (LT) models predict floods in Indian zones by analysing rainfall and river data, aiding evacuation (Mosavi et al., 2018). ML analysis of satellite data offers insights on marine plastic contamination, guiding clean-up efforts (Topouzelis et al., 2019).

11.3 BLOCKCHAIN TECHNOLOGY: A NEW PARADIGM

Bitcoin, from blockchain technology, evolved for supply chain, healthcare, and finance. The peer-to-peer network is a distributed ledger securing transactions without intermediaries (n.d.-f). Blockchain creates a trustless environment using cryptography and consensus. Blocks contain timestamp data, transaction info, and the previous block's hash. SHA-256 hash functions secure bitcoin immutability (Mauro Conti et al., 2018), supporting security and manipulation resistance (Yaga et al., 2019). Blockchain employs distributed nodes, reducing failure risk. Consensus methods, proof of work (PoW) and proof of stake (PoS), execute transactions via stake-based voting or computational conservation. PoW demands significant computing resources, enhancing security but causing power issues (de Vries, 2018). Ethereum 2.0 employs PoS for secure, energy-efficient transactions (Ethereum Roadmap, n.d.). Smart contracts automate blockchain operations. Blockchain transparency curtails banking and healthcare fraud (Casino et al., 2019). Hyperledger Fabric, a permissioned blockchain, allows controlled access while preserving decentralization (Androulaki et al., 2018). These networks offer greater scalability and privacy than public blockchains (Xu et al., 2019). Bitcoin's decentralization limits transaction throughput. Layer-2 implementations enable off-chain operations. Blockchain's pseudonymous nature complicates anti-money laundering and compliance. Governments develop regulatory systems to balance security and technology (Chen & Bellavitis, 2020). Integrating blockchain with the internet of things (IoT), AI, and quantum computing opens new possibilities. Smart contracts boost blockchain potential with predictive analytics.

A conceptual description of blockchain technology and its key features is illustrated in Figure 11.2, reflecting its significance across industries. Key features include decentralization, ensuring no single organisation owns the network, facilitating resilience and transparency. Security is paramount, with blockchain employing cryptographic techniques to build irreversible records, ensuring data integrity. Smart contracts execute set conditions automatically, enhancing efficiency and reducing costs. Blockchain, combined with IoT and AI, protects data exchanges and enables trustworthy automation among smart systems. For broader implementation, blockchain must address scalability, energy consumption, and regulation. Despite obstacles, blockchain's uses in supply chain, governance, healthcare, and finance drive innovation and transparency. The chart shows its impact, highlighting its power to secure transactions, automate, and support decentralised environments. As blockchain grows, compatibility with AI and IoT will unlock its potential, solidifying its status as future-proof technology.

Quantum computing threatens cryptographic standards, necessitating quantum-resistant encryption. Blockchain revolutionizes industries, enhancing secure

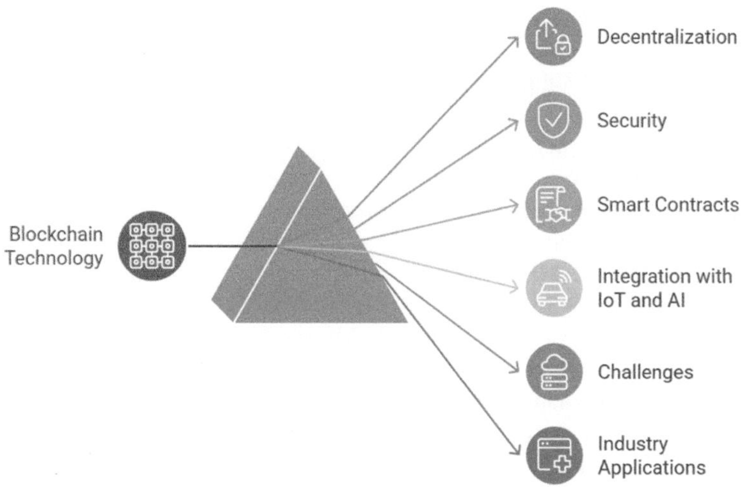

FIGURE 11.2 Unveiling the multifaceted impact of blockchain.

operations and efficiency. It helps financial institutions reduce payment processing costs, offering decentralized finance solutions and cross-border payments (Schär, 2020). In healthcare, blockchain ensures secure access to medical records, maintaining data reliability (Agbo et al., 2019). Supply chain success relies on blockchain traceability for transparency and fraud reduction (Francisco & Swanson, 2018). Environmental monitoring involves assessing natural components to detect ecosystem and human health risks (Campanella et al., 2021). Methods need improved accuracy, time, and cost. Standard environmental measurement approaches face performance restrictions, notably cost, including personnel and supply expenses (Ditria et al., 2022). Environmental programmes often have limited scope, producing incomplete condition presentations. Sample collection time and lab analysis delay emergency responses. Standard techniques suffer reduced accuracy due to human observation errors and subjective interpretation. Human involvement in lab analysis and sampling results in unreliable data.

11.4 INTEGRATION OF MACHINE LEARNING AND BLOCKCHAIN

Modern technology efficiently addresses environmental pollution. Blockchain and ML improve contamination detection, prediction, and reduction. Their integration enables distributed management systems with secure data access and automated decision-making (Mulligan et al., 2024). Combining ML pattern recognition with blockchain ensures secure sharing and trackable data set storage. Complex environmental data requires ML analysis to identify pollutants and spread patterns. Water pollutant classification uses supervised learning SVMs, neural networks, and deep learning CNNs to detect contaminants in satellite imagery and sensor information (Nkinahamira et al., 2024). ML models enhance water quality assessment by linking pollutants and environmental elements. Training predictive models with historical data enables preventive actions before pollution thresholds are reached (Bokonda et al., 2020). Anomaly-detection algorithms identify industrial incidents and chemical releases. Blockchain's ledger-based storage and verification strengthen environmental monitoring security. Traditional storage practices are susceptible to alteration, weakening regulatory compliance. Blockchain ensures data integrity with irreversible timestamps (Khan & Singh, 2022). Smart contracts impose penalties for businesses exceeding pollution limits during waste discharges.

Blockchain technology with machine learning enhances pollutant control systems. Blockchain's tamper-proof nature validates environmental data, aiding ML algorithms that need quality training data (Olawade et al., 2024). It ensures secure pollution-related data exchange. ML predictive analytics enhance blockchain environmental monitoring, enabling rapid response. Blockchain-based tokenization rewards sustainable practices, aiding pollution reduction. ML models in blockchain carbon credit systems verify emissions reduction and ensure fair incentives. Integrating blockchain with ML for contamination control faces scalability and processing challenges, compounded by regulatory demands. ML systems need large processing capacity, while blockchain requires significant storage and computing power. Full-scale environmental applications will benefit from federated learning and energy-efficient blockchain systems (Alabdali, 2025). This synergy also aids supply chain management, cybersecurity, healthcare, and finance (Mulligan et al., 2024).

Table 11.2 illustrates blockchain and ML integration into environmental management and businesses, emphasizing collaboration for efficiency, security, and transparency. Data-driven methods benefit detection, prediction, and mitigation in contamination control. Blockchain ensures safe data verification, promoting trust and compliance. Smart contrast executions, predictive analytics, and secure data transfer benefit from combining blockchain and ML. Obstacles for widespread use persist due to scalability, processing needs, and regulatory barriers. The combination exists in finance, healthcare, supply chain management, and cybersecurity. ML models benefit from blockchain protection against data-poisoning attacks Safe real-time decision-making is enabled by ML-driven sensor data in IoT and healthcare. Industry-scale ML and blockchain integration requires a framework for data management, consensus, and network interconnectivity.

Anomaly detection algorithms identify suspicious blockchain transactions (Nkinahamira et al., 2024). Reinforcement learning enhances energy-efficient consensus with proof of work and stake, reducing computational power (Quy et al., 2023). Machine learning strategies using clustering and federated learning tackle

TABLE 11.2
ML and blockchain integration in environmental management

Key Areas	Description
Role of Machine Learning in Contaminant Management	ML Aids in Detecting, Predicting, and Mitigating Environmental Contaminants through Data-Driven Decision-Making.
Blockchain for Secure Environmental Monitoring	Blockchain Ensures Secure, Tamper-Proof Storage and Verification of Pollution Data, Enhancing Transparency.
Advantages of ML-Blockchain Integration	ML Improves Predictive Analytics, While Blockchain Ensures Secure Data Sharing and Incentivizes Sustainability.
Challenges In ML-Blockchain Integration	Challenges Include Scalability, Computational Requirements, and Regulatory Hurdles in ML-Blockchain Integration.
ML and Blockchain in Various Industries	ML And Blockchain Integration Is Applied in Finance, Healthcare, Supply Chains, and Cybersecurity.
Enhancing Blockchain Efficiency with ML	Ml Optimizes Blockchain by Detecting Fraud, Enhancing Smart Contracts, and Improving Consensus Mechanisms.
Blockchain for Securing ML Models	Blockchain Mitigates ML Vulnerabilities Such as Data-Poisoning Attacks and Biased Data Manipulation.
Application in Healthcare and IoT	Blockchain Secures ML-Driven Healthcare and IoT Applications, Ensuring Trusted Decision-Making.
Framework for Efficient ML-Blockchain Integration	A Structured ML-Blockchain Framework Improves Interoperability, Scalability, and Efficiency in Various Sectors.
Key Components of ML-Blockchain Framework	Focus on Data Management, Consensus Optimization, and Cross-Chain Interoperability for Seamless Integration.

blockchain's scalability by limiting block size and minimizing delays. Machine learning-based smart contracts automatically adjust with real-time data (Alharbi, 2023; Dritsas & Trigka, 2024). Smart contracts facilitate data sharing in federated learning, safeguarding against unauthorized changes. Healthcare employs AI forecasting to diagnose diseases with blockchain-enabled medical records, ensuring privacy. Blockchain secures IoT sensor data via ML, enabling reliable decisions for industrial automation and smart cities (D. Cao et al., 2020), reducing threats, and enhancing AI analysis credibility.

A clear framework should focus on these three essential elements for effective ML-blockchain integration:

1. **Data Management and Security:** The blockchain system maintains model parameter and metadata immutability, while ML algorithms access data from external tools. Zero-knowledge proofs (ZKPs) work with encryption methods to enhance privacy.
2. **Consensus optimisation for machine learning workloads:** Machine learning demands updated consensus strategies for approval. Proof of useful work (PoUW) replaces energy-intensive proof of work (PoW), enabling nodes to perform ML tasks instead of cryptographic operations.
3. **Interoperability and standardisation:** A multichain architecture ensures communication between blockchain platforms (Ethereum, Hyperledger, Polkadot) and positively affects ML adoption on an industrial scale.

11.5 DATA COLLECTION AND MANAGEMENT

Data gathering and management have evolved with the IoT. Sensors provide data for strategic decisions, making data management vital. IoT enhances industries like smart cities, healthcare, agriculture, and automation. Research has evaluated IoT sensors and data sources. IoT sensors convert physical data into digital signals, detecting temperature, humidity, motion, pressure, light intensity, and chemical composition (Syed et al., 2022). Huber Notices improve traffic flow by assessing congestion and adjusting signals. Agriculture uses IoT soil moisture sensors for precision farming, improving fertilization and irrigation (Prakash et al., 2022). Healthcare employs IoT sensors in wearables to monitor vital signs, enabling telemedical observation. IoT sensor accuracy relies on energy efficiency, data transmission, and calibration. Wireless technologies like Zigbee, LoRa, Bluetooth, and 5G facilitate data transmission to cloud storage. IoT data systems' reliability requires addressing sensor errors, data loss, and security issues. The growth of IoT devices necessitates data management for storage, preprocessing, and analysis.

Existing database systems struggle with large IoT-originated unstructured data and diverse sources. Data management now uses cloud, edge computing, and blockchain-based distributed ledger systems (Mulligan et al., 2024). Cloud data centres process IoT data with scalable storage and computing capacity. Google Cloud IoT Core, Microsoft Azure IoT Hub, and AWS IoT platform help businesses handle data. Network bandwidth and latency hinder real-time solutions. Edge computing speeds source-based data processing, reducing latency. Organizations reduce cloud data using edge devices with processing abilities. In industrial automation, edge computing enhances sensors for quick problem detection and response (Said Mohamed et al., 2021). Security improves as sensitive data does not require internet transmission. Effective IoT data management prioritizes governance and security protocols. Data integrity, privacy, and compliance with the General Data Protection Regulation (GDPR) and the California Consumer Privacy Act (CCPA) are crucial. IoT cybersecurity relies on encryption, access control, and anomaly detection. Blockchain's secure data management is tested for IoT, providing tamper-proof storage and decentralized access control (Bellini et al., 2022). Despite its benefits, IoT issues persist. Data collection and management are crucial for organizations. Operations like information collection, storage, and processing produce decisions for better service. Growing data volumes cause privacy and security issues. New governance models are vital due to breaches and unethical use.

Figure 11.3 demonstrates how specific IoT deployment scenarios connect security constructs to computer systems and data management complexity. The analysis contains four essential elements of IoT data security and management which are organized according to their complexity alongside data volumes. The most elementary IoT set-up involving sensors stands as Basic IoT Sensor Deployment since it operates with a basic configuration and transmits minimal sensor data. The Cloud Computing for IoT Data solution manages substantial IoT data quantities effectively in basic processing conditions to support applications needing substantial storage platforms and analytics features. Edge Computing for industrial automation performs low-latency operations by keeping data processing near its sources in complex industrial sites which benefit high-volume real-time data applications. Blockchain for IoT Security objectives resolve IoT security dilemmas by safeguarding the safety of connected IoT systems that deal with minimal data yet intricate operations. The classification framework organizes IoT solutions through data volume and processing complexity

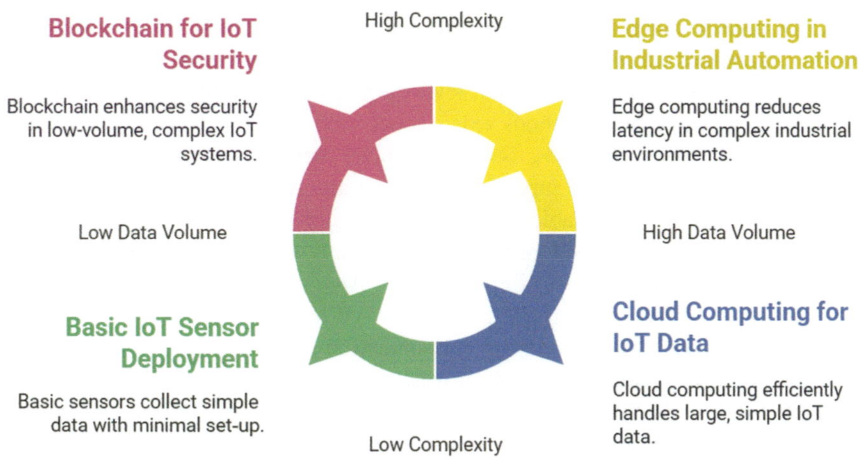

FIGURE 11.3 IOT integration in data management.

interactions while featuring security standards to achieve effective IoT application decision-making in various industries.

Information retrieval spans websites, applications, social media, and IoT devices. Organizations collect data from surveys, interactions, transactions, and machine-generated systems. Information includes structured data in databases and unstructured data like emails and videos, requiring advanced processing (McKinsey & Company, 2021). Effective data collection involves defining objectives, maintaining accuracy, and upholding ethics. Organizations use big data analytics and AI for insights to guide decisions and innovation. Data management includes storing information with protocols for accessibility and reliability, covering governance, integration, quality, and life cycle management. Proper management ensures consistency and enhances efficiency (Cost of a Data Breach 2024, n.d.). Businesses use cloud computing for scalable storage and real-time access but face security threats. Data encryption, restricted access, and governance secure information. Personal data privacy is crucial due to unauthorized theft and misuse. The European Union's GDPR and California's CCPA enforce data-privacy protections by setting standards for handling personal data. Non-compliance results in penalties and reputation damage. Organizations must create privacy policies, obtain consent, and follow transparent practices. Data security guards against cyberthreats and breaches. IBM (2023) reports organizations lost \$4.45 million on average due to security incidents in 2023, impacting finances and reputation. Businesses increasingly adopt zero-trust security models (Foote, 2022). Regular audits, employee training, and response plans mitigate risks. PETs like differential privacy, homomorphic encryption, and federated learning enable data analysis while protecting privacy (Sharma, 2023).

11.6 PREDICTIVE MODELLING OF CONTAMINANT BEHAVIOUR

Qualitative pollutant release endangers health, environmental sustainability, and production facilities. Predictive analytics uses ML for efficient contaminant behaviour modelling. Traditional modelling relies on statistics but struggles with complex, large data. ML analyses large data sets, identifying patterns for predicting contaminants' spread and degradation (Croll et al., 2025). The present study examines ML algorithms for contaminant behaviour, reviewing model types, applications, barriers, and forecasts. Three ML methods exist: supervised, unsupervised, and reinforcement learning. Decision trees, SVMs, artificial neural networks (ANNs), and random forests are common supervised learning algorithms. These models require data sets linking input variables to contaminant concentrations, dispersal routes, and breakdown speeds. ANN models predict heavy metals/pollutants in groundwater using water attributes (Cui et al., 2023). Air quality prediction uses random forest and gradient boosting models. Unsupervised learning involves clustering and dimensionality reduction, identifying contamination sources, zones, and anomalies. K-means clustering classifies industrial wastewater pollutants for remediation strategies (Zhang et al., 2025). Principal component analysis (PCA) and t-distributed stochastic neighbor embedding (t-SNE) reduce data dimensions to reveal patterns. Reinforcement learning (RL) trains agents for optimal actions by interacting with the environment, suitable for dynamic pollution control. RL ensures effective water treatment plant operation through automated adjustments (Tran & Hu, 2019). ML models control contaminant behaviour predictions in rivers, lakes, and subsurface systems. Long short-term memory (LSTM) networks excel in monitoring contaminant concentration fluctuations (Multiple Linear Regression, n.d.). ML algorithm application forecasts air pollution by analysing pollution histories, weather factors, and road conditions.

Table 11.3 examines ML applications, highlighting subjects and procedures with environmental components. ML algorithms use predictive analytics for contaminant dispersion predictions with decision trees, SVM, ANN, and random forests. Groundwater pollution monitoring and concentration prediction systems rely on ANN models with gradient boosting. K-means clustering and PCA discover pollution hotspots and analyse pollutant patterns. RL optimizes pollution control in water treatment. Edge Computing with IoT, Federated Learning, deep learning, and geospatial ML build environmental monitoring systems for risk evaluation. Data restrictions, regulatory barriers, and explanation requirements hinder ML model deployment for contaminant forecasting. Computational hybrid networks use AI processing of environmental data for model integration, aiding pollutant investigations and governance.

TABLE 11.3
ML-based predictive modelling of contaminant behaviour

Topic	ML Techniques Used	Environmental Application
Role of ML in Contaminant Behaviour Modelling	Decision trees, SVM, ANN, Random Forests	Air, water, and soil contamination modelling
Supervised Learning for Contaminant Prediction	ANN, Random Forests, Gradient Boosting	Groundwater pollution monitoring
Unsupervised Learning for Contamination Analysis	K-Means Clustering, PCA, t-SNE	Industrial wastewater analysis
Reinforcement Learning in Pollution Control	Reinforcement Learning, Policy Optimization	Adaptive water treatment strategies
Applications of ML in Environmental Domains	LSTM, Hybrid ML models, CNNs for satellite imagery	Urban and industrial emission monitoring
Challenges in ML for Contaminant Modelling	Data Imputation, Transfer Learning, XAI Techniques	Data-driven environmental decision-making
Future Directions in ML for Contaminant Prediction	Edge Computing, IoT, Federated Learning, Generative AI	Smart environmental monitoring systems
Integration of AI and Predictive Modelling	Deep Learning, Remote Sensing, Geospatial ML	Predictive environmental risk assessment
Risk Assessment and ML in Contaminant Behaviour	Monte Carlo Simulations, Bayesian Networks	Environmental regulation and policymaking
Challenges in Predictive Modelling of Contaminants	Hybrid Modelling, AI-Assisted Environmental Monitoring	Climate change adaptation in contaminant modelling

CNNs analyse satellite imagery for air pollution detection, aiding emissions monitoring. ML-based analytics predict soil contamination by heavy metals, pesticides, and hydrocarbons for land use planning and remediation. Geospatial ML models map contamination using remote sensing and soil sampling. Kriging-based geostatistical modelling with deep learning improves prediction accuracy. Challenges in ML contaminant modelling include data quality and availability, requiring large data sets. Data imputation and transfer-learning use similar conditions. Model interpretability is challenging, with many ML models resembling 'black boxes'. XAI techniques enhance transparency. Integrating domain knowledge with ML models is crucial. Hybrid approaches combining ML with physics-based models increase reliability. Edge computing and IoT devices enhance real-time data collection. Federated learning allows collaborative ML model training without raw data sharing. ML contaminant prediction supports scenario analysis and simulates pollution control. Predictive modelling assesses environmental hazards, quantifying pollutant transport, transformation, and ultimate outcome in air, water, and soil. Models understand contaminant interactions with natural systems and threats to ecosystems (Sales-Ortells et al., 2015). Advanced computational methods, AI, and ML have improved predictive model precision for environmental risks. Knowledge of contaminants' physicochemical properties is crucial for predicting environmental distribution (Pahariya et al., 2025). Contaminant movement and bioavailability are evaluated using hydrological and atmospheric models. Reactive transport modelling replicates chemical phase changes, enhancing predictions (Nkinahamira et al., 2024). ML and AI have refined contaminant behaviour models. Conventional models use equations, while ML models utilize data sets to forecast contaminant dispersion more accurately (Mackay & Paterson, 1991). Deep learning models analyse historical data to identify non-linear relationships among environmental factors, assessing industrial effects, agricultural run-off, and spills and their impact on water and soil quality. AI models integrate remote sensing, sensor networks, and geospatial data to enhance risk assessments. Risk assessment techniques have evolved with predictive modelling, including probabilistic methods for quantifying uncertainties. Monte Carlo simulations and Bayesian networks forecast contaminant concentration and health risk probabilities (Dou et al., 2018). These techniques enable policymakers to evaluate environmental management scenarios. Regulatory agencies use predictive models to set standards and exposure limits. A challenge is data availability and quality. Reliable predictions require data on contaminant sources, conditions, and pollution history. Developing regions have limited monitoring networks, affecting model validity. Scientists combine process-based and data-based methods to improve predictions (H. Chen & He, 2022). Human activity and climate change alter transport dynamics, necessitating adaptive modelling.

11.7 REAL-TIME MONITORING AND REPORTING

The integration of real-time monitoring and automated systems using ML has transformed industries by enhancing efficiency, accuracy, and decision-making. ML manages large data sets, recognizes patterns, and predicts autonomously. Real-time data processing is vital in healthcare, finance, manufacturing, and cybersecurity. This research examines the impact of real-time monitoring, ML algorithms, and automation on industries, addressing challenges and prospects. Real-time monitoring optimizes performance and addresses anomalies by observing system activities and data streams. Automated systems use ML algorithms to analyse sensor and network data, providing timely insights. It supports predictive maintenance in industry, reducing downtime and costs (Nkinahamira et al., 2024). In healthcare, ML-based systems monitor patient vitals and alert experts to patients' critical conditions. ML algorithms improve by learning from data. Supervised learning algorithms like decision trees, SVMs, and deep neural networks are used for classification and anomaly detection. In finance, ML models assess trading patterns and detect fraud in real time (Padmavathi & Kanimozhi, 2025). In cybersecurity, ML-driven intrusion detection systems (IDS) identify unauthorized access and mitigate threats. Unsupervised learning models find anomalies in unlabelled data sets. Data privacy and security are crucial due to the sensitivity of real-time data, necessitating encryption, access control, and regulation compliance (Said Mohamed et al., 2021). ML model accuracy depends on training data quality and diversity. Data bias can cause incorrect predictions, requiring validation and fairness checks. Computational complexity in real-time processing poses challenges. High-speed data streams demand significant computational resources, leading to scalability issues. Edge computing and cloud-based ML frameworks address these by distributing workloads.

Figure 11.4 presents a system with real-time monitoring coupled with ML decision functions tracking data-proficient prediction creation stages. Real-Time Monitoring Initiation enables long-term system activity recording. The system then moves into Data Collection where sensors and networks acquire necessary environmental, industrial, or system data. Data Analysis conducts ML-based algorithm assessments, producing valid information insights. Pattern Identification applies ML techniques to locate significant patterns, anomalies, and connections in gathered data. Prediction Making generates predictive forecasting models. Insight Generation converts ML-generated predictions into decision-supporting insights and recommendations that assist executives in process improvement, lowering risks, and maximizing operational efficiency. Multiple IoT-based environmental monitoring systems utilize this workflow along with industrial automation systems, financial forecasting systems, healthcare analytics tools, and various other fields. The fusion of real-time monitoring systems and ML leads to dependable, swift decisions from data that advances as a primary technology for automated predictions across diverse commercial sectors.

Integrating explainable AI (XAI) enhances transparency in automated decision-making, fostering trust in ML-powered systems. Advances in deep learning, quantum computing, and federated learning drive monitoring. Deep learning improves anomaly detection and predictive analytics, while quantum computing revolutionizes data processing. Federated learning addresses privacy and security. Blockchain enhances transparency, ensuring data security, integrity, and accessibility. Supply chain management, health, finance, and environmental observation use blockchain for accountability. The present research explores blockchain-based monitoring systems, focusing on transparency and traceability. Blockchain's distributed ledger securely records

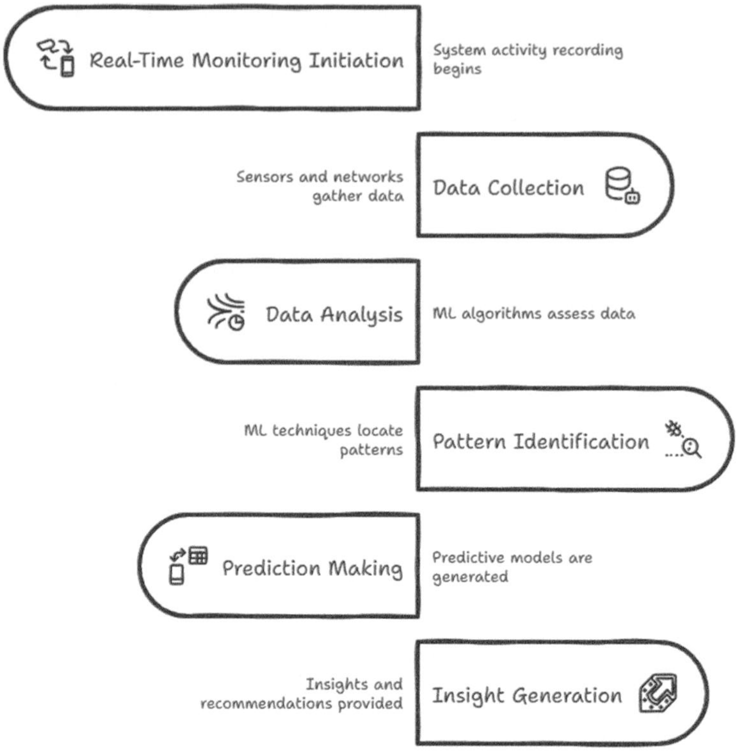

FIGURE 11.4 Integration of real-time monitoring and machine learning.

transactions, reducing fraud and tampering (Xu et al., 2019). In real-time systems, blockchains enable immediate data capture and validation. Smart contracts automate tracking (Kamilaris et al., 2019). In supply chains, blockchain tracks product movement, ensuring authenticity and reducing counterfeiting. Transactions are open, allowing stakeholders to verify origin, quality, and compliance. Blockchain ensures real-time monitoring of medical supplies, improving patient safety. Traditional systems often lack data consistency and are prone to fraud. Blockchain addresses these with a decentralized, unalterable record (Casino et al., 2019). In financial transactions, blockchain-based systems offer transparency, reducing fraud risks. Regulators can view an unalterable ledger, monitor compliance, and prevent illegal activities. Blockchain tracks carbon emissions and waste disposal, holding companies accountable. In the food industry, blockchain tracks production and transportation, enabling the tracing of contamination sources, reducing health risks and facilitating recalls (Kamilaris et al., 2019). Blockchain in real-time monitoring faces challenges. Scalability is a major issue; as networks grow, transactions become slower and costlier (Xu et al., 2019). High-frequency data updates lots of information in financial trade and IoT-driven systems exacerbate this issue. Integrating blockchain with the current IT infrastructure is challenging. Many businesses use legacy systems incompatible with blockchain, requiring investment in infrastructure, skills, and training (Kamilaris et al., 2019). Privacy concerns impede blockchain adoption. Transparency is vital, but confidentiality is essential for patient information and financial transactions. Private and permissioned blockchains offer privacy but may compromise the decentralization benefits of public chains (Casino et al., 2019). Despite challenges, blockchain has evolved, offering potential for enhanced transparency and traceability across industries. Scalability solutions like sharding and layer-2 protocols improve transaction speeds and reduce costs. Combining blockchain with AI is expected to enhance data analysis and predictive capabilities (Hernandez Aros et al., 2024). In smart cities, blockchain is explored for tracking public services, waste, and energy usage. Blockchain tracks ethical sourcing and fair labour practices, providing a record of working conditions and pay, helping hold corporations accountable (Kamilaris et al., 2019).

11.8 POLICY IMPLICATIONS AND REGULATORY FRAMEWORKS

Climate change mitigation, ecosystem degradation, and sustainability efforts are shaped by environmental policy and regulations. Governments and global bodies have established policies to regulate emissions, protect biodiversity, and promote renewable energy. These regulations affect economies and industries while providing environmental protection guidelines. Environmental policies manage activities impacting the environment. International efforts like the Paris Agreement and the Kyoto Protocol demand carbon emission cuts and support sustainable development. Nations have enacted laws to govern industrial waste, pollution, and deforestation. The Clean Air Act has improved environmental health by reducing pollutants like SO_2 and NO_x (EPA, 2020). Policies require industries to comply with emission regulations, control waste, and use sustainable materials. Businesses ignoring policies face fines, lawsuits, and reputation loss. Government grants and tax incentives support green technology investments. Laws create environmental impact assessment standards, manage resources, and control pollution. The European Environment Agency (EEA) and the US Environmental Protection Agency (EPA) ensure compliance and enforce rules. Environmental impact assessment (EIA) requires firms to evaluate environmental impacts and propose mitigation measures. The United Nations Framework Convention on Climate Change (UNFCCC) outlines an international framework for climate negotiations and emission reduction commitments. Developing countries often lack capacity, face corruption, and have inadequate resources to implement regulatory frameworks. Global cooperation, finance, and capacity development are crucial for enhancing environmental governance. The World Bank and the United Nations Environment Programme (UNEP) provide financing and assistance to enact environmental law (Glass & Newig, 2019).

Economic and industrial advancements are influenced by environmental regulations. Strict laws may raise costs but promote sustainability. Porter's theory suggests well-crafted laws enhance competitiveness. Carbon pricing encourages emissions reduction. Carbon taxes in Sweden and Canada have reduced footprints while maintaining growth (Sara, 2024). Extended producer responsibility (EPR) holds producers accountable for goods' lifespan, promoting waste minimization and recycling. A circular economy supports recycling waste into new products. Regulations like the European Green Deal and China's Circular Economy Policy promote resource efficiency (The European Green Deal, n.d.). These policies reduce damage and create opportunities. Implementing regulations faces challenges. Conflicts, financial constraints, and political opposition hinder the execution of regulations. Dependence on fossil fuels and lobbying by oil and gas companies hinder the US and China from adopting renewables. Policy differences across countries also present challenges. Developing countries face inadequate enforcement, causing carbon leakage. International collaboration and financial incentives are needed (Bates et al., 2016). The Paris Agreement emphasizes Common but Differentiated Responsibilities (CBDR), acknowledging rich countries should bear more burden in combating climate change while assisting underdeveloped countries. Future policy focuses should include climate resilience, digital environmental monitoring, and sustainable financing. The UN's Sustainable Development Goals guide environmental preservation. Technologies like blockchain, AI, and remote sensing enhance compliance and enforcement (Azevedo et al., 2024). Green finance policies include Environmental, Social, and Governance investments and sustainable bonds. Firms use climate risk disclosure and green accounting for sustainability, integrating these into financial decisions and considering environmental risks. Regulatory frameworks ensure stability and legal adherence, fostering innovation while minimizing unethical behaviour and environmental harm. As AI, fintech, healthcare, and environmental ecosystems grow more complex, frameworks require updating. Regulations must match threats, market changes, and technological advances. Cybersecurity and digital transformation affect policy. Blockchains, IoT, and AI require regulators to address privacy, transparency, and ethics. The EU's GDPR protects personal data, while the CCPA enhances consumer rights. Future regulation should focus on explainable AI, accountability, and algorithm bias. Financial regulation evolves with fintech growth, including cryptocurrencies and DeFi. Unregulated blockchain risks fraud, money laundering, and market manipulation. Bodies like the Securities and Exchange Commission and the Financial Action Task Force develop rules for cryptocurrency exchanges, initial coin offerings, and anti-money laundering procedures. Digital money's borderless nature requires international compliance. Healthcare policies focus on drug approval, telemedicine laws, and patient data protection. COVID-19 accelerated telehealth, necessitating confidentiality regulations. Ethical considerations in genetic research are crucial. Future healthcare will emphasize pandemic preparedness, digital health, and AI diagnostics. Environmental regulations are stricter as the urgency for sustainability rises. Policies like the European Green Deal and the Paris Agreement promote renewable energy, the circular economy, and emissions reduction. Future regulations will focus on emission reporting, green finance incentives, and sustainable supply chains. Biotechnology, quantum computing, and space exploration will influence regulation. Governments must create frameworks to address moral, security, and financial concerns as these frontiers develop. Stakeholder involvement, interdisciplinary research, and collaborative policymaking are essential for responsible innovation and public interest protection.

11.9 CHALLENGES AND LIMITATIONS

Despite technological advancements, ethical and technical issues like privacy, security, and equity persist. A multidisciplinary approach integrating ethics and technology is essential. Key issues include performance and scalability, as AI, IoT, and big data analytics increase computing demands. AI models need significant processing power, raising energy and environmental concerns (Smith & Jones, 2021). High costs limit access to high-performance computing. In critical applications, real-time decisions are hampered by data processing and network bandwidth limits (Brown et al., 2022). Cybersecurity and data privacy are major constraints, with rising cyberattack risks. Organizations must adopt strong security frameworks, as vulnerabilities remain despite better authentication and encryption (Davis, 2020). Many smart devices lack security, making them easy targets (Aziz Al Kabir et al., 2023). Ethical concerns include responsibility, justice, and bias. AI systems often reflect training data biases, leading to unfair outcomes and perpetuating social injustices (Johnson et al., 2023). Ensuring AI fairness and transparency is difficult due to ingrained biases. Accountability for AI errors or unethical results is unclear. Data collection and surveillance raise ethical issues. Excessive data collection poses privacy and consent challenges, with many users unaware of data use. Facial recognition and biometric monitoring enable mass surveillance without protection. Policymakers and organizations must embed ethical solutions into tech advancements. Threats can be mitigated by transparent AI models, strong cybersecurity, and clear legal guidelines. Collaboration among engineers, ethicists, and lawmakers is vital for ethical tech innovation. Technological progress has boosted industries' production, convenience, and efficiency. However, challenges like energy use, scalability, and cybersecurity hinder full potential. Overcoming these is crucial for progress.

Large data sets used by AI systems may contain biases, leading to skewed decisions in healthcare, law enforcement, and employment (Russell & Norvig, 2020). Deep learning models' 'black box' nature limits adoption and trust by making AI-generated results hard to understand. AI's potential abuse, such as privacy infringement, deepfake technologies, and surveillance, raises ethical concerns. Cybersecurity remains a major issue due to evolving cyberthreats. The attack surface for cybercriminals has expanded with increased reliance on cloud computing, IoT devices, and digital transactions. Organizations suffer financial and reputational damage from cyberattacks, exacerbated by weak encryption and users' ignorance of cybersecurity. The dynamic nature of cyberthreats complicates the creating of infallible defences. Another drawback is energy consumption, particularly in data centres, AI training, and blockchain technology. Deep learning and AI models demand substantial processing power, raising energy costs and carbon emissions (Strubell et al., 2019). Blockchain-based systems use energy-intensive proof of work techniques to validate transactions. Finding sustainable solutions remains critical as global energy demands rise. Scalability and infrastructural limitations present obstacles for 5G and quantum computing. Quantum computing can transform problem-solving in materials science, cryptography, and optimization, but current infrastructure is not ready for widespread use (Preskill, 2018). 5G networks face challenges with infrastructure deployment, spectrum allocation, and costs. In underdeveloped nations, inadequate infrastructure widens the digital divide, limiting innovation access. Ethical and regulatory issues constrain new technologies. Governments struggle to keep pace with technical advancements, lacking policies for AI, biotechnology, and driverless cars. Concerns about data privacy, algorithmic transparency, and accountability hinder implementation. The absence of international regulatory standards complicates the establishing of a consistent ethical foundation for technological development.

11.10 PROSPECTS AND INNOVATIONS

Industries have been experiencing rapid technological advancements, leading to breakthroughs in automation, AI, IoT, and quantum computing. Technology integration enhances creativity, effectiveness, and sustainability. Future prospects include blockchain for secure transactions, AI-driven automation, quantum computing for problem-solving, and IoT for better connectivity. These advancements support Industry 4.0, emphasizing cyber-physical systems, data-driven decision-making, and smart manufacturing (Said Mohamed et al., 2021). Technology integration involves AI and ML, enabling intelligent automation, process optimization, and predictive analytics. AI models boost productivity in banking, healthcare, and transportation. Robot-assisted surgery and AI-powered medical imaging improve patient outcomes and diagnostic precision (Smith & Johnson, 2022). AI-driven fraud detection analyses data sets to prevent illegal transactions and cyberthreats. AI is crucial for self-driving cars, route optimization, and for reducing human error. AI is expected to transform business structures, decision-making, and human-computer interactions. Blockchain, a transparent digital ledger, is used in smart contracts, supply chain management, and healthcare data security. Its decentralized structure ensures transaction authenticity. Future blockchain developments will employ quantum-resistant cryptographic methods to enhance security against quantum computing threats. Combining blockchain with AI can transform decentralized autonomous organizations (DAOs) and improve decision-making. Quantum computing impacts complex problems. Quantum computers use qubits for simultaneous calculations, aiding drug discovery, material research, and cryptography. Companies like IBM, Google, and D-Wave can advance by creating quantum processors to address issues beyond traditional computing's scope (Alahi et al., 2023).

Future developments in energy efficiency, climate prediction, and financial risk assessment will be driven by quantum computing. The IoT enables real-time monitoring of homes, businesses, and urban infrastructure. IoT applications like smart cities, precision farming, and industrial automation optimize resources and efficiency (Rashid & Kausik, 2024). AI and IoT enhance edge computing, enabling local data processing. Future IoT advances will focus on 5G and 6G networks for faster communication. AI and ML have transformed industries like healthcare, banking, and automation. However, AI faces challenges with interpretability, bias, and ethical decision-making. Explainable AI (XAI) is crucial for human-understandable explanations, enhancing AI's reliability. Renewable energy technologies face research gaps. Efficiency and storage remain issues despite advances in solar, wind, and hydrogen energy. Developing high-capacity, affordable batteries is key for renewable grid integration. Advances in nanomaterials and improved electrodes can boost energy density and lifespan. Hybrid systems and perovskite solar cells may address sustainable energy research gaps. Biotechnology and personalized medicine are advancing biomedical research. Disease diagnosis and treatment have improved by integrating genomes, proteomics, and AI-driven diagnostics. Legal restrictions and accessibility issues persist. CRISPR and RNAi technologies promise to treat genetic abnormalities and prevent disease, but ethical concerns require research. Quantum computing offers opportunities but is limited by error rates, decoherence, and qubit stability (Preskill, 2018). Developing fault-tolerant quantum computers and hybrid models could advance simulations, materials science, and drug discovery. Climate change demands innovative solutions. Carbon capture and storage offers mitigation alternatives, but challenges hinder implementation. Research into sustainable materials, enhanced weathering, and bioengineered carbon sequestration may address climate impacts. Interdisciplinary research is crucial for optimizing energy use, waste management, and green infrastructure in urban sustainability and smart city development (Batty, 2018). These fields hold potential for innovation and policy intervention.

11.11 CONCLUSION

The conjunction of blockchain and ML provides a good way to handle new toxins and enhance automation security and efficiency in pollution management. ML can lead to more advanced analytical tools to monitor and predict pollutants released into ecosystems at industrial chemical scales, check the chemical class level for pesticide and pharmaceutical compounds, and track microplastic pollution. ML can improve early risk detection using automated models, patterns, and decision systems. When stored using blockchain technology, data management becomes safer and more transparent through compatible information safety. Blockchain's decentralized ledgers and smart contracts reliably protect environmental monitoring data across contamination sources and treatments for regulatory compliance as unalterable and intersubjective knowledge. Blockchain can integrate with IoT mechanisms for continuous data collection from soil, water, and air sensors to prevent and track contamination. Mature deployment depends on fixing scalability obstacles, containing high costs, and meeting regulatory standards. Future research should develop blockchain management structures for big environmental data systems and enhance ML protocols for immediate pollutant identifications. These technologies advance environmental management by creating a transparent strategy for new pollutant management with data-driven approaches. ML and blockchain adoption will enable industries, researchers, and policymakers to make informed decisions for a better environment.

REFERENCES

Agbo, C. C., Mahmoud, Q. H., & Eklund, J. M. (2019). Blockchain technology in healthcare: A systematic review. *Healthcare (Basel, Switzerland)*, 7(2), 56. https://doi.org/10.3390/healthcare7020056

Alabdali, A. M. (2025). Blockchain based solid waste classification with AI powered tracking and IoT integration. *Scientific Reports*, 15(1), 15197. https://doi.org/10.1038/s41598-025-97030-2

Alahi, M. E. E., Sukkuea, A., Tina, F. W., Nag, A., Kurdthongmee, W., Suwannarat, K., & Mukhopadhyay, S. C. (2023). Integration of IoT-enabled technologies and Artificial Intelligence (AI) for smart city scenario: Recent advancements and future trends. *Sensors (Basel, Switzerland)*, 23(11), 5206. https://doi.org/10.3390/s23115206

Alharbi, M. (2023). Foreign investment attracted by Saudi Arabia. *SSRN Electronic Journal*, 26, 29. https://doi.org/10.2139/ssrn.4357486

Androulaki, E., Barger, A., Bortnikov, V., Cachin, C., Christidis, K., De Caro, A., Enyeart, D., Ferris, C., Laventman, G., Manevich, Y., Muralidharan, S., Murthy, C., Nguyen, B., Sethi, M., Singh, G., Smith, K., Sorniotti, A., Stathakopoulou, C., Vukolić, M., … & Yellick, J. (2018). Hyperledger fabric: A distributed operating system for permissioned blockchains. arXiv [Cs.DC]. https://doi.org/10.48550/ARXIV.1801.10228

Azevedo, B. F., Rocha, A. M. A. C., & Pereira, A. I. (2024). Hybrid approaches to optimization and machine learning methods: A systematic literature review. *Machine Learning*, 113(7), 4055–4097. https://doi.org/10.1007/s10994-023-06467-x

Aziz Al Kabir, M., Elmedany, W., & Sharif, M. S. (2023). Securing IoT devices against emerging security threats: Challenges and mitigation techniques. *Journal of Cyber Security Technology*, 7(4), 199–223. https://doi.org/10.1080/23742917.2023.2228053

Baklanov, A., & Zhang, Y. (2020). Advances in air quality modeling and forecasting. *Global Transitions*, 2, 261–270. https://doi.org/10.1016/j.glt.2020.11.001

Bates, M. E., Grieger, K. D., Trump, B. D., Keisler, J. M., Plourde, K. J., & Linkov, I. (2016). Emerging technologies for environmental remediation: Integrating data and judgment. *Environmental Science & Technology*, 50(1), 349–358. https://doi.org/10.1021/acs.est.5b03005

Batty, M. (2018). *Inventing future cities*. MIT Press. https://doi.org/10.7551/mitpress/11923.001.0001

Bellini, P., Nesi, P., & Pantaleo, G. (2022). IoT-enabled smart cities: A review of concepts, frameworks and key technologies. *Applied Sciences (Basel, Switzerland)*, 12(3), 1607. https://doi.org/10.3390/app12031607

Bokonda, P. L., Ouazzani-Touhami, K., & Souissi, N. (2020). Predictive analysis using machine learning: Review of trends and methods. In *2020 International Symposium on Advanced Electrical and Communication Technologies (ISAECT)*. https://doi.org/10.1109/ISAECT50560.2020.9523703

Brown, K., & Smith, J. (2022). The role of quantitative finance in financial market analysis and decision-making. *Financial Analysts Journal*, 78, 44–60.

Campanella, B., Palleschi, V., & Legnaioli, S. (2021). Introduction to vibrational spectroscopies. *ChemTexts*, 7(1), 5–21. https://doi.org/10.1007/s40828-020-00129-4

Cao, D., Hu, W., Zhao, J., Zhang, G., Zhang, B., Liu, Z., Chen, Z., & Blaabjerg, F. (2020). Reinforcement learning and its applications in modern power and energy systems: A review. *Journal of Modern Power Systems and Clean Energy*, 8(6), 1029–1042. https://doi.org/10.35833/mpce.2020.000552

Casino, F., Dasaklis, T. K., & Patsakis, C. (2019). A systematic literature review of blockchain-based applications: Status, classification and open issues. *Telematics and Informatics*, 36, 55–81. https://doi.org/10.1016/j.tele.2018.11.006

Chen, H., & He, Y. (2022). Machine learning approaches in traditional Chinese medicine: A systematic review. *The American Journal of Chinese Medicine*, 50(1), 91–131. https://doi.org/10.1142/S0192415X22500045

Chen, Y., & Bellavitis, C. (2020). Blockchain disruption and decentralized finance: The rise of decentralized business models. *Journal of Business Venturing Insights*, 13(e00151), e00151. https://doi.org/10.1016/j.jbvi.2019.e00151

Conti, Mauro, Sandeep Kumar, E., Lal, C., & Ruj, S. (2018). A survey on security and privacy issues of bitcoin. *IEEE Communications Surveys & Tutorials*, 20(4), 3416–3452. https://doi.org/10.1109/comst.2018.2842460

Cost of a Data Breach 2024. (n.d.). *Ibm.com*. Retrieved May 5, 2025, from www.ibm.com/reports/data-breach

Croll, H. C., Ikuma, K., Ong, S. K., & Sarkar, S. (2025). Reinforcement learning optimization of a water resource recovery facility: Evaluating the impact of reward function design on agent training, control optimization, and treatment risk. *Journal of Water Process Engineering*, 69(106658), 106658. https://doi.org/10.1016/j.jwpe.2024.106658

Cui, S., Gao, Y., Huang, Y., Shen, L., Zhao, Q., Pan, Y., & Zhuang, S. (2023). Advances and applications of machine learning and deep learning in environmental ecology and health. *Environmental Pollution* (*Barking, Essex: 1987*), 335(122358), 122358. https://doi.org/10.1016/j.envpol.2023.122358

Davis, J., Wolff, H., Forret, M.L. and Sullivan, S.E. (2020). Networking via Linkedin: An examination of usage and career benefits. *Journal of Vocational Behavior*, 118, Article ID: 103396. https://doi.org/10.1016/j.jvb.2020.103396

de Vries, A. (2018). Bitcoin's growing energy problem. *Joule*, 2(5), 801–805. https://doi.org/10.1016/j.joule.2018.04.016

Demestichas, K., & Daskalakis, E. (2020). Information and communication technology solutions for the circular economy. *Sustainability*, 12(18), 7272. https://doi.org/10.3390/su12187272

Dileep, G. (2020). A survey on smart grid technologies and applications. *Renewable Energy*, 146, 2589–2625. https://doi.org/10.1016/j.renene.2019.08.092

Ditria, E. M., Buelow, C. A., Gonzalez-Rivero, M., & Connolly, R. M. (2022). Artificial intelligence and automated monitoring for assisting conservation of marine ecosystems: A perspective. *Frontiers in Marine Science*, 9, Article number 918104. https://doi.org/10.3389/fmars.2022.918104

Dou, X., Zhou, Q., Chen, X., Tan, Y., He, X., Lu, P., Sui, K., Tang, B. Z., Zhang, Y., & Yuan, W. Z. (2018). Clustering-triggered emission and persistent room temperature phosphorescence of sodium alginate. *Biomacromolecules*, 19(6), 2014–2022. https://doi.org/10.1021/acs.biomac.8b00123

Dritsas, E., & Trigka, M. (2024). Machine learning for blockchain and IoT systems in smart cities: A survey. *Future Internet*, 16(9), 324. https://doi.org/10.3390/fi16090324

EPA. (2020). *The Clean Air Act and the reduction of air pollution. U.S.* Environmental Protection Agency. Retrieved from www.epa.gov

Ethereum Roadmap. (n.d.). *Ethereum.org*. Retrieved May 5, 2025, from https://ethereum.org/en/eth2/

Foote, K. D. (2022, January 11). *Data Management trends in 2022*. DATAVERSITY. http://dataversity.net/data-management-trends-in-2022/

Francisco, K., & Swanson, D. (2018). The supply chain has no clothes: Technology adoption of blockchain for supply chain transparency. *Logistics*, 2(1), 2. https://doi.org/10.3390/logistics2010002

Glass, L.-M., & Newig, J. (2019). Governance for achieving the Sustainable Development Goals: How important are participation, policy coherence, reflexivity, adaptation and democratic institutions? *Earth System Governance*, 2(100031), 100031. https://doi.org/10.1016/j.esg.2019.100031

Harris, S., Martin, M., & Diener, D. (2021). Circularity for circularity's sake? Scoping review of assessment methods for environmental performance in the circular economy. *Sustainable Production and Consumption*, 26, 172–186. https://doi.org/10.1016/j.spc.2020.09.018

Hernandez Aros, L., Bustamante Molano, L. X., Gutierrez-Portela, F., Moreno Hernandez, J. J., & Rodríguez Barrero, M. S. (2024). Financial fraud detection through the application of machine learning techniques: A literature review. *Humanities & Social Sciences Communications*, 11(1), 1–22. https://doi.org/10.1057/s41599-024-03606-0

Huntingford, C., Jeffers, E. S., Bonsall, M. B., Christensen, H. M., Lees, T., & Yang, H. (2019). Machine learning and artificial intelligence to aid climate change research and preparedness. *Environmental Research Letters*, 14(12), 124007. https://doi.org/10.1088/1748-9326/ab4e55

Johnson, A., Lee, H., & Chang, M. (2023). Bias in artificial intelligence: Ethical concerns and regulatory approaches. *AI Ethics and Society*, 7(1), 45–62.

Kamilaris, A., Fonts, A., & Prenafeta-Boldύ, F. X. (2019). The rise of blockchain technology in agriculture and food supply chains. *Trends in Food Science & Technology*, 91, 640–652. https://doi.org/10.1016/j.tifs.2019.07.034

Khan, S., & Singh, R. (2022). The role of blockchain in environmental compliance and pollution control. *Environmental Science & Technology*, 56(9), 5674–5689. https://doi.org/10.1002/bse.2882

Liakos, K. G., Busato, P., Moshou, D., Pearson, S., & Bochtis, D. (2018). Machine learning in agriculture: A review. *Sensors (Basel, Switzerland),* 18(8), 2674. https://doi.org/10.3390/s18082674

Mackay, D., & Paterson, S. (1991). Evaluating the multimedia fate of organic chemicals: A level III fugacity model. *Environmental Science & Technology*, 25(3), 427–436. https://doi.org/10.1021/es00015a008

McKinsey & Company. (2021). Big data analytics and its role in business Decisiond-making. *McKinsey Digital*, 12(4), 45–62.

Mhatre, Purva, Panchal, R., Singh, A., & Bibyan, S. (2021). A systematic literature review on the circular economy initiatives in the European Union. *Sustainable Production and Consumption*, 26, 187–202. https://doi.org/10.1016/j.spc.2020.09.008

Mosavi, A., Ozturk, P., & Chau, K.-W. (2018). Flood prediction using machine learning models: Literature review. *Water*, 10(11), 1536. https://doi.org/10.3390/w10111536

Mulligan, C., Morsfield, S., & Cheikosman, E. (2024). Blockchain for sustainability: A systematic literature review for policy impact. *Telecommunications Policy*, 48(2), 102676. https://doi.org/10.1016/j.telpol.2023.102676

Multiple Linear Regression. (n.d.). Prediction of heavy metals contamination in the groundwater of Arak region using artificial. *Yahoo.com*. Retrieved May 5, 2025, from https://r.search.yahoo.com/_ylt=Awr1SViHPxhoWAIA4IS7HAx.;_ylu=Y29sbwNzZzMEcG9zAzEEdnRpZAMEc2VjA3Ny/RV=2/RE=1747629192/RO=10/RU=https%3a%2f%2fjournals.pnu.ac.ir%2farticle_2860_f4f50e254d8d319458eb7e6dda09d229.pdf/RK=2/RS=DtDwoZ8dtvv0O9o1kHhDSxxZSYU-

Nkinahamira, F., Feng, A., Zhang, L., Rong, H., Ndagijimana, P., Guo, D., Cui, B., & Zhang, H. (2024). Machine learning approaches for monitoring environmental metal pollutants: Recent advances in source apportionment, detection, quantification, and risk assessment. *Trends in Analytical Chemistry: TRAC*, 180(117980), 117980. https://doi.org/10.1016/j.trac.2024.117980

Norouzzadeh, M. S., Nguyen, A., Kosmala, M., Swanson, A., Palmer, M. S., Packer, C., & Clune, J. (2018). Automatically identifying, counting, and describing wild animals in camera-trap images with deep learning. *Proceedings of the National Academy of*

Sciences of the United States of America, 115(25), E5716–E5725. https://doi.org/10.1073/pnas.1719367115

Ojha, Shikha, Bußler, S., & Schlüter, O. K. (2020). Food waste valorisation and circular economy concepts in insect production and processing. *Waste Management (New York, N.Y.)*, 118, 600–609. https://doi.org/10.1016/j.wasman.2020.09.010

Olawade, D. B., Wada, O. Z., Ige, A. O., Egbewole, B. I., Olojo, A., & Oladapo, B. I. (2024). Artificial intelligence in environmental monitoring: Advancements, challenges, and future directions. *Hygiene and Environmental Health Advances*, 12(100114), 100114. https://doi.org/10.1016/j.heha.2024.100114

Oliveira, M. M. de, Lago, A., & Dal' Magro, G. P. (2021). Dr. Jiří jaromír klemeš or cecília Maria villas bôas de Almeida or Yutao Wang. *Journal of Cleaner Production*, 294(126284), 126284. https://doi.org/10.1016/j.jclepro.2021.126284

Padmavathi, V., & Kanimozhi. (2025). Enhancing healthcare collaboration through blockchain-enabled digital twin integration. In Tyagi, A. K. & Tiwari, S. (eds), *Human-centric integration of 6G-enabled technologies for modern society* (pp. 315–326). Elsevier.

Pahariya, P., Shrivastav, A., Goswami, T. K., & Choudhary, R. (2025). Predictive modeling and risk assessment strategies for enhancing dairy product safety: A comprehensive review. *Food Safety and Health*, 3, 172–187. https://doi.org/10.1002/fsh3.12084

Prakash, R., Anoop, V. S., & Asharaf, S. (2022). Blockchain technology for cybersecurity: A text mining literature analysis. *International Journal of Information Management Data Insights*, 2(2), 100112. https://doi.org/10.1016/j.jjimei.2022.100112

Preskill, J. (2018). Quantum computing in the NISQ era and beyond. *Quantum*, 2(79), 79. https://doi.org/10.22331/q-2018-08-06-79

Quy, N. M., Ngoc, L. A., Ban, N. T., Van Hau, N., & Quy, V. K. (2023). Edge computing for real-time internet of things applications: Future internet revolution. *Wireless Personal Communications*, 132(2), 1423–1452. https://doi.org/10.1007/s11277-023-10669-w

Rashid, A. B., & Kausik, M. D. A. K. (2024). AI revolutionizing industries worldwide: A comprehensive overview of its diverse applications. *Hybrid Advances*, 7(100277), 100277. https://doi.org/10.1016/j.hybadv.2024.100277

Reichstein, M., Camps-Valls, G., Stevens, B., Jung, M., Denzler, J., Carvalhais, N., & Prabhat. (2019). Deep learning and process understanding for data-driven Earth system science. *Nature*, 566(7743), 195–204. https://doi.org/10.1038/s41586-019-0912-1

Russell, S., & Norvig, P. (2020). *Artificial intelligence: A modern approach* (4th ed.). Pearson.

Said Mohamed, E., Belal, A. A., Kotb Abd-Elmabod, S., El-Shirbeny, M. A., Gad, A., & Zahran, M. B. (2021). Smart farming for improving agricultural management. *Egyptian Journal of Remote Sensing and Space Sciences*, 24(3), 971–981. https://doi.org/10.1016/j.ejrs.2021.08.007

Sales-Ortells, H., Agostini, G., & Medema, G. (2015). Quantification of waterborne pathogens and associated health risks in urban water. *Environmental Science & Technology*, 49(11), 6943–6952. https://doi.org/10.1021/acs.est.5b00625

Sara, J. J. (2024, May 31). *State and trends of carbon pricing report*. World Bank Blogs. https://blogs.worldbank.org/en/climatechange/state-and-trends-of-carbon-pricing-2024--positive-progress-on-ca

Schär, F. (2020). Decentralized finance: On blockchain- and smart contract-based financial markets. *SSRN Electronic Journal*. https://doi.org/10.2139/ssrn.3571335

Sharma, H. (2023, May 19). Data privacy and cybersecurity: GDPR, CCPA, and other privacy laws, as well as concerns about data breaches. *Lawstreet Journal*. https://lawstreet.co/vantage-points/data-privacy-and-cybersecurity-gdpr-ccpa-and-other-privacy-laws-as-well-as-concerns-about-data-breaches

Smith, A. and Jones, B. (2021). Enterprise risk management: From theory to practice. *Risk Management and Insurance Review*, 24, 27–47.

Smith, J. and Johnson, P. (2022). The impact of AI on medical data analysis: A case study of IBM Watson health. *Journal of Health Informatics*, 45, 145–156.

Strubell, E., Ganesh, A., & McCallum, A. (2019). Energy and policy considerations for deep learning in NLP. In arXiv [cs.CL]. https://doi.org/10.48550/ARXIV.1906.02243

Syed, N. F., Shah, S. W., Shaghaghi, A., Anwar, A., Baig, Z., & Doss, R. (2022). Zero trust architecture (ZTA): A comprehensive survey. *IEEE Access: Practical Innovations, Open Solutions*, 10, 57143–57179. https://doi.org/10.1109/access.2022.3174679

The European Green Deal. (n.d.). *European Commission*. Retrieved May 5, 2025, from https://ec.europa.eu/info/strategy/priorities-2019-2024/european-green-deal_en

Topouzelis, K., Papakonstantinou, A., & Garaba, S. P. (2019). Detection of floating plastics from satellite and unmanned aerial systems (Plastic Litter Project 2018). *International Journal of Applied Earth Observation and Geoinformation: ITC Journal*, 79, 175–183. https://doi.org/10.1016/j.jag.2019.03.011

Tran, H.-Y., & Hu, J. (2019). Privacy-preserving big data analytics a comprehensive survey. *Journal of Parallel and Distributed Computing*, 134, 207–218. https://doi.org/10.1016/j.jpdc.2019.08.007

Xu, X., Weber, I., & Staples, M. (2019). *Architecture for blockchain applications* (2019th ed.). Springer Nature.

Yaga, D., Mell, P., Roby, N., & Scarfone, K. (2019). Blockchain Technology Overview. arXiv [Cs.CR]. https://doi.org/10.48550/ARXIV.1906.11078

Zhang, G., Wang, C., Wang, H., & Tao, Y. U. (2025). Advanced deep learning model for predicting water pollutants using spectral data and augmentation techniques: A case study of the Middle and Lower Yangtze River, China. *Process Safety and Environmental Protection: Transactions of the Institution of Chemical Engineers, Part B*, 197(107058), 107058. https://doi.org/10.1016/j.psep.2025.107058

Zounemat-Kermani, M., Batelaan, O., Fadaee, M., & Hinkelmann, R. (2021). Ensemble machine learning paradigms in hydrology: A review. *Journal of Hydrology*, 598(126266), 126266. https://doi.org/10.1016/j.jhydrol.2021.126266

12 Future Scope and Directions of Research on Emerging Contaminants and One Health

Antonino Di Bella and Filippo Fazzino

12.1 INTRODUCTION

One Health is an approach recently defined and increasingly adopted that recognizes that human, animal, plant, and environmental health are all interconnected, in particular infectious diseases that possibly spill over from animals to humans, and environmental stressors, such as ecosystem contamination, habitat destruction, and climate change. The One Health concept emphasizes how such threats can transcend individual species and ecosystems and adversely affect humans, animals, and the environment. The spread of micro- and nanoplastics represents an illustrative example of that. In fact, tiny particles of plastic litter not only contaminate soil, and especially the marine environment, but also, by oral and dermal exposure routes, can enter the food chain and adversely affect wildlife and human health. In the One Health approach, health challenges are tackled through cross-sectoral collaborations including different disciplines such as human and veterinary medicine, epidemiology, ecology, agriculture, environmental engineering, social sciences, public health, and so on [1].

In this context, pollution of air, water, and soil arising from anthropogenic activities is known to be harmful not only to humans but also to living beings and ecosystems. Particularly, threats posed by the so-called emerging contaminants (ECs) represent a critical and unprecedented global issue to be addressed within the One Health concept. ECs have been recognized for decades to enter the aquatic environment, especially via municipal, industrial, and agricultural discharges. Although knowledge about EC dynamics in the environment is still limited, various studies have found these substances worldwide, in water, soil, sediments, air, and living organisms, even in pristine areas, due to their ability to travel, transform, and persist [2].

Overall, ECs are defined as chemicals i) oftentimes present at trace levels (ng/L to μg/L), ii) only recently discovered or suspected to exert adverse effects on humans and/or the environment, iii) unaddressed by existing legislation, and iv) not routinely monitored [2] (Figure 12.1). A non-exhaustive list of ECs includes pharmaceuticals and personal care products (PPCPs), endocrine disrupting chemicals (EDCs),

DOI: 10.1201/9781032623801-12

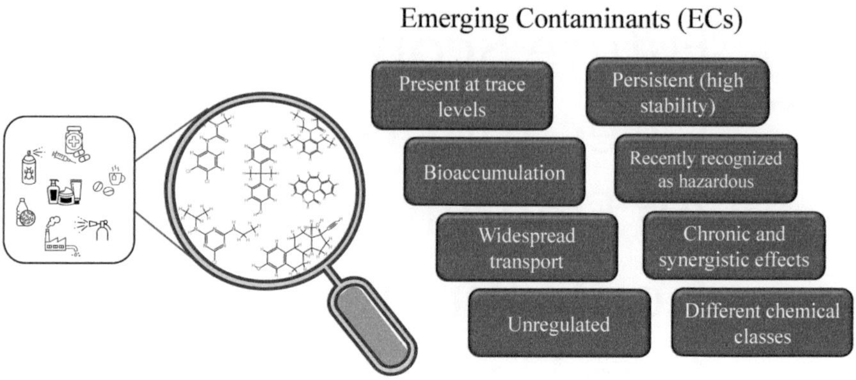

FIGURE 12.1 Schematic view of emerging contaminant (EC) features.

pesticides, per- and polyfluoroalkyl substances (PFAs) and other persistent organic pollutants (POPs), micro- and nanoplastics, and disinfection by-products (DBPs) (a more complete list is shown in Table 12.1). In this context, the definition of "emerging" can be misleading. Indeed, the vast majority of ECs have not been recently developed but have been regularly produced, used, and subsequently released into the environment for decades. For instance, POPs, such as polychlorinated biphenyls (PCBs) or chlorinated pesticides like dichloro-diphenyl-trichloroethane (DDT), have been widely used in industry and agriculture for many years. Although many have been globally banned, POPs persist in the environment and still pose a threat, as they can bioaccumulate in organisms and cause chronic toxicity, even at very low concentrations. Thus, the "emergence" is rather related to the concern about risks that such contaminants can pose, raised as a consequence of the development of new detection techniques and current awareness about their toxicity [3].

Entirely in accordance with the One Health concept, the future scope and directions of research on ECs integrate scientific, social, and economic disciplines to overcome current challenges. In this chapter, the most studied research fields are reported and the most recent strategies, regulations, and technologies dealing with the EC issue are discussed. Particularly, basic and advanced techniques for EC analysis, including their extraction from liquid and solid samples and quantification, are explored in Section 12.2, with emphasis on recent findings concerning online and real-time monitoring of EC occurrence and fate in (waste)water through surrogate parameters such as absorbance and fluorescence indexes. Then, in Section 12.3, new regulations to limit the spread of ECs, especially in a water environment, are introduced. Moreover, the most promising technologies for effective EC removal from contaminated water and soil are explored. Membrane-based systems and advanced oxidation processes can effectively treat various EC classes in urban and industrial wastewater, though limitations remain. Research is exploring surrogate monitoring and machine learning to overcome these challenges. Finally, principles and future frameworks related to risk assessment of ECs for both human and ecosystems are explored in Section 12.4, outlining the need of a more comprehensive evaluation of risk through multidisciplinary

TABLE 12.1
List of the main emerging contaminants found in the environment

ECs	Class	Sources and environmental fate	Examples
Human pharmaceuticals and veterinary drugs	Antibiotics	Excretion, hospital discharges/ disposal, accidental spills, and farmland waste WWTPs, rivers, streams, groundwater (after leaching), soil	Amoxicillin, cefazolin, chlortetracycline, ciprofloxacin, ciprofloxacin, erythromycin, doxycycline, lincomycin, sulfamethoxazole, ofloxacin, norfloxacin, oxytetracycline, penicillin, tetracycline, trimethoprim
	Analgesics		Acetaminophen, acetylsalicylic acid, codeine, dipyrone, diclofenac, indomethacin, ibuprofen, fenoprofen, fluoxetine, ketoprofen, mefenamic acid, paracetamol, naproxen, propyphenazone, salycilic acid
	Antidiabetics		Glibenclamide
	Psychiatric drugs		Carbamazepine, diazepam, gabapentin, primidone, phenytoin, salbutamol
	Blood lipid regulators		Atorvastatin, bezafibrate, clofibric acid, etofibrate, fenofibric acid, gemfibrozil, pravastatin
	X-ray contrast agents		Iopromide, iopamidol, diatrizoate
	Cardiovascular drugs (β-blockers)		Atenolol, metoprolol, propranolol, solatolol, timolol, sotalol, atenolol
	Drugs of abuse		Amphetamine, cocaine, tetrahydrocannabinol
	Veterinary drug		Flunixin
Endocrine disruptive chemicals	Steroids and hormones		Androstenedione, estradiol, estriol, estrone, diethylstilbestrol, ethinylestradiol, progesterone, testosterone

(*continued*)

TABLE 12.1 (Continued)
List of the main emerging contaminants found in the environment

ECs	Class	Sources and environmental fate	Examples
Personal care products	Antiseptics Fragrances and synthetic musks Stimulants Antihypertensive UV Filters Insect repellents	Shower waste and direct disposal of industrial waste WWTPs, terrestrial run-off freshwater, marine, estuarine environments and sediment	Triclosan, chlorophene Galaxolide, musk ketone, nitro, polycyclic, and macrocyclic musks; phthalates, tonalide Caffeine Diltiazem Benzophenone, methylbenzylidene camphor N,N-diethyltoluamide
Agriculture	Herbicide Pesticides	Agriculture waste, soil run-off Soil and water	Diuron, mecoprop, MCPA, terbuthylazine
Detergents, surfactants, and perfluorinated compounds (PFAS)	Perfluorooctane sulfonate Perfluorooctanoic acid	Industries, laundries, households, and agricultural applications in pesticides, dilutants, and dispersants	Alkylphenol ethoxylates, alkylphenols (nonylphenol and octylphenol), alkylphenol carboxylates
Disinfection by-products	Trihalomethanes and haloacetic acids	Chlorination of natural organic matter WWTPs, freshwater, and groundwater	Chloroform, bromodichloromethane, chloral hydrate, dichloroacetonitrile, and 1,1,1-trichloropropanone
Additives	Industrial	Food and municipal waste Soil, water, and air	Chelating agents (EDTA), aromatic sulfonates
	Gasoline	Disposal of exhausted engine oil, spills Soil, water, and air	Dialkyl ethers, methyl t-butyl ether
Flame retardants		Household items (furniture, electronics, appliances, baby products), and industries Wet and dry deposition on soil lead to bioaccumulation in the food web	C10–C13 chloroalkanes, hexabromocyclododecane, polybrominated diphenyl ethers, tetrabromobisphenol A, tris (2- chloroethyl) phosphate

TABLE 12.1 (Continued)
List of the main emerging contaminants found in the environment

ECs	Class	Sources and environmental fate	Examples
New classes	Nanomaterials	Research institutes Water	1,4-dioxane
	Antibiotic resistance genes	Mutations and genetic adaptation Horizontal gene transfer in microorganisms	sul (I) and sul (II) genes providing resistance to sulfonamide antibiotics, tet (W) and tet (O) genes providing resistance to tetracycline antibiotics

Source: Data compiled from: Refs. [2–5].

approaches. All these aspects are summarized in Figure 12.2. Besides EC analysis, removal, and risk evaluation, prevention of EC release into the environment and/or reduction of EC production and daily use represent the most recommended directions to pursue. To reduce EC input into the environment, strategies include using green alternatives that degrade naturally or are easily removed by treatment, and adopting sustainable processes that recycle chemicals to limit their discharge.

12.2 EC MEASUREMENT

EC presence has been observed in environmental compartments such as surface water, groundwater, soil, and air, as well as in living organisms such as plants, fish, mammals, and humans [4]. ECs can persist and bioaccumulate in organisms, making their accurate identification and quantification essential to determine their presence, understand their behavior and fate, and ultimately assess the risks they pose to the environment. As a result of detection, ECs must be distinguished, even at very low environmental concentrations, among a plethora of other substances present in the analyzed liquid or solid sample. So, high levels of sensitivity and selectivity of analytic methods have to be guaranteed. The development of new methodologies for sample preparation as well as increasingly accurate EC analytical determination methods have remarkably enhanced the capability to extract and identify EC in environmental and biological samples.

In this context, qualitative and quantitative analyses can be performed. The former is used to determine the presence of one or more contaminants in the analyzed sample without necessarily quantifying the concentration(s). Among qualitative techniques, mass spectrometry (MS) and liquid chromatography (LC) are widely employed to identify chemicals through the separation and analysis of their spectral profiles. On the other hand, quantitative analysis measures the amount of various chemical components present in the sample. High-performance liquid chromatography

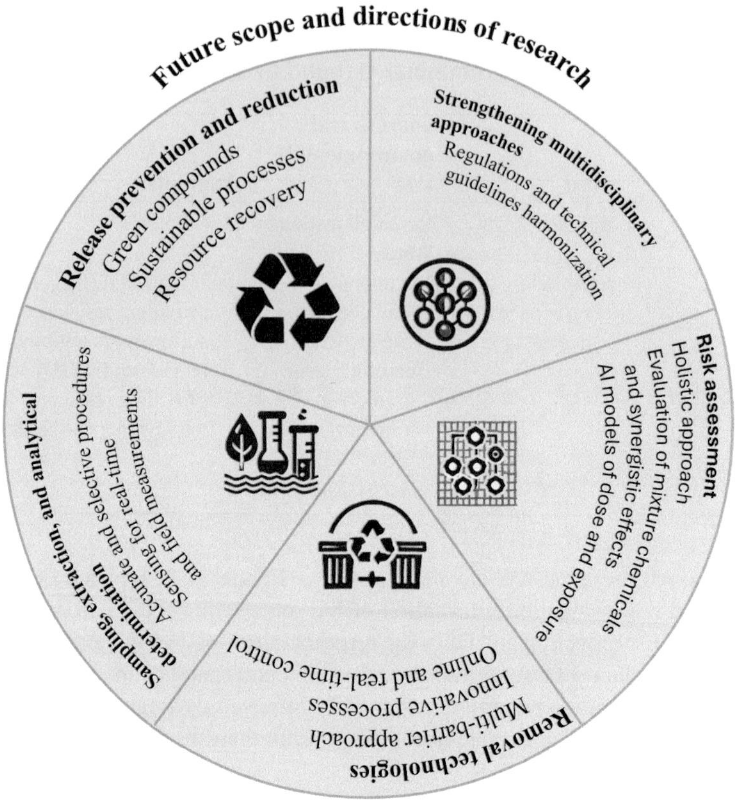

FIGURE 12.2 Pictorial summary of future scope and directions of research on EC within One Health discussed in the chapter.

(HPLC) and liquid chromatography coupled with mass spectrometry (LC-MS) are among the most widely employed methods to measure the concentration of ECs in environmental and biological matrixes. Moreover, analyses can be carried out to specifically detect ECs known to be present in the sample (target analysis) or alternatively to fully elucidate the different types of ECs including those unexpected (nontarget analysis) [5].

The overall EC detection procedure starts with sample collection and handling, followed by its preparation prior to the proper instrumental analysis and, finally, data analysis. Key challenges in EC detection include low concentration levels, matrix interferences, variability in measurements, lack of standard methods, rising numbers of new ECs, high equipment costs, long analysis times, and limited portability [6].

In the following part, traditional and especially innovative methods commonly used or recently investigated for sample preparation and analytical measurement of ECs are presented, particularly highlighting the future directions of research.

12.2.1 Sample Pretreatment Methods

The detection of trace-concentration ECs (often up to ng/L) is carried out through sensitive and advanced analytical techniques preceded by efficient extraction and enrichment methods (i.e., sample preparation). A preliminary step of water removal is performed in the case of solid samples, such as soil or sludge, or alternatively through centrifugation and decantation, air-drying (uncommon), heating (in oven at low temperatures), or lyophilization [7]. Solid-liquid extraction (SLE) is widely used to extract organic ECs like POPs from solids such as soil, sludge, or food, using solvents and mechanical agitation, including hand shaking, orbital mixing, or Soxhlet extraction. The choice of the solvent to use and the number of cycles of extraction (thus the duration of the procedure) depends on the types of ECs and the solid matrix involved in the extraction procedure. Usually, the overall procedure is time-consuming and a large amount of toxic and non-renewable solvents (e.g., methanol, acetone-hexane, cyclohexane) is needed. To address these limitations, current research focuses on greener procedures, such as replacing conventional solvents with bio-based alternatives like supercritical fluids (especially CO_2), ionic liquids, or deep eutectic solvents [8]. Further innovations in the field of SLE reside in the optimization of the process parameters by integrating unit operations. Ultrasonic-assisted extraction (UAE), microwave-assisted extraction (MAE), and pressurized liquid extraction (PLE) represent alternative procedures requiring less extraction time and solvent consumption yet that are equally effective [9]. The former is conducted in ultrasonic baths and it lies in the cavitation action of ultrasonic radiation (frequency higher than 20kHz), which induces solid particles to burst thus increasing EC mass transfer to the solvent. Similarly, MAE is carried out in an microwave oven operating at 2.45 GHz, which allows the heating up and disrupting of solid particles, thereby enhancing EC mass transfer [10]. In a PLE procedure, high pressures (10.3–20.6 MPa) are employed to reach subcritical temperature conditions for the used solvent (i.e., in liquid form at temperatures above the boiling point) to improve EC extraction kinetics [11].

Liquid-liquid extraction (LLE) and solid-phase extraction (SPE) are two classic approaches for extracting trace pollutants, especially from water matrixes. The former involves contact between the liquid matrix containing ECs and an organic solvent. Similarly to SLE, the use of greener solvents and optimal process conditions (pressure, temperature) is advisable. SPE is the most often employed technology since it is appropriate for detection of compounds with a wide variety of polarity and physicochemical characteristics. Generally, SPE consists of separating the target analytes from the liquid mobile phase to adsorbent material (stationary phase) according to their physicochemical properties. It can be conducted in small columns with disks or cartridges inside or packed with ion exchangers or other sorbents [12]. Alternatively, in the solid-phase microextraction (SPME) procedure, a fiber is directly exposed to the solution containing the analytes (direct immersion method). After a certain contact time and under optimum working conditions such as stirring, pH, and temperature, the analyte concentration in the SPME layer reaches an equilibrium level with its environment [7]. Besides extracting a wide range of compounds with diverse properties, the SPME procedure is also reported to minimize matrix effects. Moreover, especially SPME enables direct on-site sampling, and attempts of SPME have been made also for EC extraction from complex biological matrixes, in both in vitro and in vivo studies. On the other hand, solid sorbent can be dispersed in the sample solution (i.e., dispersive-micro SPE, d-μSPE) to adsorb ECs and then be separated from the solution. Afterward, a specific organic solvent is used to elute the analytes prior to instrumental analysis. By combining SPE with an LLE approach, the QuEChERS (Quick, Easy, Cheap, Effective, Rugged, and Safe) methodology has been developed and gained particular attention in the field of EC extraction. Such an approach involves two steps: extraction from homogenized samples using an organic green solvent and salt solution and, then, the supernatant is further extraction and cleaning from the supernatant using SPE techniques [13].

Extensive research focuses on developing nanosorbents suitable for various SPE applications and EC types, offering high surface area, large pore volume, fast adsorption, selectivity, reusability, and easy separation from the medium [14]. Carbonaceous nanomaterials such as carbon nanotubes (CNTs) or graphene (G)-based composites have been recently reported to act as efficient sorbents for the SPE of various ECs as well as inorganic nanoparticles (NPs) [15]. Particularly, in the case of nanosorbents doped with magnetic NPs, the separation of the adsorbent from the solution medium can be performed through a handled magnet (i.e., magnetic SPE, MSPE) instead of centrifugation or filtration methods [16]. Although these approaches have had promising results, challenges related to the widespread implementation of those novel nanosorbents in SPE procedures still remain. Further research should focus on cost-benefit analysis of large-scale sorbent production and regeneration, along with sustainable disposal strategies to manage their environmental impact at the end of their operational lifespan.

12.2.2 Instrumental Analysis of ECs

Gas and liquid chromatography (GC and LC, respectively) and mass spectrometry (MS) are the traditional methods used in analytical chemistry for separation and precise quantification of chemical species. In the last decade, either GC or LC has often been coupled with MS or tandem MS (MS/MS, namely with two separate stages of mass analysis) to increase the specificity and sensitivity of the EC detection.

Chromatography involves passing a mobile phase—either an inert gas with a vaporized mixture for GC or a liquid for LC—through a stationary phase packed into a small analytical column. The larger the molecules in the mixture, the longer the time to reach the detector at the end of the column. To shorten LC analysis time, HPLC and UHPLC are commonly used. These systems employ high-pressure pumps to force the liquid sample through smaller stationary phases, improving the speed and efficiency of separation. In the context of EC detection, GC is capable of determining volatile, nonpolar, and thermostable ECs (such as pesticides, flame retardants, and filters), whereas LC is mostly employed to detect polar, nonvolatile, and thermolabile ECs. However, very polar and nonvolatile fractions represent an issue for both techniques. A variety of chromatographic methods are being explored to fill this gap in EC detection including supercritical fluid chromatography (SFC) and hydrophilic interaction liquid chromatography (HILIC). Other chromatographic techniques are capillary electrophoresis, ion chromatography, and gel permeation chromatography [7].

In MS, molecules are converted to ions in the gas phase and separated by external electric and magnetic fields according to their mass-to-charge ratios. High-resolution mass spectrometry (HRMS) offers exceptional precision in mass-to-charge measurements by using advanced analyzers such as time-of-flight (TOF-MS), Orbitrap, or Fourier transform ion cyclotron resonance.

HRMS offers high mass accuracy and resolution, and it has been recently applied in non-target analysis due to its ability to elucidate EC structures and to transform products and metabolites.

Other advanced analytical chemistry techniques for the identification of complex organic contaminant structures and concentrations are nuclear magnetic resonance (NMR) spectroscopy and Fourier transform (near-)infrared (FTIR) spectroscopy. The former involves the reorientation of atomic nuclei with the absorption of radio-frequency electromagnetic radiation in a strong magnetic field. It enables the structural elucidation of organic EC without the need for authentic standards. However, NMR is less sensitive than the previously described MS techniques, which can result in a larger sample amount and high concentrations of target compounds for characterization. It is also less accessible than other analytical instruments [6]. FTIR is based on the absorption and transmission of infrared or near-infrared radiation by the sample, and it is reagent-free, nondestructive, quickly responsive, and it allows sample reuse after measurement. As a limitation, FTIR primarily provides information about the functional groups in molecules rather than detailed structural information, and it suffers from interferences from water and other compounds commonly present in samples [10].

The relative complexity of all the above reported techniques often requires highly trained and skilled personnel to deal with the equipment in terms of measurement procedure, regular calibration, maintenance, and precise data interpretation. These features make the overall EC analysis usually time-consuming and costly. Additionally, the current high cost of equipment restricts its availability to highly specialized laboratories. Such limitations curb the widespread and ordinary use of these technologies and affect their practical effectiveness in environmental monitoring.

In this context, absorbance and fluorescence spectroscopy is being explored as a more promising, cheaper, quicker, and more reliable method to qualitatively detect EC. Generally, spectroscopy makes it possible to characterize the dissolved organic matter (DOM) in aquatic environments through molecule excitation by incident light at different wavelengths. Absorbance is the quantity of incident light absorbed by the molecule (defined as chromophore), while fluorescence is the radiation (at a larger wavelength than incident light) emitted by the excited molecule (defined as fluorophore) [17]. As a result, water samples can be characterized by distinct signatures (i.e., absorbance and/or fluorescence spectra, respectively) depending on DOM quality and quantity (Figure 12.3). Fluorescence spectroscopy is typically more sensitive than UV absorbance with the possibility of detecting single molecules and discriminating different sources of organic matter.

In order to decode information present in absorbance and/or fluorescence spectra, different methods have been developed and a plethora of indexes have been

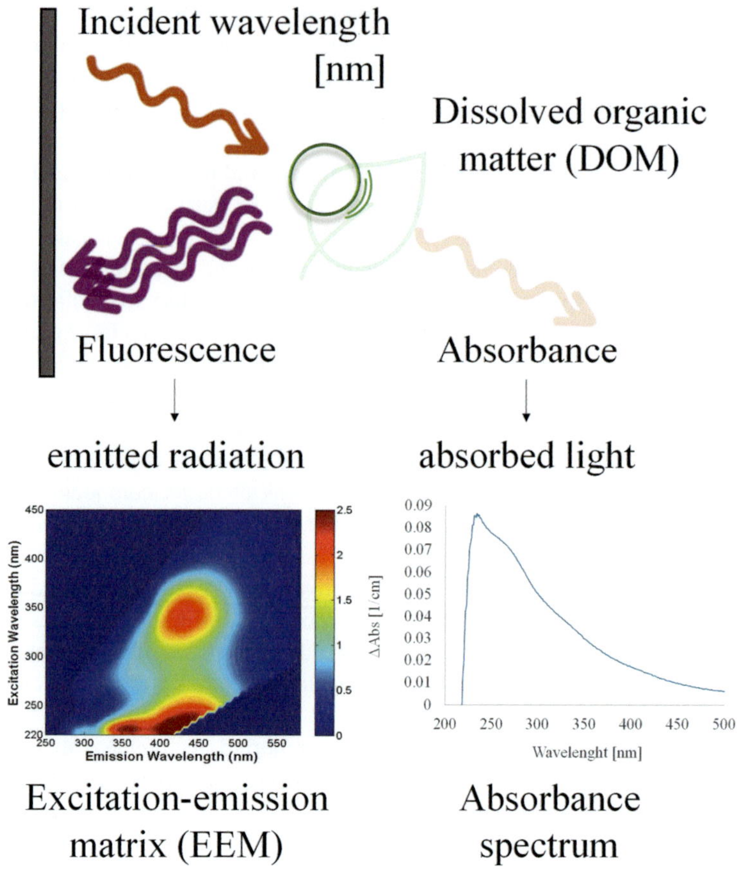

FIGURE 12.3 Principles of fluorescence and absorbance spectroscopy.

generated. Absorbance can be reported as: i) single-wavelength values (e.g., A_{254}), ii) wavelength ratios (e.g., A_{254}/A_{202}), iii) pre/post-treatment differences ($\Delta A\lambda$), and iv) slope across specific wavelengths, known as the absorbance slope index (ASI). On the other hand, fluorescence spectroscopy generates three-dimensional excitation (ex)/emission (em) matrices (EEMs) that can be interpreted through the peak picking method (PP), fluorescence regional integration (FRI), and/or parallel factor analysis (PARAFAC). These methods generate various indexes, including peak intensities at specific excitation/emission wavelengths, distinct EEM regions, PARAFAC components, and fluorescence-based metrics like FI, TF, HIX, BIX, and others [17]. Recent research discovered significant correlations between some of the aforementioned spectroscopic indexes and concentrations of ECs (analytically determined by conventional techniques) in both natural water samples and effluents of advanced wastewater treatment processes [18]. As a result, spectroscopic indexes are being increasingly investigated as surrogate parameters for "indirect" yet reliable and rapid EC estimation.

The vast majority of these studies relied on benchtop instrumentation designed to be used inside a laboratory environment. On the one hand, laboratory-based analysis allows to accomplish complete DOM characterization through data processing so that stronger and more significant correlations with EC contents are established. On the other hand, benchtop equipment requires high current power supplies and special handling. Moreover, improper sample collection, storage, and preparation can adversely affect the correctness of the measurement [18]. Sample transport and spectral interpretation extend analysis time. To address this, portable absorbance- and fluorescence-based devices are increasingly being developed, enabling quicker and more practical in situ environmental measurements. Particularly, fluorescence-based sensors can detect a unique couple of ex/em wavelengths (i.e., one-dimensional data compared to the three-dimensional EEMs), specific to the target analytes, instead of the entire EEM. Many fluorescence sensors use LED technology as it ensures compact size, relatively low price, and a long lifetime. As a consequence, such an approach further reduces instrument size and power consumption, and it allows even quicker EC estimation (actual real-time monitoring) by measuring more specific fluorescence peaks [19]. However, regular maintenance of these devices has to be guaranteed if run continuously as fouling of optical sensors adversely affects fluorescence measurements as well as water turbidity.

In terms of sensing methods for EC detection, biosensors and microfluidic sensors are also gaining increasing attention. Biosensors are innovative devices combining a biological element (e.g., enzyme, antibody, DNA) with a physicochemical detector to convert changes in mass, fluorescence, charge, or refractive index into an electronic signal [20]. Biosensors are selective, sensitive, and suitable for real-time monitoring, but environmental conditions like temperature and pH can affect the biological component's stability, potentially leading to unreliable or inconsistent results. Microfluidic sensors analyze nanoliter to microliter volumes of water samples by integrating pretreatment, reaction, separation, and detection stages into a single chip, typically built on substrates like silicon, glass, or polymers. Fluid flows inside the microchannels of the chip due to driving forces including electrical drive, pressure drive, centrifugal force drive, and capillary drive. The detection of ECs can occur through optical, electrochemical, or biological microsystems or by connecting laboratory testing instruments to the chips. Recent and promising attempts at EC detection by microfluidic sensors have been reported [21].

Despite their potential, sensing methods for EC detection are not yet widely applied in real-world settings, as most studies are lab-based and prototypes often fail to perform reliably under field conditions. Similarly, the majority of reported studies involve the use of standard solutions or EC-spiked aqueous samples, and thus do not consider the complexity of the composition of actual environmental and biological matrices. Finally, sensor measurements are rarely compared and validated with standard analytical methods and in accordance with accuracy and precision features requested by quality assurance/quality control (QA/QC) standards.

Table 12.2 summarizes the reported technologies for EC measurements with their respective most common applications, limitations, and estimation of costs.

TABLE 12.2
Traditional and innovative technologies for EC measurements

Technology	Application	Limitation(s)	Cost	Ref.
LC-MS/MS (Liquid Chromatography-Tandem Mass Spectrometry)	Routine analysis and precise quantification in water and soil	Requires reference standards and prepared samples; expensive	$$$$	[22]
HRMS (High-Resolution Mass Spectrometry)	Identification of known and unknown ECs, non-targeted analysis	High cost and slow analysis; requires experts for data interpretation	$$$$$	[23]
GC-MS (Gas Chromatography-Mass Spectrometry)	Especially applied for volatile ECs	Limited to volatile compounds; requires derivatization for non-volatile compounds	$$$	[24]
TOF-MS (Time-of-Flight Mass Spectrometry)	Non-targeted identification and screening of low-concentration ECs	Limited precision compared to Orbitrap for targeted analysis; high cost	$$$$	[25]
Orbitrap (Orbitrap Mass Spectrometry)	Quantification of known compounds and discovery of new non-target contaminants	Slow analysis and high cost; requires advanced skills for data interpretation	$$$$$	[25]
IC (Ion Chromatography)	Monitoring of specific compounds, less common for detailed analysis	Less versatile for complex analysis, limited to anionic ECs	$$	[26]
Spectroscopic Techniques (Absorbance, Fluorescence)	Preliminary surrogate screening; less effective for complex ECs like PFAS	Poor sensitivity and specificity; needs robust correlations to track EC occurrence	$$	[27]
Biosensors	Rapid, low-cost environmental screening, useful for field monitoring	Limited sensitivity; not suitable for precise quantitative analysis	$	[28]

12.3 EC REMOVAL TECHNOLOGIES

In order to control the spread of ECs in the environment, methods for their removal are being explored. Research on effective and convenient treatment techniques is continuously evolving and represents one of the main challenges in the overall EC management. The adopted technologies rely on processes ranging from physical adsorption to chemical degradation, as well as thermal and biological systems. The effectiveness of each technology depends mostly on the specific characteristics of the contaminants to be treated. In general, the development of efficient, green, and cost-effective materials to be used as adsorbents, photocatalysts, membranes, or membrane fillers, is one of the main research trends [29].

In terms of research direction, work is mainly devoted to water and wastewater treatment technologies. Indeed, EC occurrence in aqueous environment is due to wastewater discharges from municipal, industrial, and agricultural sources. Increasing attention has been paid to such issues since conventional wastewater treatment plants (WWTPs) have been shown to be largely ineffective for EC abatement [30]. WWTPs generally include preliminary and primary treatments to remove solids, secondary treatment to degrade biodegradable matter, and tertiary processes to reduce nutrients and pathogens before discharging the treated effluent. At the European scale, in December 2022, the Urban Waste Water Treatment Directive (UWWTD) was updated with the objective of increasing the protection of human health and the environment from harmful discharges. The revised directive particularly focuses on the improvement of wastewater treatment processes. Secondary treatment should cover wastewater from agglomerations of 1,000 p.e. or more, while tertiary treatment is required for facilities treating over 100,000 p.e., or 10,000–100,000 p.e. in eutrophication-sensitive areas. Quaternary treatment will target EC removal in urban WWTPs over 100,000 p.e., and 10,000–100,000 p.e. in sensitive areas, such as low-dilution water bodies or drinking water sources, by December 31, 2040.

The improved quality of WWTP effluent requested by the revised Directive can lead to possible reuse scenarios from irrigation scopes to potable reuse. Especially regarding the latter, ensuring the effectiveness, selectivity, and reliability of EC abatement appears of paramount importance. For this reason, current academic and corporate research agree on the development of multi-barrier systems, namely the combination of different advanced (quaternary) processes [31]. Among others, reverse osmosis (RO) and advanced oxidation processes (AOPs, O_3- and/or UV-based processes) have been determined as the most suitable treatment technologies to remove ECs from (waste)water.

The former is included within the membrane-based technologies regularly used in the (waste)water treatment sector to implement the physical separation of pollutants from the influent when it passes through them. As a result, permeate (i.e., treated effluent) and concentrate or retentate (stream where pollutants are concentrated) are distinguished. Membranes can be made up of synthetic polymers (polyethylene, polytetrafluoroethylene, polypropylene, cellulose acetate, poly vinyl alcohol, polyimide, polysulfone, polyethersulfone) or inorganic materials (ceramics such as Al_2O_3, TiO_2, ZrO_2, and zeolites) [32]. Besides RO, the other technologies involving the use of membranes are filtration (including micro-, ultra-, and nano-, MF, UF, and NF, respectively), and membrane biological reactor (MBR). In particular, MF and UF are mostly recommended as pre-treatment prior to other membrane processes since they are effective in removing suspended matter, and large colloidal, microorganisms, preventing fouling of further NF and/or RO. Concerning ECs, NF and RO have been observed to remove ECs with low to high efficiency [33], while negligible removal is generally recorded for MF and UF. MBR represents the integration of conventional biological removal with filtration and enhancement of biomass activity attached to pores in order to improve abatement of recalcitrant compounds including some ECs. The main drawbacks of RO (and membranes in general) are represented by membrane fouling (thus correlated with the need for their replacement with additional costs) and concentrate disposal. While the former can be limited by appropriately pretreating the influent, the latter is as critical as large the membrane efficiency is. Traditional methods for concentrate treatment involve its discharge into WWTPs or evaporation (solar) ponds, but they are only slightly effective for EC reduction, whereas thermal-based technologies and UV-based AOPs can lead to successful EC degradation [34].

AOP-driven degradation of ECs relies on highly reactive radicals that can effectively oxidize a broad range of ECs. Indeed, AOPs involve the generation of highly reactive radical species such as hydroxyl ($HO^{\bullet}$), superoxide ($O_2^{\bullet -}$), hydroperoxyl ($HO_2^{\bullet}$), alkoxyl ($RO^{\bullet}$), sulfate ($SO_4^{\bullet -}$), and chlorine ($Cl^{\bullet}$) radicals [35]. Especially hydroxyl radical is known to be a strong oxidant, so it is able to oxidize almost all reduced material present in wastewater. The hydroxyl radical reacts with the dissolved compounds and, after a series of oxidation reactions, causes their complete mineralization. Processes that generate $HO^{\bullet}$ radicals involve the use of ozone (O_3) and/or hydrogen peroxide (H_2O_2), possibly along with ultra-violet radiation (UV), solar light, ultrasound (US), or a catalyst (e.g., Fe^{2+}, Fe^{3+} and TiO_2). Even though AOP efficacy in EC abatement has been demonstrated, their application at full-scale, however, is still missing. Such difficulty in scaling them up is related to challenges that need to be overcome. First, influent quality strongly affects the performance of AOPs related to EC removal. Various scavengers in wastewater, such as humic acids, amino acids, sulfide, and bicarbonate, can react with hydroxyl radicals more rapidly than ECs, thereby reducing AOP effectiveness and hindering contaminant degradation [30]. Second, the selectivity of AOPs related to target ECs depends on their different reactivities toward the oxidizing agents. Indeed, some ECs are known to have high reactivity with both O_3 and $HO^{\bullet}$ (e.g., carbamazepine, naproxen, bisphenol A), whereas others are mainly photo-susceptible (e.g., fluoxetine, simazine) [36]. As a consequence, depending on the type of ECs targeted to be removed, one or more AOPs should be accordingly implemented. The formation of known (and still unknown) harmful oxidation by-products represents another significant challenge for AOP applications. For instance, ozonation may result in the formation of N-nitrosodimethylamine (NDMA) and bromate when natural organic matter (NOM) and bromide occur in treated (waste)water . Finally, AOPs are usually characterized by high energy demand and substantial use of chemicals as well as expensive equipment and monitoring tools. Implementing quaternary treatment will involve significant additional costs for existing plants, which are expected to be partially covered by industries responsible for EC emissions, in line with the "polluter-pays" principle.

In this context, the optimization of AOP conditions through real-time control of removal performance would allow considerable savings of energy and materials. Marino et al. [19] successfully correlated the removal of target ECs, including PFAS and pharmaceuticals, from WWTP effluent using pilot-scale O_3 and UV-based AOPs with real-time fluorescence signals from a field sensor. Results indicate that fluorescence is a highly sensitive surrogate indicator for AOP control, allowing rapid detection of changes in water quality, including the oxidation of emerging contaminants by hydroxyl radicals. Other similar attempts have been made, but mostly under laboratory conditions, highlighting the need to translate these promising results into real-world applications for consistent and effective EC monitoring. Real-time sensor data are crucial for adjusting AOP process conditions, such as reagent dosage and contact time, to enhance overall treatment efficiency and ensure consistent degradation of emerging contaminants. Sensors can be placed at AOP unit inlets or outlets, with a set EC removal threshold. If fluorescence exceeds the limit, an alarm is triggered and process parameters are adjusted to restore removal efficiency.

Despite promising results, challenges remain in applying fluorescence sensors to contaminant concentrations in the real world. For instance, periodic calibrations seem to be necessary to check the sensitivity of the site-specific correlations with removals of target ECs, especially in the case of variation of influent/effluent quality. Moreover, it is still not clear how continuous operation can alter fluorescence detection and what is the suggested frequency of probe cleaning. Future applications of these sensors could include integration with machine learning models for predictive monitoring and optimization of AOPs, further enhancing their utility in managing water quality, energy consumption, and costs. Additionally, research into integrating fluorescence with other sensor technologies, such as Raman or UV absorbance sensors, could provide even more robust and multi-parametric monitoring systems for comprehensive tracking of contaminants.

With regard to soil remediation by ECs, heating and adsorption technologies are the most promising. This method involves heating soil using sources like combustion chambers, microwaves, or electrical resistance to induce desorption and volatilization of organic ECs, such as POPs, which are then treated in the gas phase. The high cost of energy and equipment is the main limitation of such treatment, but performance in organic EC removal is excellent. Recent early-stage research focuses on developing numerical models to predict thermal desorption rates of ECs from soil grains during heating, aiming to better understand contaminant behavior under different treatment conditions. These models help define optimal process settings before pilot or full-scale applications and support reliable estimations of total treatment time and energy demand, improving the efficiency and planning of remediation strategies. Adsorption of ECs occurs physically and chemically, relying on the interaction of hydrogen bonds and functional groups, electrostatic interaction, ionic exchange, and pore filling. Among the adsorbents, activated carbon (in granular, powdered, and also recently investigated liquid form, GAC, PAC, LAC, respectively) and biochar are the most widely used. Especially the latter has attracted great interest since it is generally considered a sustainable material when obtained by the chemical conversion of biomass (e.g., sewage sludge, agricultural or livestock waste) based on a thermal treatment [37]. In order to improve selectivity and regeneration capability, and trigger EC catalytic degradation, traditional sorbents are recently being enhanced by including in their structure several nanomaterials. Successful results in terms of EC (mainly pesticides, and micro- and nano-plastics) have been reported [38]. Nanoparticles have been also demonstrated to amplify pollutant uptake by plants since they have been observed to bind pollutants, thereby increasing their accessibility to plants, and promoting the production of phyto-enzymes and plant growth hormones. However, applications of nanomaterials for enhanced EC removal from soil are still at laboratory scale, and thus under controlled conditions (pH, temperature, soil composition, nutrient levels) and modest soil sample amounts.

12.4 EC ENVIRONMENTAL AND HEALTH RISK ASSESSMENT

Environmental and human health risk assessment (RA) evaluates the likely or actual harm of one or more threats, such as chemical pollutants, pathogens, or anthropogenic activities, on the environment and human health, respectively. Such evaluation is useful to guide policy, regulation, and management strategies when evaluating and comparing alternative scenarios and cost-effective actions to minimize the risks.

As research on EC identification/quantification and removal is ongoing and the number of newly recognized ECs is ever growing, assessment of the risks that EC occurrence poses to both human and environmental health is crucial. Accordingly, in the case of direct reuse in the irrigation of reclaimed wastewater, the European Commission has recently established the development of preventive RA.

Conventionally, the overall RA procedure consists of four steps: i) problem formulation, ii) exposure assessment, iii) hazard assessment, and iv) risk characterization.

The problem formulation step generally involves the selection of the target endpoints and contaminants based on which the RA is conducted. The initial choice of endpoints (such as environmental or human health) determines the direction and type of RA procedure to be followed. Indeed, as endpoints, environment and human health can be considered. The former includes, as exposed population (receptors), aquatic organisms and terrestrial wildlife, and the target contaminants are usually chemicals such as ECs that can be ingested by animals through contaminated food (i.e., crops) and water. Human health risks involve chemical (e.g., ECs) and biological (e.g., pathogens) threats that may cause acute or chronic effects through oral or dermal exposure to contaminated food and water. To date, basically all the studies dealing with RA have addressed separately the impacts of ECs on the environment and human health [39]. However, within the One Health concept, further investigation is needed to comprehensively evaluate the effects of ECs on human and ecosystems following a holistic approach. In particular, a recent preliminary study highlights that considering only the human health risk would result in the underestimation of the overall risk associated with some ECs (i.e., pharmaceuticals). Indeed, while human health risk results were negligible for pharmaceutical exposure in a wastewater indirect reuse scenario, environmental RA exceeded the warning threshold in the same case [39]. As a result, the authors outlined the need for further regulations and technical guidelines that standardize the joint protection of human and environmental health as well as further studies in this regard.

Depending on endpoints and receptors, further research is also needed to designate the most appropriate ECs to consider as target compounds for any effective and accurate RAs. Indeed, due to the large variety of ECs and their different compositions/characteristics, it seems essential to select only a restricted number of ECs that can be representative of EC classes. In this regard, Verlicchi et al. (2023) [40] recently suggested a methodology based on EC occurrence, persistence, bioaccumulation, and toxicity to define EC indicators in the case of wastewater reuse for irrigation. Moreover, ECs to be selected as reference compounds should be those included in the list reported in the revision of the European Drinking Water Directive and in the periodically updated watch list of substances of concern in aquatic environments. Finally, the choice of reference ECs should take into account the availability of concentration data in the existing literature and databases and/or from site-specific measurements thereby ensuring the robustness of the results of the assessment.

The second step aims at the identification of exposure pathways and levels, namely EC concentration in environmental matrixes (i.e., measured environmental concentration, MEC) and in crops (C_{CROP}). A vital role in this step is played by appropriate sampling techniques, precise extraction methods, and accurate instrumental measurements. For this reason, the development and advancement of EC detection procedures that are as reliable and cost-effective as possible is advisable, as discussed in Section 12.2. However, when field measurements or historical data are not available or not sufficient to support a quantitative analysis, a qualitative risk analysis methodology can be adopted [41]. Particularly, it relies on mathematical models describing the fate and behavior of ECs from their release into the environment, from one or different sources, to the target receptors. Such numerical models are able to simulate the migration, transformation, and (bio)accumulation of ECs through different compartments such as air, water, soil, and biota. Although several models exist, further research is needed to address key challenges in simulating EC behavior, including limited training data and poor understanding of how environmental factors affect fate processes. Additional complexity arises from multi-pollutant interactions and their combined effects within environmental matrices, such as precipitation or adsorption, which remain difficult to capture accurately in current predictive modeling approaches.

On the other hand, the use of AI technologies, such as machine learning and deep learning, has been proved and increasingly encouraged to effectively predict the chemical reactivity, toxicity, and transport of various ECs. Nevertheless, the qualitative approach is not suitable for risk evaluation in water reuse scenarios for potable purposes, where quantitative chemical and microbial risk assessments (QCRA and QMRA, respectively) are recommended [41].

In the hazardous assessment step, the dose-response relationship between target pollutants and receptors is established and toxicological reference concentrations are determined. Predicted no-effect concentration (PNEC) and reference dose (RfD) indicate safe exposure levels for environmental and human health risk assessments, respectively, where no adverse effects are expected over a defined exposure period [40]. Both the values are determined as a result of (eco)toxicological studies. However, due to the cost of experimentation and legislative restrictions, defining the effects of all hazardous chemicals in relation to organisms is challenging. For this reason, prediction of ecotoxicity through advanced statistical methods as well as machine-learning techniques is being explored [42]. Furthermore, traditional ecotoxicological studies focused on the effects of individual ECs but in in vitro and in vivo studies could be conducted to evaluate the potential effects provided by the exposure to multiple pollutants simultaneously.

As a last step of the overall RA procedure, risk characterization provides an estimation of the magnitude of the risk through calculated risk indexes. Risk characterization can be usually performed through three distinct approaches: qualitative, deterministic, and probabilistic. The former delivers risk matrices as outputs to categorize the risks according to a scale of likelihood and/or severity of the impacts [41]. The probabilistic approach allows to account for the uncertainty of the overall RA procedure since it quantifies the risk in terms of statistical distribution of the benchmark quotient (BQ) and disability-adjusted life year (DALY) for QCRA and QMRA, respectively. Finally, in the deterministic approach, a single precautionary value is determined for each target EC. Specifically, in the case of environmental RA, the risk quotient (RQ) is calculated by dividing the maximum MEC by PNEC, whereas, for human health RA, the hazard quotient (HQ) is the ratio of the maximum C_{CROP} to RfD. When assessing the risk posed by mixtures of chemicals, the concentration addition (CA) model and the hazard index (HI) are respectively adopted. Both the parameters are obtained as a sum of each RQ and HQ, respectively, describing the risk associated with the single target EC. This approach relies on the precautionary assumption that chemicals exert additive effects [43]. However, overall risk may be underestimated, as pollutants from different sources can interact with each other or with naturally occurring compounds (e.g., organic matter), potentially leading to synergistic and more harmful effects. Therefore, further studies are necessary in this regard.

12.5 CONCLUSIONS

Emerging contaminants (ECs) pose a threat to human, animal, and environmental health due to their widespread occurrence, environmental persistence, and bioaccumulation potential. The One Health approach allows for effectively tackling the EC issue at a global level through cross-sectoral collaborations. Within this perspective, the future scope and directions of research regarding EC have been discussed in this chapter.

Traditional analytical techniques, such as chromatography coupled with mass spectrometry, have been shown to offer high sensitivity and specificity, but their widespread use is often limited by their significant costs, long-term analyses, and the need for high-skilled personnel. To address these limitations, research is advancing toward innovative approaches such as biosensors, surrogate parameter-based spectroscopy, and machine learning models to enhance real-time monitoring and predictive capabilities. Additionally, the development of low-cost, portable sensors capable of real-time EC detection in diverse matrices should be prioritized.

With regard to EC remediation, advanced oxidation processes and membrane-based technologies have demonstrated remarkable efficacy in removing ECs from wastewater. However, challenges such as high operational costs, membrane fouling, and the generation of harmful by-products need to be overcome. As a solution, online and real-time process control through field-deployable sensors would allow for adjusting process conditions, such as reagent doses and contact times, to optimize EC removal but saving costs. Similarly, adsorption technologies using materials like activated carbon, biochar, and nanomaterials offer versatile solutions, especially for soil remediation, but scalability and cost-effectiveness remain barriers to widespread implementation. In this case, the use of sustainable and recyclable adsorbents, such as biochar enhanced with nanomaterials, should be explored further to balance effectiveness with environmental impact.

Finally, a holistic approach to environmental and health risk assessment has been shown to be crucial in order to account for the cumulative and synergistic effects of ECs on a global scale in alignment with the One Health paradigm. Such an approach requires that regulatory guidelines should be updated to standardize the assessment of both chemical and biological threats posed by ECs.

REFERENCES

[1] European Centre for Disease Prevention and Control., *ECDC One Health Framework: May 2024.*, Publications Office, LU, 2024. https://data.europa.eu/doi/10.2900/634973 (accessed November 25, 2024).

[2] V. Geissen, H. Mol, E. Klumpp, G. Umlauf, M. Nadal, M. van der Ploeg, S.E.A.T.M. van de Zee, C.J. Ritsema, Emerging pollutants in the environment: A challenge for water resource management, *International Soil and Water Conservation Research* 3 (2015) 57–65. https://doi.org/10.1016/j.iswcr.2015.03.002.

[3] V. Dulio, B. van Bavel, E. Brorström-Lundén, J. Harmsen, J. Hollender, M. Schlabach, J. Slobodnik, K. Thomas, J. Koschorreck, Emerging pollutants in the EU: 10 years of NORMAN in support of environmental policies and regulations, *Environmental Sciences Europe* 30 (2018) 5. https://doi.org/10.1186/s12302-018-0135-3.

[4] P. Biswas, B.P. Vellanki, A.A. Kazmi, Investigating a broad range of emerging contaminants in a set of anthropogenically impacted environmental compartments, *Science of The Total Environment* 824 (2022) 153757. https://doi.org/10.1016/j.scitotenv.2022.153757.

[5] M. Paszkiewicz, K. Godlewska, H. Lis, M. Caban, A. Białk-Bielińska, P. Stepnowski, Advances in suspect screening and non-target analysis of polar emerging contaminants in the environmental monitoring, *TrAC Trends in Analytical Chemistry* 154 (2022) 116671. https://doi.org/10.1016/j.trac.2022.116671.

[6] X. Li, X. Shen, W. Jiang, Y. Xi, S. Li, Comprehensive review of emerging contaminants: Detection technologies, environmental impact, and management strategies, *Ecotoxicology and Environmental Safety* 278 (2024) 116420. https://doi.org/10.1016/j.ecoenv.2024.116420.

[7] S. Shyamalagowri, N. Shanthi, J. Manjunathan, M. Kamaraj, A. Manikandan, J. Aravind, Techniques for the detection and quantification of emerging contaminants, *Physical Sciences Reviews* 8 (2023) 2191–2218. https://doi.org/10.1515/psr-2021-0055.

[8] P. Ishtaweera, G.A. Baker, Progress in the application of ionic liquids and deep eutectic solvents for the separation and quantification of per- and polyfluoroalkyl substances,

Journal of Hazardous Materials 465 (2024) 132959. https://doi.org/10.1016/j.jhazmat.2023.132959.

[9] H. Qin, H. Liu, Y. Liu, S. Di, Y. Bao, Y. Zhai, S. Zhu, Recent advances in sample preparation and chromatographic analysis of pharmaceuticals and personal care products in environment, *TrAC Trends in Analytical Chemistry* 164 (2023) 117112. https://doi.org/10.1016/j.trac.2023.117112.

[10] M. Llompart, M. Celeiro, T. Dagnac, Microwave-assisted extraction of pharmaceuticals, personal care products and industrial contaminants in the environment, *TrAC Trends in Analytical Chemistry* 116 (2019) 136–150. https://doi.org/10.1016/j.trac.2019.04.029.

[11] Y. Soriano, V. Andreu, Y. Picó, Pressurized liquid extraction of organic contaminants in environmental and food samples, *TrAC Trends in Analytical Chemistry* 173 (2024) 117624. https://doi.org/10.1016/j.trac.2024.117624.

[12] M.E.I. Badawy, M.A.M. El-Nouby, P.K. Kimani, L.W. Lim, E.I. Rabea, A review of the modern principles and applications of solid-phase extraction techniques in chromatographic analysis, *Analytical Sciences* 38 (2022) 1457–1487. https://doi.org/10.1007/s44211-022-00190-8.

[13] P. Khodayari, H. Ebrahimzadeh, A green QuEChERS syringe filter based micro-solid phase extraction using hydrophobic natural deep eutectic solvent as immobilized sorbent for simultaneous analysis of five anti-diabetic drugs by HPLC-UV, *Analytica Chimica Acta* 1279 (2023) 341765. https://doi.org/10.1016/j.aca.2023.341765.

[14] M.N.L. Suseela, M.K. Viswanadh, A.K. Mehata, V. Priya, Vikas, A. Setia, A.K. Malik, P. Gokul, J. Selvin, M.S. Muthu, Advances in solid-phase extraction techniques: Role of nanosorbents for the enrichment of antibiotics for analytical quantification, *Journal of Chromatography A* 1695 (2023) 463937. https://doi.org/10.1016/j.chroma.2023.463937.

[15] Z.S. Korrani, E. Khalili, H. Kamyab, W.A. Wan Ibrahim, H. Hashim, A new solid phase extraction sorbent developed based on cyanopropyl functionalized silica nanoparticles for organophosphorus pesticides determination, *Environmental Research* 238 (2023) 117167. https://doi.org/10.1016/j.envres.2023.117167.

[16] M. Kalaboka, V. Sakkas, Magnetic solid-phase extraction based on silica and graphene materials for sensitive analysis of emerging contaminants in wastewater with the aid of UHPLC-Orbitrap-MS, *Molecules* 28 (2023) 2277. https://doi.org/10.3390/molecules28052277.

[17] G.V. Korshin, M. Sgroi, H. Ratnaweera, Spectroscopic surrogates for real time monitoring of water quality in wastewater treatment and water reuse, *Current Opinion in Environmental Science & Health* 2 (2018) 12–19. https://doi.org/10.1016/j.coesh.2017.11.003.

[18] M. Sgroi, T. Anumol, F.G.A. Vagliasindi, S.A. Snyder, P. Roccaro, Comparison of the new Cl2/O3/UV process with different ozone- and UV-based AOPs for wastewater treatment at pilot scale: Removal of pharmaceuticals and changes in fluorescing organic matter, *Science of The Total Environment* 765 (2021) 142720. https://doi.org/10.1016/j.scitotenv.2020.142720.

[19] L. Marino, E. Gagliano, D. Santoro, P. Roccaro, Fluorescence sensor enabled control of contaminants of emerging concern in reclaimed wastewater using ozone-based treatment processes, *Water Research* 268 (2025) 122616. https://doi.org/10.1016/j.watres.2024.122616.

[20] N.M. Kulshreshtha, D. Shrivastava, P.S. Bisen, 14 – Contaminant sensors: nanotechnology-based contaminant sensors, in: A.M. Grumezescu (Ed.), *Nanobiosensors*, Academic Press, 2017: pp. 573–628. https://doi.org/10.1016/B978-0-12-804301-1.00014-X.

[21] Y. Zhang, J. Li, S. Jiao, Y. Li, Y. Zhou, X. Zhang, B. Maryam, X. Liu, Microfluidic sensors for the detection of emerging contaminants in water: A review, *Science of The Total Environment* 929 (2024) 172734. https://doi.org/10.1016/j.scitotenv.2024.172734.

[22] C.P. Shelor, C. Warren, C.V. Odinaka, K. Dumre, Comprehensive review of combustion ion chromatography for the analysis of total, adsorbable, and extractable organic fluorine, *Journal of Separation Science* 47 (2024) 2400235. https://doi.org/10.1002/jssc.202400235.

[23] J.M. Partington, S. Rana, D. Szabo, T. Anumol, B.O. Clarke, Comparison of high-resolution mass spectrometry acquisition methods for the simultaneous quantification and identification of per- and polyfluoroalkyl substances (PFAS), *Analytical and Bioanalytical Chemistry* 416 (2024) 895–912. https://doi.org/10.1007/s00216-023-05075-x.

[24] C.A. McDonough, J.L. Guelfo, C.P. Higgins, Measuring total PFASs in water: The tradeoff between selectivity and inclusivity, *Current Opinion in Environmental Science & Health* 7 (2019) 13–18. https://doi.org/10.1016/j.coesh.2018.08.005.

[25] E. Pasecnaja, D. Zacs, Determination of Perfluorinated Substances (PFAS) Using LC-ORBITRAP-MS in Certain foodstuffs of Animal Origin According to newly established EU legislation, *Food Analytical Methods* 17 (2024) 640–649. https://doi.org/10.1007/s12161-024-02603-y.

[26] K. Chen, S. Li, H. Yang, J. Zou, L. Yang, J. Li, L. Ma, Feasibility of using gas chromatography-ion mobility spectrometry to identify characteristic volatile compounds related to brandy aging, *Journal of Food Composition and Analysis* 98 (2021) 103812. https://doi.org/10.1016/j.jfca.2021.103812.

[27] M.T. Aung, K.K. Shimabuku, N. Soares-Quinete, J.P. Kearns, Leveraging DOM UV absorbance and fluorescence to accurately predict and monitor short-chain PFAS removal by fixed-bed carbon adsorbers, *Water Research* 213 (2022) 118146. https://doi.org/10.1016/j.watres.2022.118146.

[28] F. Zhao, L. Wang, M. Li, M. Wang, G. Liu, J. Ping, Nanozyme-based biosensor for organophosphorus pesticide monitoring: Functional design, biosensing strategy, and detection application, *TrAC Trends in Analytical Chemistry* 165 (2023) 117152. https://doi.org/10.1016/j.trac.2023.117152.

[29] E.K. Radwan, H.H. Abdel Ghafar, M. b. m. Ibrahim, A.S. Moursy, Recent trends in treatment technologies of emerging contaminants, *Environmental Quality Management* 32 (2023) 7–25. https://doi.org/10.1002/tqem.21877.

[30] L. Rizzo, S. Malato, D. Antakyali, V.G. Beretsou, M.B. Đolić, W. Gernjak, E. Heath, I. Ivancev-Tumbas, P. Karaolia, A.R. Lado Ribeiro, G. Mascolo, C.S. McArdell, H. Schaar, A.M.T. Silva, D. Fatta-Kassinos, Consolidated vs new advanced treatment methods for the removal of contaminants of emerging concern from urban wastewater, *Science of The Total Environment* 655 (2019) 986–1008. https://doi.org/10.1016/j.scitotenv.2018.11.265.

[31] N. Voulvoulis, Water reuse from a circular economy perspective and potential risks from an unregulated approach, *Current Opinion in Environmental Science & Health* 2 (2018) 32–45. https://doi.org/10.1016/j.coesh.2018.01.005.

[32] A. Lee, J.W. Elam, S.B. Darling, Membrane materials for water purification: design, development, and application, *Environmental Science: Water Research Technology* 2 (2016) 17–42. https://doi.org/10.1039/C5EW00159E.

[33] L. Yang, C. Xia, J. Jiang, X. Chen, Y. Zhou, C. Yuan, L. Bai, S. Meng, G. Cao, Removal of antibiotics and estrogens by nanofiltration and reverse osmosis membranes, *Journal of Hazardous Materials* 461 (2024) 132628. https://doi.org/10.1016/j.jhazmat.2023.132628.

[34] K. Mangalgiri, Z. Cheng, H. Liu, Development of dissolved organic matter-based indicators to understand the degradation of organic contaminants in reverse osmosis concentrate from potable reuse systems, *Journal of Hazardous Materials* 470 (2024) 134060. https://doi.org/10.1016/j.jhazmat.2024.134060.

[35] W. Hamd, E.A. Daher, T.S. Tofa, J. Dutta, Recent advances in photocatalytic removal of microplastics: Mechanisms, kinetic degradation, and reactor design, *Frontiers in Marine Science* 9 (2022) 885614. https://doi.org/10.3389/fmars.2022.885614.

[36] H.-W. Yu, T. Anumol, M. Park, I. Pepper, J. Scheideler, S.A. Snyder, On-line sensor monitoring for chemical contaminant attenuation during UV/H2O2 advanced oxidation process, *Water Research* 81 (2015) 250–260. https://doi.org/10.1016/j.watres.2015.05.064.

[37] M. Coccia, E. Bontempi, New trajectories of technologies for the removal of pollutants and emerging contaminants in the environment, *Environmental Research* 229 (2023) 115938. https://doi.org/10.1016/j.envres.2023.115938.

[38] P. Rajput, P. Kumar, A.K. Priya, S. Kumari, S.R.G. Shiade, V.D. Rajput, A. Fathi, A. Pradhan, R. Sarfraz, S. Sushkova, S. Mandzhieva, T. Minkina, A. Soldatov, M.H. Wong, C. Rensing, Nanomaterials and biochar mediated remediation of emerging contaminants, *Science of The Total Environment* 916 (2024) 170064. https://doi.org/10.1016/j.scitotenv.2024.170064.

[39] L. Penserini, B. Cantoni, M. Antonelli, Modelling the impacts generated by reclaimed wastewater reuse in agriculture: From literature gaps to an integrated risk assessment in a One Health perspective, *Journal of Environmental Management* 371 (2024) 122715. https://doi.org/10.1016/j.jenvman.2024.122715.

[40] P. Verlicchi, V. Grillini, E. Lacasa, E. Archer, P. Krzeminski, A.I. Gomes, V.J.P. Vilar, M.A. Rodrigo, J. Gäbler, L. Schäfer, Selection of indicator contaminants of emerging concern when reusing reclaimed water for irrigation — A proposed methodology, *Science of The Total Environment* 873 (2023) 162359. https://doi.org/10.1016/j.scitotenv.2023.162359.

[41] D.B.M. Vidotti, P.F. Soares, J.A. Lukiantchuki, C.T. Benatti, A qualitative risk assessment model for water reuse: Risks related to agricultural irrigation in Brazil, *Science of The Total Environment* 931 (2024) 172965. https://doi.org/10.1016/j.scitotenv.2024.172965.

[42] G. Gini, F. Zanoli, Machine learning and deep learning methods in ecotoxicological QSAR modeling, in: K. Roy (Ed.), *Ecotoxicological QSARs*, Springer US, 2020: pp. 111–149. https://doi.org/10.1007/978-1-0716-0150-1_6.

[43] F. Spilsbury, V. Kisielius, K. Bester, T. Backhaus, Ecotoxicological mixture risk assessment of 35 pharmaceuticals in wastewater effluents following post-treatment with ozone and/or granulated activated carbon, *Science of The Total Environment* 906 (2024) 167440. https://doi.org/10.1016/j.scitotenv.2023.167440.

Index

B

C

Q

For Product Safety Concerns and Information please contact our EU representative GPSR@taylorandfrancis.com Taylor & Francis Verlag GmbH, Kaufingerstraße 24, 80331 München, Germany